MATERIALS SCIENCE AND TECHNOLOGIES

ADVANCED MATERIALS AND SYSTEMS FOR ELECTROCHEMICAL TECHNOLOGIES

MATERIALS SCIENCE AND TECHNOLOGIES

Additional books in this series can be found on Nova's website
under the Series tab.

Additional e-books in this series can be found on Nova's website
under the e-books tab.

MATERIALS SCIENCE AND TECHNOLOGIES

ADVANCED MATERIALS AND SYSTEMS FOR ELECTROCHEMICAL TECHNOLOGIES

MAURO COELHO DOS SANTOS
EDITOR

New York

Library of Congress Cataloging-in-Publication Data

ISBN: 978-1-53613-852-8

Published by Nova Science Publishers, Inc. † New York

This book is dedicated to:

My Father: José Coelho dos Santos
My Mother: Terezinha Ferreira dos Santos
My Father in Law: José Franco Honório

In Memoriam

My Wife: Káthia Maria Honório
My Daughter: Isadora Honório Coelho dos Santos

In Life

CONTENTS

ACKNOWLEDGMENTS

A special thanks to:

My Mother in Law: Maria de Lourdes Honório

My Sister: Maria Aparecida dos Santos

My Brothers: José Aparecido Coelho dos Santos, Luis Coelho dos Santos and Antonio Coelho dos Santos Sobrinho

My Nephews: Caio Rocha Coelho dos Santos, Leonardo Martins Honório

My Nieces: Fabiana Aparecida Matias dos Santos, Juliana Aparecida Matias dos Santos, Marininha and Manu.

My Brother in Law: Thiago Arlindo Honório

My Sisters in Law: Gracinda, Marlene, Joana, Karina and Paula

My Friends Forever and Ever: Mario Martins Honório, Marcelo Calegaro, Marcos Gaião, Christiane, Manu and Helena.

Both our family and our friends give reason and feelings to our lives.

Knowledge is better than judgment.

God bless us all!

FOREWORD

Over the last decade, the electrochemical technologies applied to energy conversion, electroanalysis and wastewater treatment have experienced a great development thanks to the preparation of novel advanced materials with enhanced electrocatalytic properties. These environmentally friendly technologies are crucial for developing new human activities in a sustainable world. Despite the excellent results achieved so far, more fundamental research is needed for continuous innovation and future use at industrial scale. This book describes some remarkable materials used to carry out these electrochemical processes, with a critical revision and future perspectives of their applicability to different fields.

The first two chapters of the book are devoted to the description of efficient electrocatalysts for the electrochemical energy conversion in a direct ethanol fuel cell. On the one hand, Chapter 1 emphasizes the use of anode materials such as PtSn with different Sn content and Pt_3Sn to enhance the fuel cell performance yielding a maximum power density. On the other hand, Chapter 2 presents the main goals of carbon-based anode materials as nanoparticles support, like carbon nanotubes and more recently, graphene, along with their improved properties by doping with nitrogen or metal oxides. The development of cathodic electrocatalysts with enhanced activity for oxygen reduction is also analyzed. Next, Chapter 3 reviews the most recent trends in the development of promising enzymatic biofuel cells containing nanoscale materials. The design of techniques to immobilize enzymes onto the electrode surface and the incorporation of carbon nanotubes, nanofibers and nanocomposites to the bioelectrode are described.

A second section of the book is dedicated to the recent advances in electroanalysis. First, a detailed discussion on the preparation, properties and applications of chemically modified electrodes in the electrochemical (bio)sensing field is presented in Chapter 4. The interesting characteristics of the reductive activation of nitroaromatic redox mediators that form the reversible couple hydroxylamine/nitroso in multi-walled carbon nanotubes is here emphasized. Chapter 5 discusses the performance of nanoplasmonic sensors, which offer good chemical stability, high surface area and excellent electronic properties with competitive levels of sensitivity, robustness and integration. More general synthetic routes, properties and applications in the electrochemical biosensors are treated in Chapter 6, highlighting the most recent materials used such as graphene and its derivatives, nanoparticles (metallic, magnetic, and quantum dots), dendrimers and aptamers. An interesting overview of the current state and future trends of these materials is provided.

The last part of the book is related to environmental applications. The morphology of TiO_2 nanotubes and their modification with metals, oxides and wall decoration are first considered in Chapter 7. Examples on the applications of TiO_2 nanotubes in photocatalysis, photoelectrocatalysis, solar cells, semiconductors and biomedical areas are further presented and the main characteristics required for these applications are discussed. Chapter 8 considers the recent development of electrochemical advanced oxidation processes for organic destruction in synthetic and simulated wastewater. Special emphasis is made on hydrogen peroxide electrogeneration to produce the strong oxidant hydroxyl radical to attack organic pollutants. The best cathodic materials for the formation of hydrogen peroxide from oxygen reduction and combined processes with various materials to treat wastewater are highlighted.

The book Advanced Materials and Systems for Electrochemical Technologies that presents the recognized Prof. Dr. Mauro Coelho dos Santos is then an excellent contribution to know the state-of-art of advanced materials for relevant electrochemical technologies. It is highly recommended to chemists, electrochemists and chemical engineers interested in the development of the materials required for such technologies and their future applications.

Prof. Dr. Enric Brillas
University of Barcelona, Barcelona, Spain

FUEL CELLS

In: Advanced Materials and Systems …
Editor: Mauro Coelho dos Santos
ISBN: 978-1-53613-852-8

Chapter 1

PtSn Electrocatalysts for Ethanol Oxidation Reaction: The Effect of Metal Ratio and Thermal Treatment

Flavio Colmati[1,]* and Ermete Antolini[2]
[1]Intituto de Química, Universidade Federal de Goiás, Goiás, Brasil
[2]Scuola di Scienza dei Material, Cogoleto, Genova, Italy

Abstract

The ethanol oxidation reaction has a technological application for electrochemical energy conversion in a direct ethanol fuel cell. The ethanol fuel cell is a system that converts chemical energy into electrical energy by generating protons and electrons from ethanol molecules; the protons permeate across the electrolyte to the cathode, while the electrons flow through an external circuit. Ethanol adsorbs in the electrode and the oxidation reaction proceeds with deprotonation. Therefore, the anode material characteristics, specifically PtSn with different Sn content in the catalysts presets, alter the ethanol oxidation reaction pathways by changing the direct ethanol fuel cell performance and enhancing the fuel cell performance. In addition, Pt_3Sn subjected to a thermal treatment at reductive atmosphere presented an even higher increase in the direct ethanol fuel cell performance, having obtained the maximum power density with Pt_3Sn thermally treated at 200°C in H_2 atmosphere.

Keywords: DEFC, PEMFC, PtSn, ethanol electrocatalysts, ethanol electro-oxidation

---

* Corresponding Author Email: colmati@ufg.br.

INTRODUCTION

The ethanol oxidation reaction (EOR) has many applications, including a technological application of electrochemical ethanol oxidation reaction for electrochemical energy conversion through a system named direct ethanol fuel cell (DEFC).

DIRECT ETHANOL FUEL CELLS

Direct ethanol fuel cells are based on Proton Exchange Membrane Fuel Cells (PEMFCs). PEMFCs are a fuel cell system that spontaneously converts chemical energy into electrical energy from hydrogen oxidation and oxygen reduction reactions. The anode is supplied with hydrogen, which is adsorbed in Pt-based electrodes and broken down into protons and electrons; subsequently, protons permeate across the electrolyte to the cathode, while the electrons flow through an external circuit. The oxygen molecules capture the free electrons in the cathode and combine them with protons, resulting in water production. Figure 1 shows a schematic of the PEMFC system.

This system operates at low temperatures (from 60 to 110°C) because of the use of Nafion; note that temperature has a strong influence on kinetic chemical reactions. The current generated from an electrochemical system depends on the kinetics of the anode and cathode reactions. Considering these operational conditions for such reactions, electro-catalysts must have very high electrocatalytic activity. Platinum is the best metal to be used in an electrocatalyst for reactions of hydrogen oxidation reaction and oxygen reduction at low temperatures; however, it has limitations and depends on the hydrogen purity. Further details can be found in the references [1-4].

Replacing hydrogen gas with liquid fuel has some advantages; alcohol of low molecular structure is very attractive, especially ethanol, because of its low toxicity and possibility of production from agricultural products.

By maintaining the same structure illustrated in Figure 1, but substituting hydrogen fuel with ethanol solution, a system called a Direct Ethanol Fuel Cell (DEFC) is produced that involves the following: an anode directly supplied with ethanol [5, 6]; adsorption of ethanol by the electrode; an oxidation reaction toward ethanol molecule deprotonation, resulting in the generation of proton ions and electrons, followed by their separation from each other; protons permeating across the electrolyte to the cathode; and electrons flowing through an external circuit. Similar to a PEMFC, oxygen molecules capture free electrons in the cathode and combine them with protons, resulting in water production.

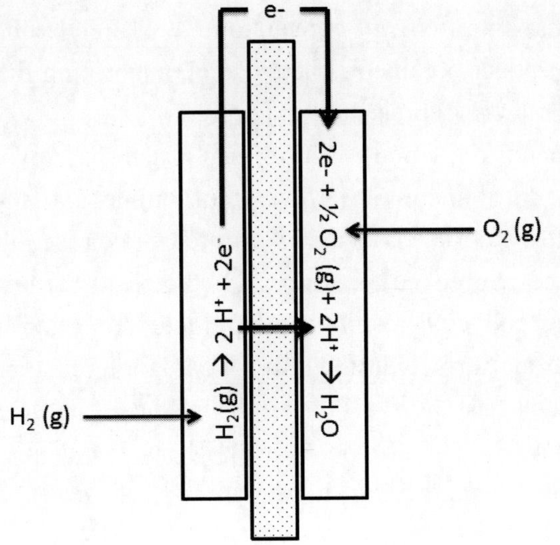

Figure 1. Schematic of a Proton Exchange Membrane Fuel Cell (PEMFC).

Furthermore, when used as an electrocatalyst in an ethanol oxidation reaction, platinum has low activity and requires some modifications in the electrochemical surface. Platinum has limitations regarding a DEFC since ethanol oxidation reaction follows parallel pathways that reduce energy density and form products strongly adsorbed by platinum, thereby drastically reducing the electrochemical activity of Pt toward ethanol oxidation reaction (as discussed next).

When ethanol enters the DEFC in an electrochemical alcohol oxidation reaction, the ethanol molecule is adsorbed on the electrocatalytic surface and the molecules are broken down with the removal of H^+ electrons from the ethanol molecule, resulting in intermediates of the reaction that are strongly adsorbed by platinum.

Many studies have reported electrochemical ethanol oxidation reaction pathways [7-10]; currently, the pathways with better acceptance are represented in Figure 2 below.

Ethanol is adsorbed by platinum (1). The resulting reaction (2) is the oxidation of ethanol into acetaldehyde, which can be oxidized into acetic acid (3). Acetic acid is the final product. Reactions (4) and (5) represent the routes involving the C–C bond cleavage in the molecule to generate adsorbed CO and other fragments with a single carbon atom. Finally, reaction (6) is the oxidation of CO adsorbed into CO_2.

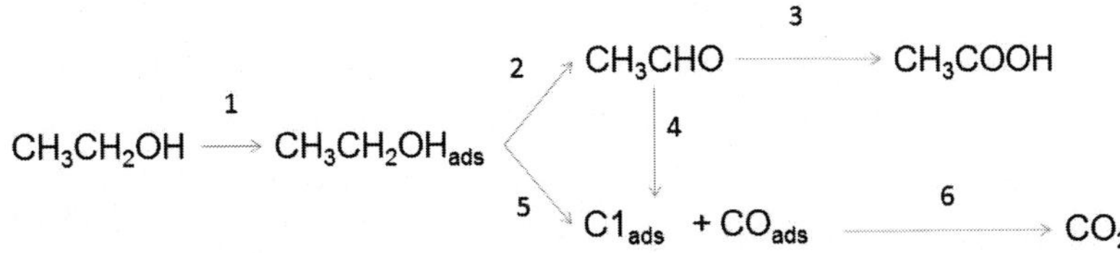

Figure 2. Scheme for the oxidation of ethanol on platinum.

Desirable pathways for ethanol oxidation reaction encompass steps 1-5-6 or steps 1-2-4-6, resulting in maximum energy conversion, i.e., 12 electrons per ethanol molecule; however, the main product obtained is acetic acid, suggesting that the ethanol oxidation reaction follows steps 1-2-3, represented in Figure 2.

Study of the ethanol oxidation reaction mechanism has involved the application of some non-electrochemical techniques *In situ* or online and the screening of several reaction intermediates. Fourier Transform Infrared Spectroscopy (FTIR) is among the main *In situ* techniques employed, in which [11] the electrode potential is controlled and the FTIR spectra are collected resulting in characteristic bands from the formation of products on the electrode surface during ethanol oxidation reaction.

Figueiredo et al. [12] showed FTIR results of the ethanol oxidation reaction on Pt modified with Sb and Bi, and Rocha et al. [13] showed FTIR results of the ethanol oxidation reaction on Nb modifying Pt catalyst; in both cases, acid acetic is the main product.

We also used differential electrochemical mass spectrometry (DEMS), in which the products of ethanol oxidation reaction are transported to an MS chamber and detected through mass/charge ratio after ionization. Rizo et al. [14] used Sn to modify Pt single crystals and obtained a different perspective regarding ethanol oxidation reaction with DEMS analysis.

Alternatively, the fuel cell output can be directly connected to a high-performance liquid chromatography (HPLC) system, in which the products can be separated, identified, and quantified. Lamy et al. [15] presented the results of an analysis using a DEFC connected with an HPLC system and observed the formation of acetaldehyde and acetic acid.

Other experiments, such as nuclear magnetic resonance (NMR) [16], have also been employed less frequently; all the results indicate a similar pathway for the ethanol oxidation reaction.

As abovementioned, acetic acid is the main product obtained from the ethanol oxidation reaction, in addition to the formation of acetaldehyde and a low amount of CO_2; consequently, ethanol is not completely oxidized in platinum-based electrodes. In addition, replacing palladium with platinum has also been considered, for example, Antolini [17] reported the Pd electrochemical activity toward ethanol oxidation reaction in a direct ethanol fuel cell. More recently, Geraldes et al. [18] examined Pd and PdSn, with the FTIR analysis showing acetate and acetaldehyde as the main products once again.

Furthermore, for electrochemical energy conversion, the ethanol oxidation reaction should occur with the lowest potential possible to obtain a high potential difference in the DEFC. The oxygen reduction reaction occurs at approximately 0.8 V vs. RHE under practical conditions (theoretical potential value of 1.23 V) in the cathode electrode. Hence, the ethanol oxidation reaction should occur with a potential below 0.6 V vs. RHE;

this characteristic is the main platinum limitation, as some intermediates of ethanol oxidation reaction adsorbed by platinum have oxidation above 0.8 V.

As pure platinum is limited as a catalyst for ethanol oxidation reaction, other transition metal elements are used to modify platinum properties. As observed from the ethanol oxidation reaction mechanism, an intermediate CO is maintained strongly and adsorbed by platinum; Gastaiger et al. [19] in 1995 studied CO oxidation in Pt and observed the need of a high potential to oxidize CO into CO_2 in Pt. Introducing Ru in the Pt lattice led to a strong decrease in the potential of CO oxidation in PtRu materials; in addition, the best atomic ratio observed was 1:1 [20].

Subsequently, we observed that PtRu did not present high electro-catalytic activity for the ethanol oxidation reaction. Lamy et al. [21] reported their studies with PtSn at different atomic ratios. In the beginning, this material does not show chemical stability since the tin is oxidized at the onset potential of the ethanol oxidation reaction.

There are studies with other metals modifying Pt toward ethanol oxidation reaction, such as rhodium (Rh) [22], molybdenum (Mo) [23], and cerium (Ce) [24], among others [25-28], all contributing to ethanol oxidation reaction in platinum with either bifunctional or electronic effect. The activity of the bimetallic electrocatalyst is dependent on the modification of platinum by the second metal. Some studies showed that Rh can break the C-C bond [29]; however, the more active material for the ethanol oxidation reaction is Pt_3Sn [30], even with the limitation of breaking C-C bonds and the major product being acetic acid, the current from the ethanol oxidation reaction is higher when Pt_3Sn is used as electrocatalyst [31].

METHODS

PtSn Nanomaterials

The PtSn nanoparticles for ethanol oxidation reaction are usually prepared from the chemical reduction of ions of Sn and Pt in aqueous solutions. As reductors, formic acid [32], borohydride [33], ethanol reflux [34] and poly-alcohols (called polyol) [35] can be used. The metallic ions are reduced and deposited on high surface area carbon, (Vulcan XC-72R, 240 m^2 g^{-1}, from Cabot Corporation, Boston, MA, USA), but other supports have also been used, such as carbon nanotubes [36] and graphene [37]. The atomic ratio of Pt/Sn is determined from the amounts of Pt ions and Sn ions added into the reactor; the change in atomic ratio modifies the electrocatalytic properties. The data shown in the work were obtained with PtSn prepared with formic acid as a reductor [32]. An aqueous suspension with the carbon powder, Pt ions, and Sn ions were heated to 80°C, and then 2 mol L^{-1} of formic acid was slowly added. Once metal ions were reduced, the catalysts were filtered and dried [38].

Morphological and Chemical Characterization

Because the nanoparticles are obtained, the materials should be characterized morphologically using transmission electron microscopy (TEM), X-ray diffraction (DRX) and EDX.

Transmission electron microscopy (TEM) or high-resolution transmission electron microscopy (HRTEM) shows the particle size distribution on the carbon support and the shape of the nanoparticles, and HRTEM allows for determining the interatomic plane distances and surface crystallographic orientation of the surface; moreover, the distribution of nanoparticle size (d_n) can be obtained.

Figure 3 shows an image obtained using transmission electron microscopy of the Pt on carbon Vulcan XC-72R. The black points are metal nanoparticles, and the green points are the carbon support. Note that the inset of Figure 3 shows an image obtained using high-resolution transmission electron microscopy; this image allows the interatomic planes distances to be obtained. Figure 3 illustrates that the nanoparticles are well dispersed on the carbon support with low agglomerate formation.

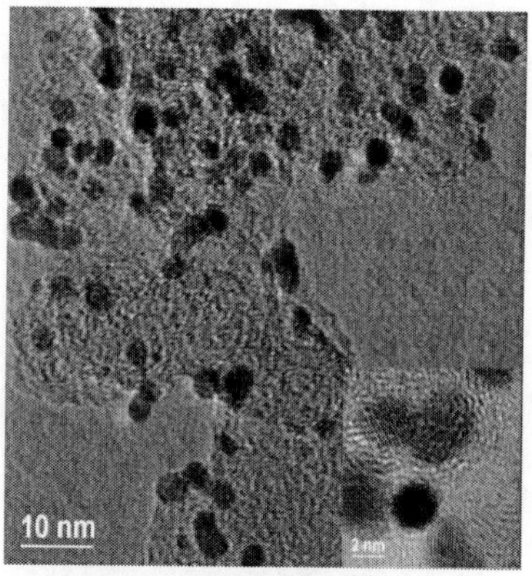

Pt_9Sn/C

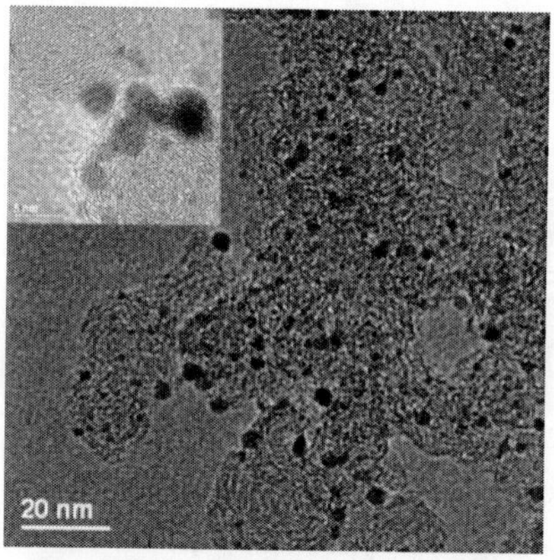

Pt_4Sn/C

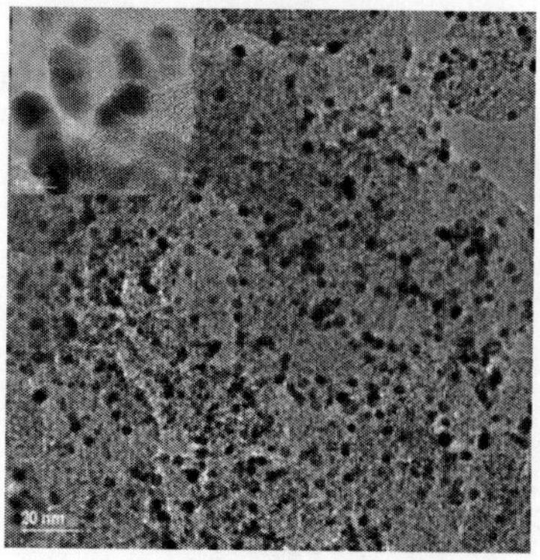

Pt_3Sn/C

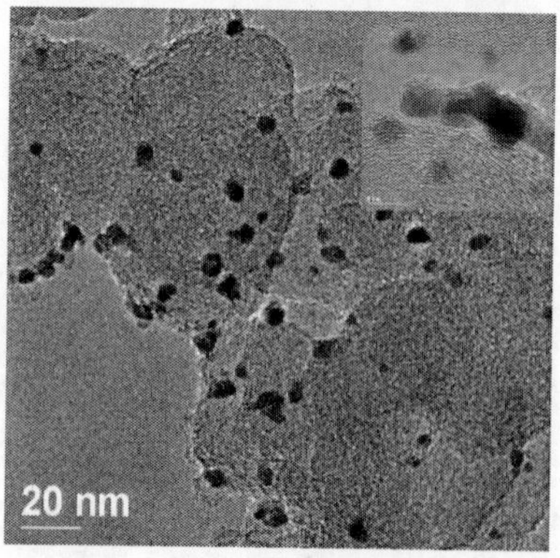

Pt_2Sn/C

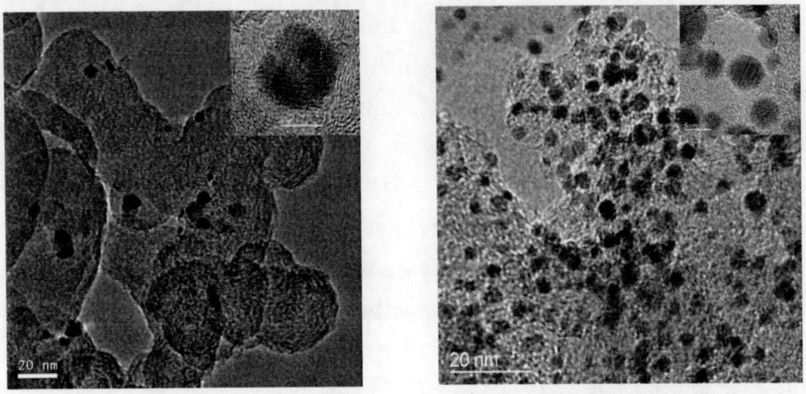

PtSn/C Pt$_3$Sn/C E-TEK

Figure 3. TEM images of PtSn materials with different amounts of Sn. Inset: HRTEM images.

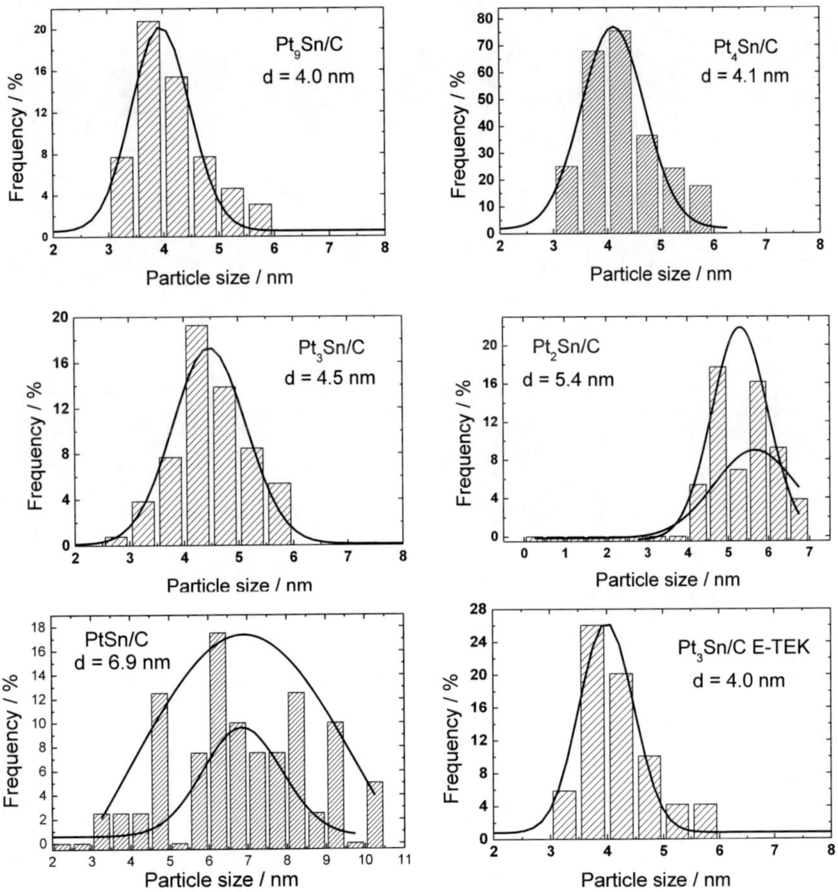

Figure 4. Histogram of the PtSn particle size with different Sn amounts.

The particle size distribution can be obtained by counting the number of particles of each size range. Equation 3 generates the histogram shown in Figure 4. The histogram images should be obtained at a low resolution, and the number of particles should be above 300 nanoparticles in several areas of the TEM images. In this manner, it is possible to observe a Gaussian distribution of the particle size; the maximum of the Gaussian distribution was assumed as the average particle size (d). In addition, the prepared materials rich in Sn present more than one Gaussian distribution (see Pt_2Sn and PtSn) for high and low frequencies; however, the maximum values for both Gaussian distributions are not very different, suggesting they have the same average particle size. Figure 4 indicates the average particle size.

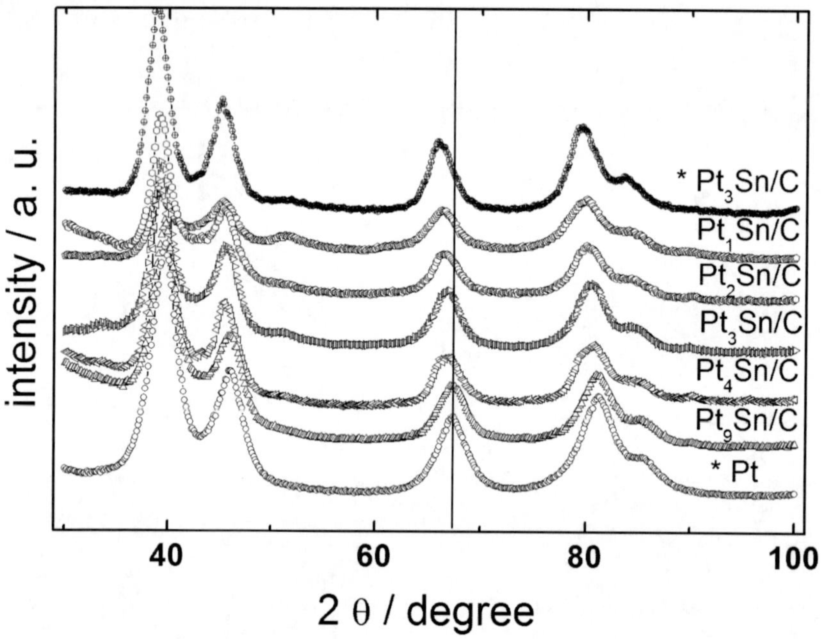

Figure 5. XRD patterns of the carbon supported 20 wt % PtSn with different amounts of tin. * Commercial material from E-TEK co.

X-ray diffraction (XRD) shows the crystallographic structure of the nanoparticles, and the shift in the peak positions indicate the formation of solid solution or phase segregation. Figure 5 shows the XRD pattern of PtSn prepared with different Sn contents. Figure 5 also shows the XRD pattern of Pt and Pt_3Sn from E-TEK Co. All materials exhibit peaks of the fcc structure, typical of platinum; peaks of cassiterite can also be observed with low intensity. A substantial observation is the shift of the diffraction peaks to lower degree with an increase in the tin content in the catalysts. These shifted peak position can be assigned to the insertion of tin in the platinum lattice. A vertical black line is shown in Figure 5 to facilitate this observation.

As shown in Figure 5, the XRD results reveal diffraction peaks typical for platinum (face cubic centered structure, fcc); the peak positions are shown in Table 1 (note that PtSn peak positions are lower, depending on the tin content). Moreover, Table 1 shows the assigned peak and the degree of the diffraction.

Table 1. XRD peak position of the platinum (face cubic centered structure, fcc)

Pt (fcc) peaks	**Angle degree/2θ**
(111)	40
(200)	47
(220)	67
(311)	82
(222)	87

Table 2. Peak 220 position for Pt and PtSn at different Sn contents

Material	**220 Peak position/2θ degree^{-1}**
Pt (E-TEK)	67.46
Pt_9Sn	67.20
Pt_4Sn	67.06
Pt_3Sn	66.93
Pt_2Sn	66.53
PtSn	66.26
Pt_3Sn (E-TEK)*	65.86

* Considered 100% alloy.

Instead of (111) showing high intensity, the peak (220) is often used for calculations to avoid carbon support backscattering. The crystal size and lattice parameters are calculated according to Bragg's law.

Figure 5 shows a shift of the peaks to lower 2θ degrees with tin content in the catalyst that originates from the insertion of Sn into Pt lattice, as well as an expansion of the distance between Pt – Pt atoms. The vertical black line at 2θ = 82 degrees facilitates observation of the peak shift. Table 2 indicates the (220) peak position of Pt and PtSn with different Sn contents in the electrocatalysts.

This shift corresponds to solid solution formation; it is possible to establish the amount of solid solution by using the lattice parameters of both the alloy and the sample. Equation (1) measures the formation of solid solution amount.

$$x = \frac{a - a_o}{a_s - a_o} x_s \tag{1}$$

where a is the lattice parameter of the alloy (assumed Pt_3Sn E-TEK, a – 4.00 nm); a_o is the reference lattice parameter (Pt, 0.399 nm); a_s is the sample lattice parameter and X_s is the amount of tin in the material.

The materials prepared result in different solid solutions, demonstrating an increase in the amount of tin of the material and an increase in the solid solution formation; however, this method of preparation of the electrocatalyst does not result in 100% of solid solution, as some tin is in the oxide form. A full discussion can be found in [39].

Energy-Dispersive X-ray Spectroscopy (EDS or EDX) is used for the elemental analysis or chemical characterization of a sample; this technique shows the amounts of elements presented in the sample by analyzing the X-rays emitted from the sample. The EDX results indicate the semi-quantitative composition of the material because the amount of the metal cannot be entirely present in the final material, as it can be partially lost during the preparation process. Thus, the EDX results are very important. As discussed earlier, the composition of the material influences the mechanism of ethanol oxidation reaction. All synthesized materials were analyzed by EDX; an atomic ratio close to nominal composition was observed in all Pt:Sn materials [38].

X-Ray Photoelectron Spectroscopy (XPS)

Focusing on PtSn, many papers report that the best catalyst for the ethanol oxidation reaction is PtSn. The high affinity of Sn with Pt permits solid solution formation; thus, Sn can modify the 5d orbital electrons of platinum and modify the adsorption and reaction of the ethanol on the surface [40, 41]. Moreover, Sn can activate the water of the electrolyte and produce hydroxide adsorbed on Sn (OH^-_{ads}) called the hydroxide species, which can oxidize the ethanol or intermediate of the reaction via a bifunctional mechanism [42]. It is known that Sn inhibits C – C bond cleavage when it is not alloyed with Pt [43]; in contrast, when Pt_3Sn forms a solid solution, it shows very high electrocatalytic activity for ethanol oxidation reaction via the bifunctional mechanisms and the electronic effect.

X-ray Absorption Spectroscopy (XAS) near the edge (XANES) provides information on orbital filling of the central atom. The Pt edge transition energy of electron from 2p to 5d orbitals, called the LIII edge is 11564 eV. Each sample is scanned from -20 eV before and 30 eV after the edge to characterize the peak of adsorption, the intensity of which indicates the filling of the 5d Pt orbital. There are two methods to obtain this information, the Mansour method [44] and the Shukla method [45]. The Mansour method is more complete and is necessary to obtain the XAS spectra at the L_{III} and L_{II} edges; with a mathematical treatment with these curves it is possible to obtain the 5d Pt orbital filling. The Shukla method [45] is easier to use and is necessary to use only for the L_{III} XAS spectrum; a baseline can be obtained from an arctan fitting and with mathematical treatment to obtain the 5d Pt orbital filling. From XANES analysis, it was observed that

Sn causes a partial filling of the Pt d-band vacancies [46]. Figure 6 shows the X-ray adsorption curves at the Pt LIII edge (11564 eV) for PtSn materials with different amounts of Sn on the electrocatalysts registered with electrode polarized at 0.8 V vs. ERH. These data indicated that the increase in Sn content on electrocatalysts increases the Pt d-band vacancies filling; however, when the experiments were conducted at the low potential of ca. 0.3 V vs. ERH, the modification in the d-band was not adequate to produce a visible effect [47]

Figure 6 shows the similar intensity of the curves; however, as the amount of tin increases, the curve intensities decrease because the Sn inserts electrons into platinum and the 5d orbital has further filling with the increased tin content at the catalysts [48]. This behavior modifies the energy of ethanol adsorption on the surface. These experiments have a singular advantage of being conducted *In situ*, with the electrode at 0.8 V vs. RHE.

Electrochemical Characterization

A single fuel cell can be used as a half electrochemical cell and register cyclic voltammetry and linear sweep voltammetry. In this case, the working electrode should be supplied with N_2 (g), water, or ethanol solution, and the other electrode should be

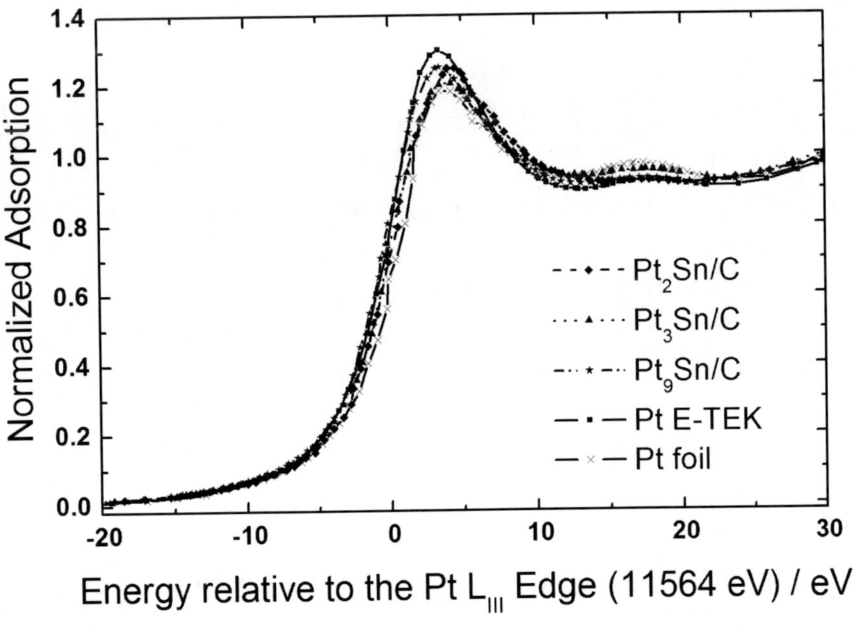

Figure 6. X-ray Adsorption Near Edge Spectra at the Pt L_{III} edge at 0.8 V vs. ERH; Pt foil as a reference (ex situ).

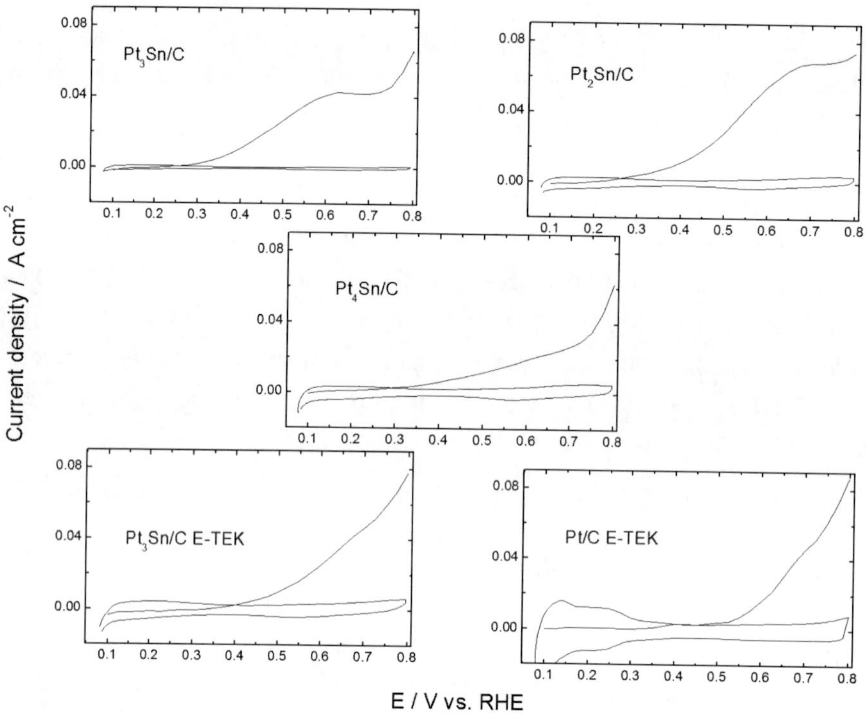

Figure 7. Cyclic voltammetry and linear sweep voltammetry of the PtSn material with different Sn contents. Cyclic voltammetry registered with diffusion gas electrodes, anode supplied with N_2 and cathode (dynamic reference) supplied with H2. Linear sweep voltammetry anode supplied with ethanol solution and cathode (dynamic reference) supplied with H2(g).

supplied with H_2(g); the latter electrode works as a counter and reference electrode simultaneously, called the dynamic reference electrode [49]. The main difference between this system and the half electrochemical cell with three electrodes is the dynamic reference electrode operates as a reference and the auxiliary electrode at the same time. Figure 7 shows cyclic voltammetry and linear sweep voltammetry data registered using a single cell. The cathode side was supplied with H_2(g), where the catalysts were Pt/C, and the anode side, where the catalysts were PtSn with different Sn contents, was supplied with ethanol solution. By performing these experiments, it is possible to determine the onset potential of the ethanol oxidation reaction, as shown in Table 3.

Pt cyclic voltammetry has a profile similar to Pt in H_2SO_4 acid, i.e., the voltammograms show the hydrogen region with hydrogen adsorption in electrocatalysts between 0.05 and 0.3 V vs. RHE, a double layer region between 0.3 and 0.6 V vs. RHE, and an oxide region above 0.6 V vs. RHE, where OH^- species are adsorbed on the electrocatalyst surface. Moreover, when ethanol solution is supplied on the anode side and the linear sweep voltammetry is registered, it is possible to determine the onset potential of the ethanol oxidation reaction in the electrocatalysts. The onset potential is determined where the current of ethanol oxidation is 5% above the current registered for cyclic voltammetry (without ethanol, capacitive current). The onset potential is a very important characteristic for electrochemical cells, with the lower the onset potential, the higher is the potential difference in a fuel cell. The values of the onset potential for different PtSn materials are shown in Table 3.

Table 3. Onset potential of ethanol oxidation reaction in PtSn materials with different Sn contents

Material	Onset potential/mV vs. RHE
Pt	500
Pt_3Sn E-TEK	410
Pt_4Sn	360
Pt_3Sn	300
Pt_2Sn	280

Direct Ethanol Fuel Cell

The performance of the direct ethanol fuel cell was studied with gas diffusion electrodes, with the cathode prepared with 20 wt % Pt/C from E-TEK and the anode prepared with a PtSn electrocatalyst. All electrodes were prepared at the same loading, 1 mg Pt cm^{-2}, and the electrodes have a geometric area of 4.62 cm^2. The electrolyte used was a Nafion® 115 membrane, and the membrane electrode assembly was prepared through a hot press with 50 kg cm^{-2}.

The amounts of tin in the Pt lattice in the anode catalysts modify the maximum power density through DEFC; such modification in the DEFC performance can be attributed to the properties of PtSn previously reported. Figure 8 shows the current density vs. cell potential, and the inset shows power densities vs. cell potential curves with different tin contents in the anode material. Figure 8 demonstrates that an increase in Sn increases the cell performance and provides better performance with the use of commercial Pt_3Sn in the anode.

The fuel cell curves indicated improved performance with an increased amount of tin in the anode electrode. These results may be associated with the action of Sn modifying the Pt electronic structure and contributing to ethanolic species adsorbed by electrocatalyst through a bifunctional mechanism along with having acetic acid and acetaldehyde as the main reaction products [50]. Further studies are required to achieve an electrocatalyst able to lead ethanol to a full oxidation (i.e., $CO_2 + H_2O$) with high reaction kinetics [51, 52].

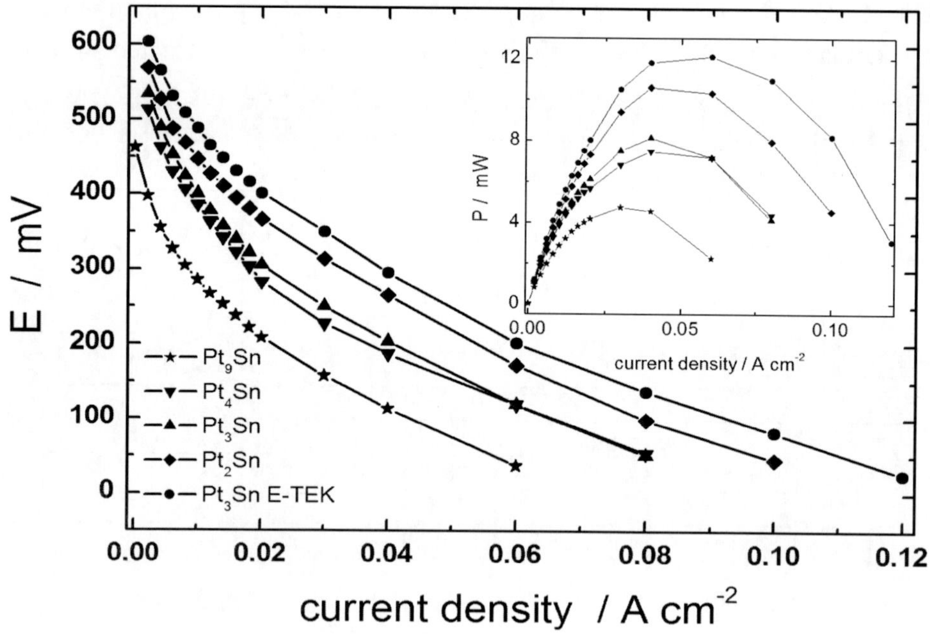

Figure 8. Current density vs. cell potential curves from a direct ethanol fuel cell operating at 90°C with the anode supplied with ethanol 1 mol L^{-1} and the cathode supplied with 3 atm of O_2; the electrodes have 1 mg Pt cm^{-2}, Pt/C E-TEK in the cathode and PtSn with different Sn contents in the anode.

Effect of Thermal Treatment on Pt_3Sn Material

As commercial material, Pt_3Sn E-TEK had the best performance for the direct ethanol fuel cell. A material with the same atomic ratio was selected to study the thermal treatment effect and the commercial catalyst was used for comparison.

The XRD data shown in Figure 5 enabled the observation of an increase in Sn content through Pt electrocatalyst with a shift in the peak position via formation of a solid solution. In addition, an improvement in cell performance, as shown in Figure 8, as well as a decrease in the onset potential of the ethanol oxidation reaction, illustrated in Figure 7, were observed.

Furthermore, the solid solution formation proved not very high with the preparation of the PtSn material. In addition to the experimental atomic ratio close to nominal atomic ratio, the tin present in the catalysts is not fully alloyed with platinum; it is partially in the oxide form [53, 54]. Thus, to improve the solid solution formation, the Pt_3Sn material was subjected to a thermal treatment at 200 and 500°C in a reductive atmosphere (H_2) to reduce tin oxides present in the electrocatalyst [39, 55] and increase the electrochemical activity regarding ethanol oxidation reaction.

Figure 9 shows the XRD of Pt_3Sn according to its preparation and subjected to thermal treatment; the XRD of Pt/C and Pt_3Sn/C commercial materials from E-TEK are also shown for comparison purposes. All materials have the same amounts of Pt and Sn and were subjected only to thermal treatment.

Figure 9 illustrates a shift from the peak position to lower 2θ degrees after the Pt_3Sn material was subject to a thermal treatment; this shift may be attributed to tin oxide reduction and formation of a solid solution with platinum. In addition, because all materials have the same amount of Sn, the Pt d-band was filled [55, 56]. The vertical black line is added to assist in the observation of the peak position shift. Moreover, a thermal treatment at very high temperatures (ca. 500°C) may also be observed with a segregation phase and other tin phases. The strong increase in crystal size is thought to correspond to both narrow and strong peaks [39, 55].

The analysis of this material was performed using the images obtained via transmission electron microscopy, as shown in Figure 10.

The transmission electron microscopy analysis of Pt_3Sn material indicates some very large particles, possibly because of clusters resulting from thermal treatment; a previous study provides further details on this process [55]. Briefly, the thermal treatment at 200°C results in a small alteration in the particle size (from 4 to 8 nm); however, the alloy degree was strongly increased, as verified in the XRD analysis [55].

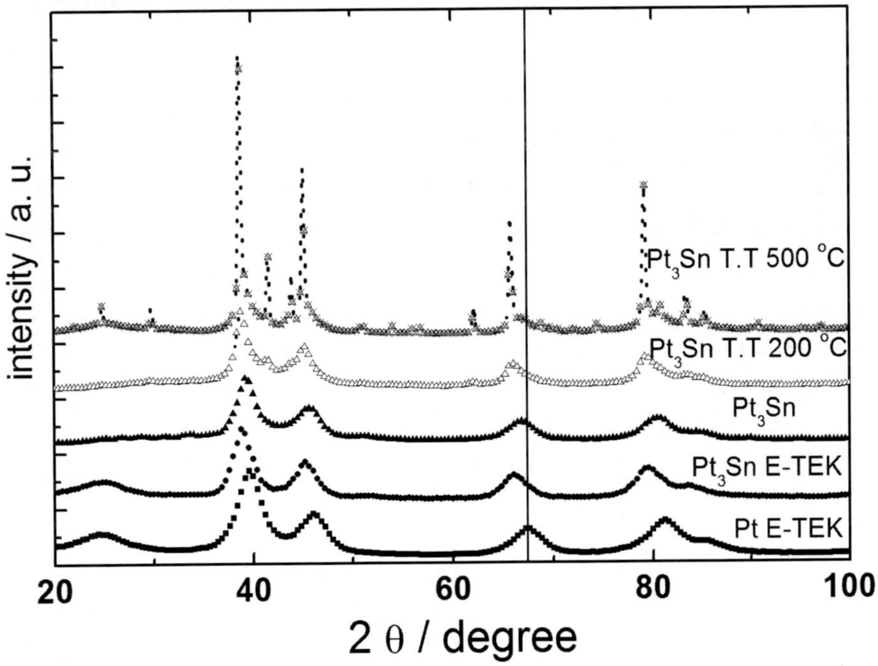

Figure 9. XRD peak patterns of carbon supported 20 wt % Pt_3Sn subjected to thermal treatment in a reductive atmosphere at different temperatures. * Commercial material from E-TEK co, not subjected to any treatment.

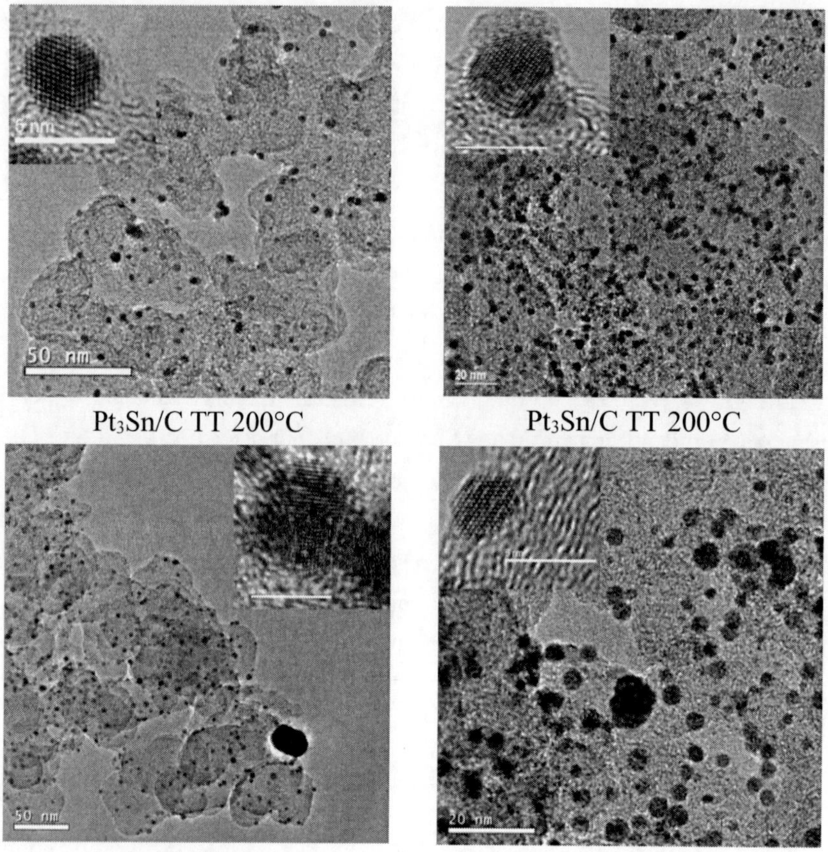

Pt$_3$Sn/C TT 200°C   Pt$_3$Sn/C TT 200°C

Pt$_3$Sn/C TT 500°C   Pt$_3$Sn/C TT 500°C

Figure 10. TEM images of Pt$_3$Sn material subjected to thermal treatment in a reductive atmosphere at different temperatures.

Figure 11 shows the histograms of the treated material, revealing an increase in the particle size, with particles larger than 70 nm for material treated at 500°C. Once more, two Gaussian adjustments occurred: one with low size frequencies and another with high size frequencies; however, the maximum size for both adjustments is almost the same and thus was assumed as the average particle size, as shown in the histogram figures. Pt$_3$Sn TT at 500°C presents particles larger than 30 nm that are not presented in the histograms but are considered in the Gaussian adjustment.

Pt$_3$Sn material subjected to a thermal treatment also shows different electrochemical characteristics. The following will describe how the solid solution degree increases the electrochemical characteristic.

Figure 12 illustrates an example of cyclic voltammetry and linear sweep voltammetry data of Pt$_3$Sn according to its preparation and subjected to a thermal treatment at different temperatures.

Figure 13 illustrates the linear sweep voltammetry of Pt$_3$Sn subjected to a thermal treatment. Thermal catalysts treated at 200°C had a high current density compared with

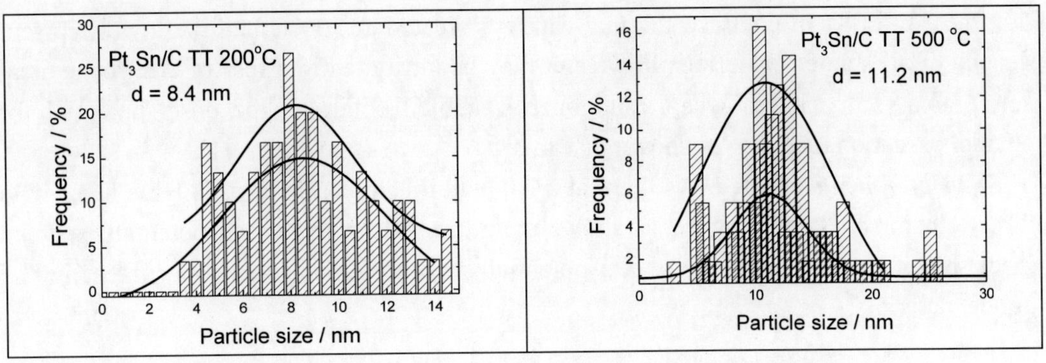

Figure 11. Histogram of Pt_3Sn particle size subjected to thermal treatment at 200 and 500°C in an H_2 atmosphere.

Table 4. Onset potential of Pt_3Sn subjected to thermal treatment at different temperatures

Material	Onset potential/mV vs. RHE
PtSn TT 500 °C	300
PtSn TT 200 °C	200
PtSn as prepared	200
*PtSn E-TEK	260

* Commercial material was not subjected to any treatment but used as a reference material.

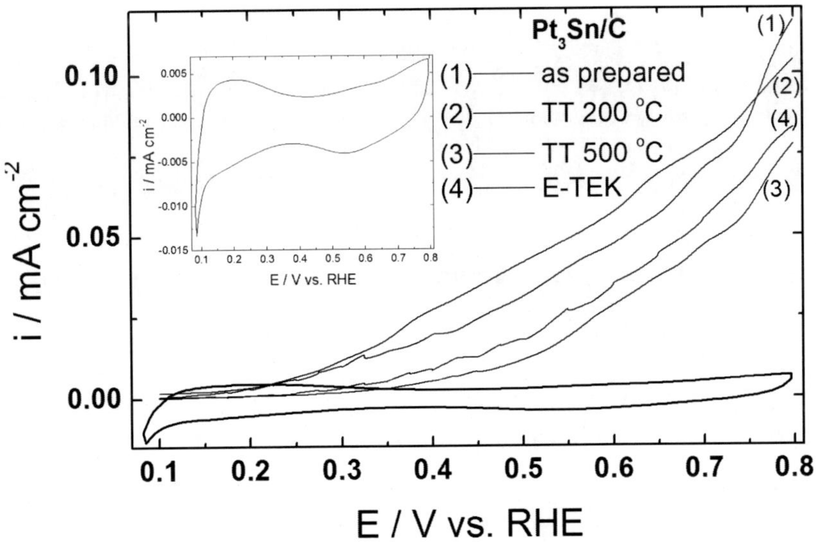

Figure 13. Cyclic voltammetry of the Pt_3Sn E-TEK (as a baseline) and linear sweep voltammetry of the Pt_3Sn materials subjected to thermal treatment. Cyclic voltammetry registered with the diffusion of gas electrodes, anode supplied with N_2 and cathode (dynamic reference) supplied with H_2. Linear sweep voltammetry anode supplied with ethanol solution and cathode (dynamic reference) supplied with H_2(g), in an experiment carried out at 90°C. Inset: cyclic voltammetry of Pt_3Sn E-TEK.

Pt_3Sn E-TEK. In contrast, thermal catalysts treated at 500°C had the lowest current density of all studied materials; this result may be attributed to a loss of electrochemical active area [55]. Linear sweep voltammetry enabled obtaining the onset potential for ethanol oxidation reaction, as shown in Table 4.

Finally, when subjected to a thermal treatment, this material was available in a single direct ethanol fuel cell. Figure 14 shows the current density vs. cell potential, and the inset shows power densities vs. cell potential curves with different tin content in the anode material.

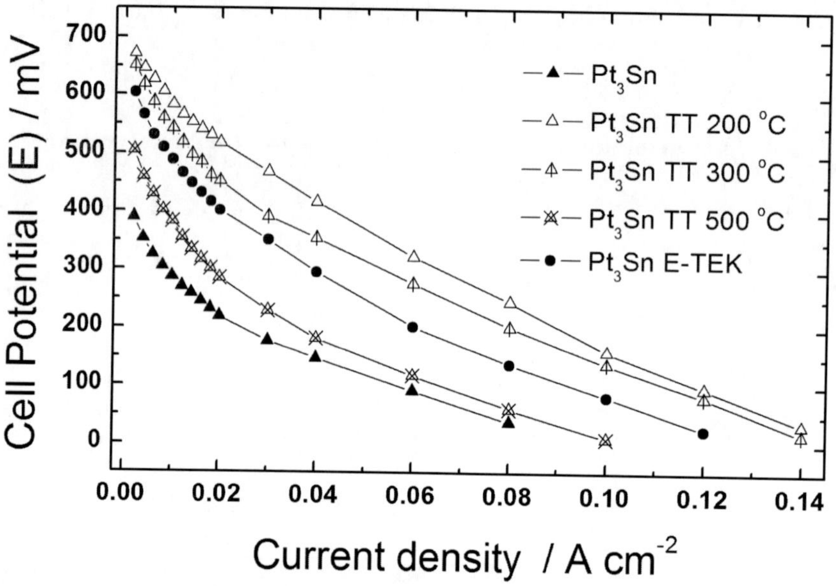

Figure 14. Current density vs. cell potential of the Pt_3Sn subjected to thermal treatment at different temperatures. Commercial material was not subjected to any treatment, as it was used as the reference material.

Figure 14 demonstrates that the higher the alloy formation, the greater the cell performance, as well as a better performance observed when commercial Pt_3Sn was subjected to 200°C. Above this temperature, nanoparticle clusters are formed that reduce the electrochemically active surface area and fuel cell performance. Thermal treatment in a reductive atmosphere increases the electrochemical activity of Pt_3Sn toward ethanol oxidation reaction in a direct ethanol fuel cell; the main products are acetaldehyde and acetic acid.

There are other studies to enhance the ethanol oxidation reaction with trimetallic materials for the anode. Magalhães and Colmati [57] showed that PtSnCu reduces the onset potential. Antolini et al. [58-60] reported PtSnPd [58] and PtSnRh [59, 60] as the anode in a direct ethanol fuel cell, but the catalysts for the anode in a DEFC were not sufficiently developed.

CONCLUSION

Ethanol electrooxidation reaction is a very important electrochemical energy conversion reaction that is applied in direct ethanol fuel cell systems; however, electrocatalysts are currently not able to convert ethanol into carbon dioxide.

PtSn has currently proved to have the best cell performance for catalysts of ethanol oxidation reaction and the best fuel cell performance; however, the major products of the reaction are acetaldehyde and acetic acid (main product).

The increase in solid solution between Pt and Sn improves the fuel cell performance; moreover, tin oxides on electrocatalyst are active and contribute to the electrocatalysis of the ethanol oxidation reaction.

It is important to conduct many further studies to elucidate the ethanol oxidation reaction based on the platinum surface and to develop electrocatalysts able to fully oxidize ethanol and with high reaction kinetics.

REFERENCES

[1] Renau, J., Barroso, J., Lozano, A., Nueno, A., Sánchez, F., Martín, J., Barreras, F. (2016). Design and manufacture of a high-temperature PEMFC and its cooling system to power a lightweight UAV for a high altitude mission, *International Journal of Hydrogen Energy 41 (43):*19702-19712.

[2] Jeon, S. W., Cha, D., Kim, H. S., Kim, Y. (2016). Analysis of the system efficiency of an intermediate temperature proton exchange membrane fuel cell at elevated temperature and relative humidity conditions, *Applied Energy 166:*165-173.

[3] Rosli, R. E., Sulong, A. B., Daud, W. R. W., Zulkifley, M. A., Husaini, T., Rosli, M. I., Majlan, E. H., Haque, M. A. (2016). A review of high-temperature proton exchange membrane fuel cell (HT-PEMFC) system, *International Journal of Hydrogen Energy In Press.*

[4] Decoopman, B, Vincent, R., Rosini, S., Paganelli, G., Thivel, P.-X. (2016). Proton exchange membrane fuel cell reversible performance loss induced by carbon monoxide produced during operation, *Journal of Power Sources 324:*492-498.

[5] Kamarudin, M. Z. F., Kamarudin, S. K., Masdar, M. S., Daud, W. R. W. (2013). Review: Direct ethanol fuel cells, *International Journal of Hydrogen Energy 38:*9438-9453.

[6] Badwal, S. P. S., Giddey, S., Kulkarni, A., Goel, J., Basu, S. (2015). Direct ethanol fuel cells for transport and stationary applications - A comprehensive review, *Applied Energy 145:* 80-103.

[7] Vigier, F., Coutanceau, F., Hahn, F., Belgsir, E. M., Lamy, C. (2004). On the mechanism of ethanol electro-oxidation on Pt and PtSn catalysts: electrochemical

and *In situ* IR reflectance spectroscopy studies, *Journal of Electroanalytical Chemistry 563:*81-89.

[8] Jiang, L., Zhou, Z., Li, W., Zhou, W., Song, S., Li, H., Sun, G., Xin, Q. (2004). Effects of treatment in different atmosphere on Pt_3Sn/C electrocatalysts for ethanol electro-oxidation, *Energy & Fuels 18:*866-871.

[9] Wang, Q, Sun, G. Q., Jiang, L. H., Xin, Q., Sun, S. G., Jiang, Y. X., Chen, S. P., Jusys, Z., Behm, R. J. (2007). Adsorption and oxidation of ethanol on colloid-based Pt/C, PtRu/C and Pt_3Sn/C catalysts: *In situ* FTIR spectroscopy and on-line DEMS studies, *Physical Chemistry Chemical Physics 9:*2686-2696.

[10] Soares, L. A., Purgato, F. L. S., Morais, C. Napporn, T. W. Kokoh, K. B., Olivi, P. (2016). Effect of Acetic Acid on Carbon Monoxide Electrooxidation over Tin Oxide and Rhodium-Modified Platinum Electrode Materials, *Electrocatalysis 8:*1-5.

[11] Beyhan, S., Léger, J-M., Kadırgan, F. (2014). *In situ* FTIR investigation of acetic acid electrooxidation on carbon supported Pt-Sn based trimetallic catalysts: Influence of the nature of the third metal, *Applied Surface Science 321:*426-431.

[12] Figueiredo, M. C., Santasalo-Aarnio, A., Vidal-Iglesias, J. Solla-Gullón F, Feliu, J. M. Kontturi, K., Kallio, T. (2013). Tailoring properties of platinum supported catalysts by irreversible adsorbed adatoms toward ethanol oxidation for direct ethanol fuel cells, *Applied Catalysis: B-Environmental 140-141:*378-385.

[13] Rocha, T. A., Linares, J. J., Colmati, F., Ciapina, E. G., Gonzalez, E. R. (2012). Electrocatalytic activity of platinum-niobium nanoparticles for ethanol oxidation, *Journal of the Electrochemical Society 159:*F650-F658.

[14] Rizo, R., Lázaro, M. J., Pastor, E., Koper, M. T. M. (2016). Ethanol oxidation on Sn-modified Pt single-crystal electrodes: new mechanistic insights from on-line electrochemical mass spectro-scopy, *Chem Electro Chem 3:*2196-2201.

[15] Rousseau, S., Coutanceau, C., Lamy, C., Léger, J-M., (2006). Direct ethanol fuel cell (DEFC): electrical performances and reaction products distribution under operating conditions with different platinum-based anodes, *Journal of Power Sources 158:*18-24.

[16] Hwang, R. Y., Byun, Y. S., Han, O. H. (2016). Electrochemical phenomena in direct methanol fuel cells investigated by multiple NMR analyses of cathode and anode exhausts, *Journal of Electroanalytical Chemistry 775:*320-324.

[17] Antolini, E. (2009). Palladium in fuel cell catalysis, *Energy & Environmental Science 2:*915-931.

[18] Geraldes, A. N., da Silva, D. F., da Silva, J. C. M., de Sá, O. A., Spinacé, E. V., Neto, A. O., dos Santos, M. C. (2015). Palladium and palladium-tin supported on multi wall carbon nanotubes or carbon for alkaline direct ethanol fuel cell, *Journal of Power Sources, 275:*189-199.

[19] Gasteiger, H. A., Markovic, N. M., Ross-Jr, P. N. (1995). H2 and CO Electrooxidation on Well-Characterized Pt, Ru, and Pt-Ru. 1. Rotating Disk Electrode Studies of the Pure Gases Including Temperature Effects, *The Journal of Physical Chemistry 99:* 8292-8301.

[20] Igarashi, H., Fujino, T., Zhu, Y., Uchida, H., Watanabe, M. (2001). CO tolerance of Pt alloy electrocatalysts for polymer electrolyte fuel cells and the detoxification mechanism, *Physical Chemistry Chemistry Physics 3:* 306-314.

[21] Delime, F., Léger, J-M., Lamy, C. (1999). Enhancement of the electrooxidation of ethanol on a Pt-PEM electrode modified by tin. Part I: Half-cell study, *Journal of Applied Electrochemistry 29 (11):*1249-1254.

[22] Lima, F. H. B., Gonzalez, E. R. (2008). Ethanol electro-oxidation on carbon-supported Pt-Ru, Pt-Rh and Pt-Ru-Rh nanoparticles, *Electrochimica Acta 53(6):*2963-2971.

[23] Santiago, E. I., Camara, G. A., Ticianelli, E. A. (2003). CO tolerance on PtMo/C electrocatalysts prepared by the formic acid method, *Electrochimica Acta 48:*3527-3534.

[24] Fiala, R., Vaclavu, M., Rednyk, A., Khalakhan, I., Vorokhta, M., Lavkova, J., Potin, V., Matolinova, I., Matolin, V. (2015). Pt-CeOx thin film catalysts for PEMFC, *Catalysis Today 240:*236-241.

[25] Boucher, H., Shevchenki, E. V., Robert, A., Mekis, I., Kornowski, A., Grübel, G., Weller, H. (2005). Determination of nanocrystal sizes: A comparison of TEM, SAXS and DRX studies of highly monodisperse $CoPt_3$ Particles, *Langmuir 21:*1931-1936.

[26] Queiroz, A. C., Silva, W. O., Rodrigues, I. A., Lima, F. H. B. (2014). Identification of bimetallic electrocatalysts for ethanol and acetaldehyde oxidation: Probing C2-pathway and activity for hydrogen oxidation for indirect hydrogen fuel cells, *Applied Catalysis B: Environmental 160-161:*423-435.

[27] Akhairi, M. A. F., Kamarudin, S. K. (2016). Catalysts in direct ethanol fuel cell (DEFC): An overview, *International Journal of Hydrogen Energy 41(7):*4214-4228.

[28] dos Reis, R. G. C. S., Colmati, F. (2016). Electrochemical alcohol oxidation: a comparative study of the behaviour of methanol, ethanol, propanol, and butanol on carbon-supported PtSn, PtCu and Pt nanoparticles, *Journal of Solid State Electrochemistry 20:*2559-2567.

[29] Cantane, D. A., Ambrosio, W. F., Chatenet, M., Lima, F. H. B. (2012). Electro-oxidation of ethanol on Pt/C, Rh/C, and Pt/Rh/C-based electrocatalysts investigated by on-line DEMS, *Journal of Electroanalytical Chemistry 681:*56-65.

[30] Antolini, E. (2007). Catalysts for direct ethanol fuel cells, *Journal of Power Sources, 170 (1):*1-12.

[31] Friedl, J., Stimming, U. (2013). Model catalyst studies on hydrogen and ethanol oxidation for fuel cells, *Electrochimica Acta 101:*41- 58.

[32] Pinheiro, A. L. N., Oliveira-Neto, A., Souza, E. C., Perez, J., Paganin, V. A., Ticianelli, E. A., Gonzalez, E. R. (2003). Electrocaatalysis on noble metal and noble metal alloys dispersed on high surface area carbon, *Journal of New Materials for Electrchemical Systems 6:*1-8.

[33] Hidayati, N., Scott, K. (2016) Electro-oxidation of Ethanol on Carbon Supported PtSn and PtSnNi Catalysts, *Bulletin of Chemical Reaction Engineering & Catalysis 11(1):*10-20.

[34] Akhairi, M. A. F., Kamarudin, S. K. (2016). Catalysts in direct ethanol fuel cell (DEFC): An overview, *International Journal of Hydrogen Energy 41:*4214-4228.

[35] González-Quijano, D., Pech-Rodríguez, W. J., Escalante-García, J. I., Vargas-Gutiérrez, G., Rodríguez-Varela, F. J. Highly Active Pt-Sn/C. (2014). Catalysts for Ethanol Electro-Oxidation Prepared by a Polyol-Alcohol Reduction Process, *ECS Transactions 61 (29):*1-9.

[36] Gupta, C., Maheshwari, P. H., Dhakate, S. R. (2015). Development of multiwalled carbon nanotubes platinum nanocomposite as efficient PEM fuel cell catalyst, *Mater Renew Sustain Energy (2016) 5(2)* 1 -11.

[37] Pinithchaisakula, A., Themsirimongkon, S., Promsawan, N., Weankeaw, P., Ounnunkad, K., Saipanya, S. (2017). An Investigation of a Polydopamine-Graphene Oxide Composite as a Support for an Anode Fuel Cell Catalyst, *Electrocatalysis 8:*36-45.

[38] Colmati, F., Antolini, E., Gonzalez, E. R. (2005). Pt-Sn electrocatalysts for methanol oxidation synthesized by reduction with formic acid. *Electrochimica Acta 50:*5496-5503.

[39] Colmati, F., Antolini, E., Gonzalez, E. R. (2008). Effect of thermal treatment on phase composition and ethanol oxidation activity of a carbon supported Pt50Sn50 alloy catalysts. *Journal of Solid State Electrochemistry 12:*591-599.

[40] Ishitobi, H., Ino, Y., Nakagawa, N. (2017). Anode catalysts with enhanced ethanol electrooxidation activity by effective interaction between Pt-Sn-SiO2 for a direct ethanol fuel cell. *International Journal of Hydrogen Energy* 42:26897-26904.

[41] Mukerjee, S., McBree, J. (1999). An *In situ* X-ray adsoption spectroscopy investigation of the effect of Sn additions to carbon-supported Pt electrocatalysts. Part I, *Journal of the Electrochemical Society, 146 (2):* 600-606.

[42] Kheradmandinia, S., Khandan, N., Eikani, M. H. (2016). Synthesis and evaluation of CO electro-oxidation activity of carbon supported SnO2, CoO and Ni nano catalysts for a PEM fuel cell anode, *International Journal of Hydrogen Energy 41(42):* 19070-19080.

[43] Antolini, E. (2011). An empirical model to evaluate the contribution of alloyed and non-alloyed tin to the ethanol oxidation reaction on Pt-Sn/C catalysts based on the

presence of SnO_2 and a $Pt_{(1-x)}Sn_x$ solid solution: Application to DEFC performance, *International Journal of Hydrogen Energy 36:* 11043-11047.

[44] Mansour, A. N., Cook, J. W., Sayers, D. E. (1984). Quantitative technique for the determination of the number of unoccupled d-electron states in a platinum catalyst using the L2,3 X-ray absorption edge spectra, *Journal of Physical Chemistry 88:*2330-2334.

[45] Shukla, A. K., Raman, R. K., Choudhury, N. A., Priolkar, K. R., Sarede, P. R., Emura, S., Kumashiro, R. (2004). Carbon-supported Pt-Fe alloy as a methanol-resistant oxygen-reduction catalyst for direct methanol fuel cell, *Journal of Electroanalytical Chemistry 563*:181-190.

[46] Sousa Jr. R., Colmati, F., Ciapina, E. G., Conzalez, E. R. (2007). An analysis of X-ray absorption spectra in the XANES region of platinum-based electrocatalysts for low-temperature fuel cell, *Journal of Solid State Electrochemistry 11:* 1549-1557.

[47] Colmati, F., Antolini, E., Gonzalez, E. R. (2007). Ethanol oxidation on carbon supported Pt-Sn electrocatalysts prepared by reduction with formic acid, *Journal of the Electrochemical Society 154:*B39-B47.

[48] Gonzalez, E. R., Colmati, F., Antolini, E. (2006). Ethanol Oxidation on Pt-Sn Electrocatalysts Supported on Carbon Prepared by Reduction with Formic Acid, *ECS Transactions 3 (1):* 1307-1316.

[49] Küver, A., Vogel, I., Vielstich, W. (1994). Distinct performance evaluation of a direct methanol SPE fuel cell. A new method using a dynamic hydrogen reference electrode, *J Power Sources 52 (1):* 77-80.

[50] Sánchez-Monreal, J., García-Salaberri, P. A., Vera, M. (2017). A genetically optimized kinetic model for ethanol electro-oxidation on Pt-based binary catalysts used in direct ethanol fuel cells, *Journal of Power Sources 363:*341-355.

[51] Antoniassi, R. M., Silva, J. C. M., Oliveira Neto, A., Spinacé, E. V. (2017). Synthesis of Pt+SnO2/C electrocatalysts containing Pt nanoparticles with preferential (100) orientation for direct ethanol fuel cell, *Applied Catalysis B: Environmental 218:*91-100.

[52] Song, S., Wang, Y., Shen, P. (2007). Thermodynamic and kinetic considerations for ethanol electrooxidation in direct ethanol fuel cells, *Chinese Journal of Catalysis 28:*752-754.

[53] Cordeiro, G. L., Ussui, V., Messias, N. A., Piasentin, R. M. De Lima, N. B., Neto, A. O., Lazar, D. R. R. (2017). Effect of Sn loading on the characteristics of Pt electrocatalysts supported on reduced graphene oxide for application as direct ethanol fuel cell anode, *International Journal of Electrochemical Science 12:*3795-3813.

[54] Rizo, R., Sebastián, D., Lázaro, M. J., Pastor, E. (2017). On the design of Pt-Sn efficient catalyst for carbon monoxide and ethanol oxidation in acid and alkaline media, *Applied Catalysis B: Environmental 200:*246-254.

[55] Colmati, F., Antolini, E., Gonzalez, E. R. (2007). Etanol oxidation on carbon-supported $Pt_{75}Sn_{25}$ electrocatalysts prepared by reduction with formic acid: effect of thermal treatment, *Applied Catalysis B: Environmental 73:* 106-115.

[56] Colmati, F. E., Gonzalez, E. R. (2007). Electronic effects in Pt-Sn alloy as anodes in direct ethanol fuel cell, *ECS Transactions 11(1):* 1425-1435.

[57] Magalhães, M. M., Colmati, F. (2014). Carbon-Supported PtSnCu, PtCu and PtSn electrocatalysts for ethanol oxidation in acid media, *Journal of the Brazilian Chemical Society 25:*1317-1325.

[58] Antolini, E., Colmati, F., Gonzalez, E. R. (2009). Ethanol oxidation on carbon supported (PtSn)alloy/SnO2 and (PtSnPd)alloy/SnO2 catalysts with a fixed Pt/SnO2 ratio: Effect of the alloy phase characteristics, *Journal of Power Sources 193:*555-561.

[59] Colmati, F., Antolini, E., Gonzalez, E. R. (2008). Preparation, structural characterization and activity for ethanol oxidation of carbon supported ternaty Pt-Sn-Rh catalysts, *Journal of Alloys and Compounds 456:*264-270.

[60] de Souza, E. A., Giz, M. J., Camara, G. A., Antolini, E., Passos, R. R. (2014). Ethanol electro-oxidation on partially alloyed Pt-Sn-Rh/C catalysts, *Electrochimica Acta 147:*483-489.

In: Advanced Materials and Systems … ISBN: 978-1-53613-852-8
Editor: Mauro Coelho dos Santos

Chapter 2

ELECTROCATALYSTS FOR DIRECT ETHANOL FUEL CELLS

Júlio César Martins da Silva, Thiago Lopes, Almir Oliveira Neto and Estevam Vitorio Spinacé
Fuel Cell and Hydrogen Center, Nuclear and Energy Research Institute, IPEN-CNEN/SP, São Paulo, Brazil

ABSTRACT

In this chapter, an overview of electro-catalysts for direct ethanol fuel cells based on polymer electrolyte membranes is presented, with special attention given to the alkaline anion exchange membrane-based devices. Considering the anodic reaction, it is evidenced that a substantial effort has been dedicated towards increasing the fuel cell efficiency by developing electrocatalysts with higher catalytic activity. In this sense, palladium is pointed out as a promising substitute to platinum with regards to electrocatalytic activity, price and element availability. Furthermore, palladium-based nanoparticles are pointed out as the electrocatalysts with the highest electrocatalytic activity for ethanol electro-oxidation in alkaline media. Taking into account the support materials for the electrocatalysts nanoparticles, it was possible to observe an evolution in carbon-based materials, first carbon nanotubes overcame the carbon black, where more recently, graphene is addressed as the best support choice for catalysts nanoparticles. Different efforts have been made to improve the properties of the support materials, such as doping with nitrogen or with metal oxides like TiO_2, for example. Considering the cathode materials, the main goal is to develop catalysts with enhanced activity to the oxygen reduction reaction as well as tolerance to ethanol molecules that crossover the electrolyte membrane.

Keywords: electrocatalysts, ethanol electro-oxidation, oxygen reduction reaction, alkaline media, polymer electrolyte membranes fuel cell

1. INTRODUCTION

Fuel cells are devices which convert chemical energy into electrical energy with high efficiency and low emission of pollutants [1]. Among different kinds of fuel cells, alkaline direct fuel cells (ADFC) have been attracting a great attention since the development of alkaline anion exchange membranes (AAEMs) [2-7]. The AAEMs solve the problem related to the carbonate precipitation on the traditional alkaline fuel cells which uses a KOH solution as electrolyte [2, 4, 5]. It is reported that alkaline anion exchange membrane fuel cell (AAEMFC) has some advantages over proton exchange membrane fuel cell (PEMFC), such as faster kinetics of oxygen reduction reaction and less corrosive environment which allows the utilization of less expensive electrocatalysts than in acid media [1-3, 6].

The AAEMFCs are operated at low temperature, usually at 80 °C and the OH^- produced at the cathode are conducted through the polymeric membrane to the anode producing water, which reduces the anode dryness and the problem related to the crossover of the fuels, like ethanol for example [2, 7].

These kind of devices can be fed with different fuel molecules, such as methanol, ethanol, formic acid, etc. [8-10]. Among them, ethanol has been recognized as a promising candidate for AAEMFCs, since it can be produced directly from the fermentation of biomass, which makes it a renewable fuel, leaving the natural balance of carbon dioxide in the atmosphere unaltered [11-13]. Additionally, ethanol presents low toxicity, high energy density (8.0 kW kg^{-1}) and it is a low cost product [4, 5, 14-17]. A simplified scheme of the alkaline anion exchange membrane direct ethanol fuel cells (AAEMDEFCs) can be seen in Figure 1. In this scheme, it was considered the complete ethanol oxidation to CO_2 at the anode and the direct four-electron pathway for the oxygen reduction reaction at the cathode.

In Figure 1, the molecules of ethanol, water, carbon dioxide and hydroxide ions are not to scale, in fact the molecules are much smaller than the catalyst particle sizes, among these molecules, ethanol is the largest and present a diameter of 0.44 nm, while the catalyst nanoparticles usually range from 2 to 10 nm [18-20]. Due to the small size of ethanol and water molecules (~ 0.28 nm) [18], they can diffuse through the catalytic layer where the reactions take place. A zoom in the catalyst layer of the cathode, which ideally should have the same porosity of the anode, can be seen in the Figure 1. Equations 1 and 2 represent the electrochemical reactions at the anode and cathode, respectively [21]. It is important to point out that in this representation, it was considered the complete ethanol electro-oxidation to carbon dioxide.

$$\text{Anode: } H_3CCH_2OH + 12\ OH^- \rightarrow 2\ CO_2 + 9\ H_2O + 12\ e^- \quad (1)$$

$$\text{Cathode: } 3\ O_2 + 6\ H_2O + 12\ e^- \rightarrow 12\ OH^- \quad (2)$$

The complete oxidation of ethanol to CO_2 involves 12 electrons per ethanol molecule [22], however, to convert ethanol to CO_2, the C—C bond has to be cleaved, which is difficult at low temperatures [17, 23, 24]. As a consequence, acetaldehyde and acetic acid (acetate in alkaline media) are dominant products [17, 25], which represent a loss in the Faradic efficiency to the process [26, 27]. A simplified scheme of ethanol electro-oxidation is presented in the Figure 2.

The oxidation of ethanol to acetate involves an exchange of 4 electrons per ethanol molecule, while the ethanol oxidation to acetaldehyde involves only 2 electrons [14, 22, 28], which corresponds to a Faradic efficiency of 33.3% and 16.67%, respectively.

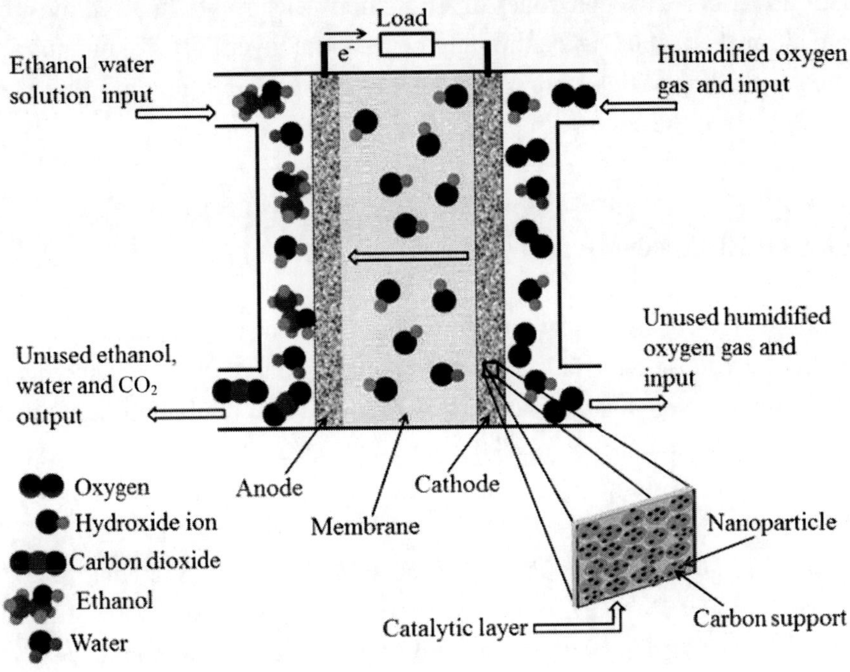

Figure 1. A simplified schematic view of a AAEMEDFC.

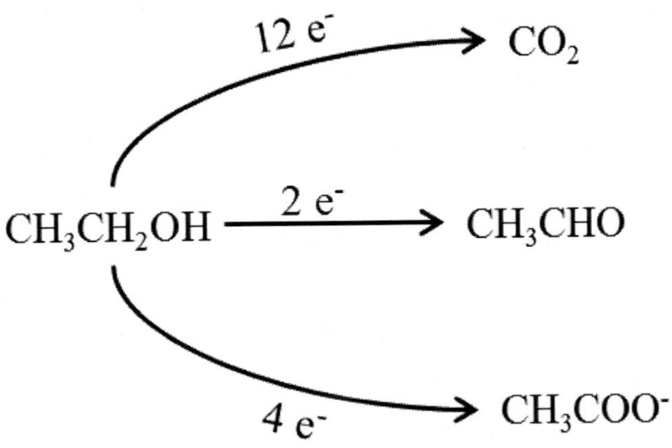

Figure 2. Scheme of ethanol electro-oxidation in alkaline media.

2. Catalysts for Ethanol Electro-Oxidation in Alkaline Media

Electrocatalysts play a key role in accelerating and directing the ethanol electro-oxidation. In this sense, many efforts and improvements have been made along the years. It is well known that catalyst nanoparticles increase the catalytic reaction rate compared to bulk materials [29]. Some properties of the materials can change with their size, as a result of different causes, like quantum confinement for example. Thus, nanomaterials can present different properties from the bulk materials [29, 30] and also a higher surface area [31]. Furthermore, the electrodes of a polymeric electrolyte fuel cell must be porous in order to allow the fuel diffusion through the catalytic layer [32]. Taking this aspect into account, catalyst nanoparticles are ideal and can be supported on cheap porous support materials (in most of the cases, carbon black).

2.1. Platinum Electrocatalysts

Considering the electro-oxidation of ethanol in acid media, platinum is the metal with the highest catalytic activity [13, 33], where platinum presents outstanding corrosion resistant in both acid and alkaline media. However, the cost of platinum is extremely high [34], which represents about 50% of the fuel cell costs [33]. Additionally, platinum is only found at 0.003-5 ppm in the earth's crust and it is considered the second most expensive metal after rhodium [35].

Considering the electro-oxidation of ethanol in alkaline media, platinum and platinum-based materials have been proposed as anode catalysts [8, 12, 36-38]. In contrast, as in acid media, platinum electrocatalyst does not promote the C—C bond cleavage with high efficiency [39]. The main products from ethanol electro-oxidation in alkaline media on Pt/C are acetaldehyde and acetate [39]. In addition, considering C—C bond splitting, carbon monoxide is an intermediate to the carbon dioxide formation, and platinum is rapidly poisoned by CO [15, 40]. In this sense, the catalyst surface is blocked by CO and further ethanol molecules cannot adsorb on catalytic sites to be oxidized. In Figure 3, a simplified scheme is given illustrating the block of the catalyst surface by CO.

In order to improve the catalytic activity of platinum toward ethanol electro-oxidation, binary e.g., PtSn [8, 37, 39] or multimetallic i.e., PtPdAu [41], Ru@PtPd/C [13] and platinum-based electrocatalysts have been proposed [8, 12, 36-38].

Rubén Rizo et al. [8] investigated the ethanol electro-oxidation in acid and alkaline media on PtSn electrocatalysts. From the author's results, it is possible to observe that the current density from ethanol electro-oxidation in alkaline media was higher than that in acid media. Baranova et al. [37] showed that bi-phase PtSn alloy + SnOx presents a higher catalytic activity toward ethanol electro-oxidation in alkaline media than PtSn alloy. The mean role of tin in the bimetallic PtSn catalysts is attributed to the bifunctional mechanism, where tin provides OH at a lower overpotential than platinum, favoring the oxidation of intermediate species, and also to the electronic effect, which consists in tailoring the electronic state of platinum atoms, weakening the strength of adsorbed intermediates during the ethanol electro-oxidation reaction [8, 37]. In Figure 4, it can be seen a simplified scheme that represents the CO adsorbed on the platinum nanoparticle surface and OH on SnO_x surface. In this way, SnO_x supplies oxygen to oxidize CO to CO_2 as represented in Equations 3 and 4 [42, 43].

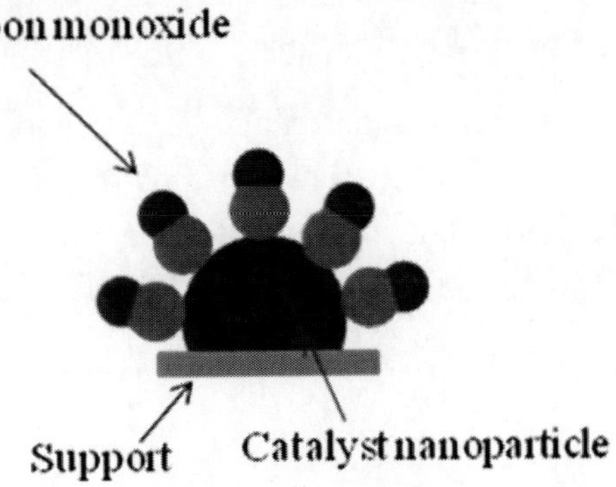

Figure 3. A simplified schematic of the nanoparticle with the surface blocked by CO.

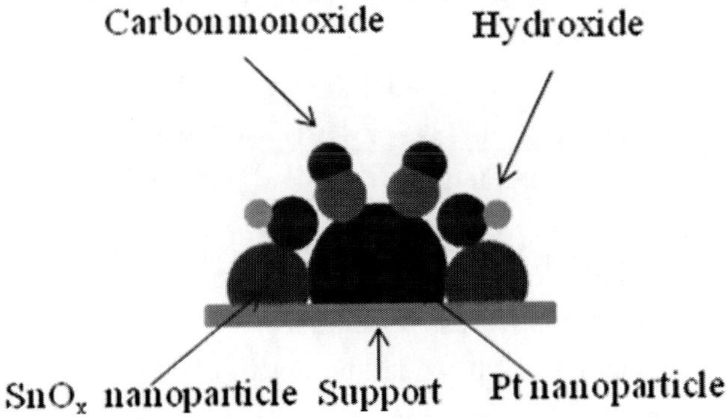

Figure 4. A simplified scheme of the platinum nanoparticle with the CO adsorbed on the surface and tin nanoparticle with OH adsorbed on the surface.

$$SnOx + H_2O \rightarrow SnOxOHads + H+ + e- \quad (3)$$

$$Pt—CO + SnOxOH \rightarrow Pt + SnOx + CO_2 + H+ + e- \quad (4)$$

Silva et al. [36] reported the results from an alkaline direct ethanol fuel cell using Pt/C, Au/C and PtAu/C at different atomic ratios (Pt:Au 90:10, 70:30 and 50:50). The Figure 5a corresponds to the maximum current density and Figure 5b corresponds to the maximum power density obtained during the experiments.

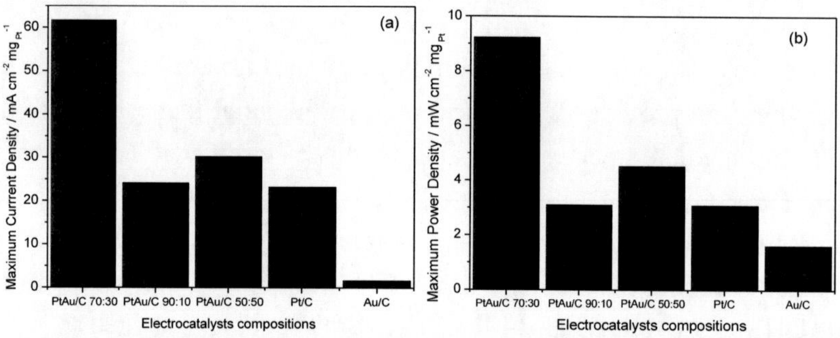

Figure 5. Maximum current density (a) and maximum power density curves (b) of a single direct ethanol fuel cell at operated 75 °C fed with 2 M ethanol in 2 M KOH. Anode and cathode 1 mg of Pt per cm^2 excepted for Au/C anode which contained 1 mg of Au per cm^2. Data from Ref. [36].

The best result was obtained using PtAu/C 70:30 as anode, as it can be seen in Figure 5. Showing that atomic ratio between platinum and gold is very important to enhance the current density from the fuel cell device fed with ethanol.

2.2. Palladium Electrocatalysts

As mentioned earlier, the high price and limited supply of platinum limits its application in large scale production of fuel cells. Palladium is considered an alternative substitute for platinum since it is more abundant and less expensive [3, 44, 45]. Considering the electro-oxidation of ethanol in alkaline media, palladium shows a higher catalytic activity than platinum [1, 4, 45-49]. However, to fulfill the commercial application of the AAMFCs, the activity of palladium for ethanol electro-oxidation in alkaline media needs to be enhanced [4, 46, 47].

It has been reported that acetate is the main product from ethanol electro-oxidation on palladium [1, 45, 50]. In other words, palladium is not effective to promote the C—C bond splitting. The ethanol electro-oxidation reaction pathway in alkaline media for acetate and acetaldehyde involves multiple steps, which are given in Equations 5 - 14 [1, 22, 51].

Pathway to acetate:

$$CH_3CH_2OHsol \rightarrow CH_3CH_2OHad \quad (5)$$

$$OH^- \rightarrow OHad + e^- \quad (6)$$

$$CH_3CH_2OHad + 3\ OHad \rightarrow CH_3COad + 3\ H_2O + 3\ e^- \quad (7)$$

$$CH_3COad + OHad \rightarrow CH_3COOH \quad (8)$$

$$CH_3COOH + OH^- \rightarrow CH_3COO^- + H_2O \tag{9}$$

Pathway to acetaldehyde:

$$2\,OH^- \rightarrow 2\,OHad + 2\,e^- \tag{10}$$

$$CH_3CH_2OH + OHad \rightarrow CH_3CH_2O\bullet + H_2O \tag{11}$$

$$CH_3CH_2O\bullet + OHad \rightarrow CH_3CHO + H_2O \tag{12}$$

However, it is reported that the electro-oxidation of acetaldehyde is a main reaction to form acetate in alkaline media (steps 13 and 14).

$$CH_3CHO + 2\,OH^- \rightarrow CH_3COOH + H_2O + 2\,e- \tag{13}$$

$$CH_3COOH + OH^- \rightarrow CH_3COO^- + H_2O \tag{14}$$

The pathway of ethanol electro-oxidation to form CO_2 involves multiples steps that are not clear in the literature [22].

To improve the catalytic activity of palladium toward ethanol electro-oxidation, different elements have been used to produce bimetallic or multimetallic Pd-based electrocatalysts, such as, PdTb/C, PdCuPb/C, PdAg/C, PdTe/C, PdPb/C, PdSb/C, PdRu/C, PdSn/C and PdAu/C [1, 4, 6, 46, 47, 52-57]. The palladium electronic structure can be modified by the introduction of another element in the catalyst. Thus, palladium alloys could result in a shift of the d-band center of palladium, which modifies the adsorption capacity of adsorbates onto the catalyst surface, which may enhance the catalytic activity toward ethanol electro-oxidation [54, 58].

Among different elements used to prepare Pd-based materials, gold has attracted particular attention, probably due to its reactivity for CO and hydrocarbons electro-oxidation in alkaline media. Additionally, gold promotes the modification of palladium electronic configuration by withdrawing electrons, which stabilize the catalyst during the ethanol electro-oxidation [56, 57].

Taking into account the catalyst composition, it is well known that the atomic ratio of the elements in the catalysts have a profound influence in the catalytic activity. Xu et al. [59] reported that among different Pd:Au composition (Pd_7Au/C, Pd_3Au/C and PdAu/C) Pd_3Au/C shows the best catalytic activity. Also, Qin et al. [56] showed that PtAu with low alloy degree presents a higher catalytic activity than PtAu with high alloy degree. As it can be seen, besides the composition, the crystalline structure of the catalysts also promotes influences on the catalytic activity. Additionally, the morphology of the catalysts plays an important role in the catalytic activity of the materials [3, 60].

2.3. Support Materials for Nanoparticles Electrocatalysts

Support materials for catalysts nanoparticles play a very important role in the electro-oxidation of ethanol. A suitable support must attend some requirements, such as high electrical conductivity, large surface area and adequate stability in operating conditions [61]. Carbon black is a common choice as support for electrocatalysts nanoparticles in PEMFCs and AAEMFCs because of its large surface area, high electrical conductivity, pore structures and low cost [34, 62]. However, carbon blacks are usually prone to undergo corrosion under the operation conditions of PEMFC [63, 64]. In the last few years, carbon nanotubes have attracted significant attention as support for catalysts nanoparticles for ethanol electro-oxidation [65-67] because of the peculiar properties such as, high surface area, high electrical conductivity and stability. The effect of carbon nanotubes as a catalyst support can be different depending on the structure of the nanotube. Single wallet carbon nanotubes (SWCNT) (see Figure 6) are usually a mixture of semiconductor and metallic materials in terms of electrical conductivity, whereas multi wallet carbon nanotubes (MWCNT) was considered as a metallic electrical conductor [67]. Biuck et al. [66] reported that the catalytic activity of palladium nanoparticles towards ethanol electro-oxidation is influenced by the carbon support. They supported palladium nanoparticles on carbon black, multi wall carbon nanotubes (MWCNT) and single wall carbon nanotubes (SWCNT), and according to the authors, the Pd/SWCNT showed greater catalytic activity compared to the Pd/C and Pd/MWCNT.

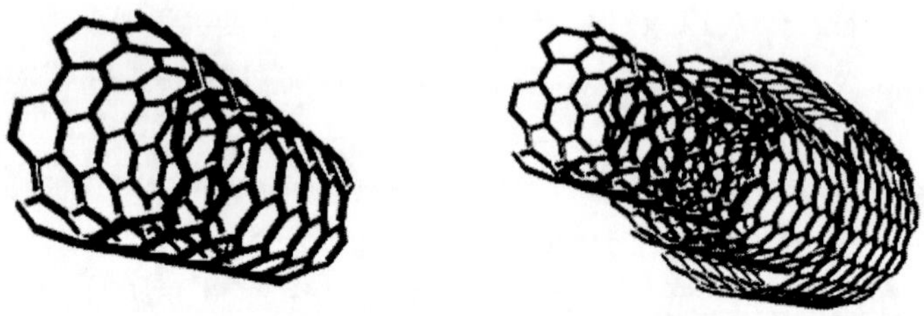

Figure 6. Schematic diagrams of single wall carbon nanotubes (left side) and multi-wall carbon nanotube (right side).

However, problems like, poor dispersibility, consequently high agglomeration, and also poor binding sites to anchor the precursor of the catalyst nanoparticles affects the catalytic activity of the materials supported on carbon nanotubes [68]. Thus, carbon nanotubes have been functionalized by introducing acid groups in order to improve the hydrophilicity and solve the problems related to the agglomerations of the nanotubes and also improve the nanoparticles dispersion [67, 68]. The functionalization of carbon nanotubes is usually performed by covalent and non-covalent functionalization. The covalent functionalization is usually carried out in HNO_3 and non-covalent performed by adsorption of functional groups onto the tubes [67].

Shuwen Li et al. [68] supported palladium nanoparticles on MWCNTs, MWCNTs covalent functionalized with nitric acid and non-covalent functionalized using ionic liquid molecules. According to the authors, the palladium nanoparticles supported on non-covalent functionalized MWCNTs showed higher catalytic activity toward ethanol electro-oxidation in alkaline media than the palladium nanoparticles supported on covalent functionalized MWCNTs and on MWCNTs. The superior catalytic activity was attributed to the high distribution, small main particle size and an excellent conductivity.

In the recently years, graphene (Figure 7) has been considered as a suitable option as support for catalyst nanoparticles due to some important properties such as large surface area, high stability, high electrical and thermal conductivity [69, 70]. The catalytic activity of palladium nanoparticles supported on reduced graphene oxide toward ethanol electro-oxidation in alkaline media is significantly improved compared to palladium nanoparticles supported on carbon black or carbon nanotubes [71].

In order to enhance the catalytic activity of the materials towards ethanol electro-oxidation, efforts have been devoted to the modification of graphene, such as the incorporation of nitrogen atoms to graphene, which promotes a positive effect of the nanoparticles dispersion and also the modification of the surface structure of the carbons, enhancing electron-donor of the carbon, which promotes the metal support interaction owing the enhancement of the catalytic activity toward ethanol electro-oxidation [72]. Peng Wu et al. [72] reported that the catalytic activity of palladium nanoparticles supported on nitrogen-doped graphene shows a greater catalytic activity than palladium nanoparticles supported on graphene and carbon black.

Many efforts have been made in order to improve the efficiency of the support materials towards ethanol electro-oxidation in alkaline media. In this sense, Qi Liu et al. [73] deposited manganese dioxide (MnO_2) on the graphene nanoribbons, therefore, the authors combined the excellent electronic conductivity of graphene nanoribbons and also the oxygen groups which facilitate the dispersion of the material in aqueous solvents with the MnO_2 that promotes the formation of hydroxide at low overpotential, which improves the tolerance of strongly adsorbed intermediates, like CO. Additionally, MnO_2 presents suitable electrochemical stability and it is of low-cost. According to the authors, palladium nanoparticles supported on graphene nanoribbons coated with MnO_2 showed a higher catalytic activity towards ethanol electro-oxidation than palladium nanoparticles supported on reduced graphene oxide, on multi wall carbon nanotube and on carbon black.

Metal oxides like titanium dioxide (TiO_2) have also been used as catalyst support in the electro-oxidation of ethanol in alkaline media [74]. TiO_2 is considered an appropriate choice as support material due to its high corrosion resistance, as well as, high electrochemical stability [61]. However, TiO_2 present low conductivity at low temperature (below 200°C), which is a disadvantage for fuel cell application [61]. In order to overcome this drawback, efforts have been made, for instance, Yuan-Hang Qin

[74] has used carbon-doped TiO_2 as support for palladium nanoparticles and reported that Pd/TiO_2-C showed higher catalytic activity and stability towards ethanol electro-oxidation in alkaline media than Pd/C.

Other strategy to mitigate the problem related to the low electronic conductivity of TiO_2 is doping it with a donor-type element [61, 75]. Son Truong Nguyen et al. [61] have synthesized Nb-doped TiO_2 because niobium is a suitable donor dopant for TiO_2. According to the authors, the addition of niobium into TiO_2 structure improved the electronic conductivity of TiO_2. They reported that PdAg nanoparticles supported on Nb-TiO_2 showed a higher catalytic activity and durability towards ethanol electro-oxidation in alkaline media than the PdAg supported on carbon and PdAg supported on TiO_2.

Besides TiO_2, other elements, like nickel, have been used as material support. Maksudul Hasan et al. [76] have synthesized nickel nanowires to be used as support for palladium nanoparticles with different morphologies. According to the authors, the material could be used as an alternative to carbon nanotubes.

As it can be seen, there are different catalysts materials with diverse composition and morphology and also support materials proposed in the literature related to the ethanol electro-oxidation in alkaline media. Certainly, there are excellent works that were not mentioned in this chapter. However, the intention was to present an overview of the electrocatalysts for ethanol electro-oxidation in alkaline media.

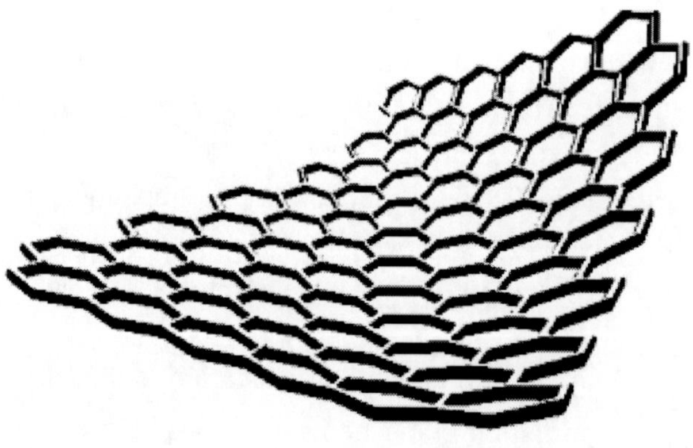

Figure 7. Schematic diagram of graphene sheet.

It is important to point out that in fuel cells, the reactions that occur at the cathode compartment are very important. Actually, the oxygen reduction reaction is considered as one of the most important electrocatalytic reactions in fuel cells and will be addressed in the following section.

3. Cathode Catalysts for Direct Ethanol Fuel Cells Based on Polymer Electrolyte Membranes

Following the development of Direct Methanol Fuel Cells (DMFCs), Direct Ethanol Fuel Cells (DEFCs) are primarily based on polymer electrolytes, both acid and alkaline membranes. However, the use of such polymer electrolytes presents a significant challenge due to the permeability of ethanol through these membranes [2, 77]. This means that fuel will crossover from the anode compartment to the cathode compartment of the fuel cell through the electrolyte. Therefore, the ethanol fuel cell will be associated with a "chemical short", as part of the ethanol provided for the anodic process diffuses, instead, through the membrane and is oxidized at the cathode due to the high potential of this electrode. This results in a mixed electrode potential, as both oxygen reduction and ethanol oxidation occur at the same electrode, which decreases the cathode electrode potential and therefore the overall voltage efficiency of such a fuel cell. As an example, Lamy et al. [77] have shown for Nafion™ membranes that a 50% decrease in ethanol crossover results in an improvement of about one order of magnitude in the power density of PEMDEFCs, where the open circuit potential increased by more than 400 mV. As a logical consequence, much of the scientific attention has been devoted to the development of novel polymer electrolytes presenting lower permeabilities for ethanol and other alchols [78-83]. Relatively later, attention has also been given progressively towards developing new electrocatalysts insensitive to the presence of the alcohol in the cathode, or selective towards the reduction of oxygen when ethanol is also present in the cathodic electrode [84-86]. This seems a suitable option as polymer electrolytes and its composites are expensive, while alcohol tolerante oxygen reduction catalysts are usually cheaper than its counter part, i.e., Pt [2, 84-90].

The oxygen reduction reaction (ORR) is considered as one of the most important electrocatalytic reactions given its role in electrochemical energy conversion, industrial process and corrosion. Despite notable progress has been achieved over the years on developing novel ORR catalysts, the electrocatalytic activity together with stability and cost still need to be improved towards implementation in practical electrochemical energy conversion devices [91, 92]. The kinetics of the ORR are sluggish both in acid and in alkaline electrolytes, with acceptable reaction rates achieved only with suitable catalysts [91-94]. This sluggishness is severely complicated in alcohol fuel cells, given the crossover of alcohol along with water from the liquid feed anode electrode to the gas feed cathode electrode, which is driven primarily by concentration gradients. This liquid crossover causes several effects at the cathode, such as electrode flooding and catalyst active area diminishment for the ORR due to the oxidation of the alcohol. The former effect results from the high cathode electrode potential and the presence of oxygen, which favor the oxidation of the liquid fuel at the reducing electrode, generating a leaking current and a mixed electrode potential. Therefore, depending on the alcohol crossover rate, a significant fraction of the total electrocatalytically active surface area, otherwise available for the oxygen reduction reaction, is "stolen" by the alcohol oxidation reaction, which severely impacts both the voltage and fuel conversion efficiencies of the fuel cell [77, 83].

However, pioneer studies revealed that alloys of platinum-group-metals (PGMs) with non-PGM tend to result in catalysts that promote higher performances in DEFCs upon catalyzing the oxygen reduction reaction at the cathode [85, 86]. The forwarding discussion and review on this type of catalysts is subdivided into platinum group metal (PGM) and non-PGM based catalysts under a chronological manner with a primary focus on materials with potential to reach practical application requirements.

3.1. Platinum Group Metal based Ethanol Tolerant ORR Electrocatalysts

As early as 1967, the effect of ethanol as an interference over the oxygen reduction reaction was studied (dropping mercury electrode) [95]. However, it was not until much later that studies on the ethanol tolerance of oxygen reduction catalysts were reported in the literature focused on energy conversion applications. In 2006, Savadogo and Varela [96] reported on the impact of ethanol on the oxygen reduction on-set potential and on the PEMDEFC performance employing Pt/C, PtRu/C and Ru/C as cathode catalysts. Even though Ru/C presented suitable tolerance to ethanol, its poor catalytic activity towards the ORR rendered lower DEFC performances than the standard Pt/C cathode catalyst. The same authors also investigated the impact of ethanol on the on-set potential of the ORR, in liquid electrolyte (0.5 M H_2SO_4) at room temperature, catalyzed by sputtered Pd and PdCo catalysts [97]. Even though these materials presented some tolerance to ethanol upon catalysis of the ORR, the electroactivity of these catalysts towards reducing oxygen was inferior to that of Pt/C, making it unsuitable for practical applications. The first work to present ORR catalysts capable of effectively producing higher

PEMDEFCs performances than the standard Pt/C was that of Lopes et al. in 2007 [85]. In this work, these authors studied a commercial (ETek) $Pt_{75}Co_{25}$ alloy catalyst in a PEMDEFC at different operating temperatures, Figure 8. Based on the experimental results, these researchers concluded that due to the same ethanol tolerance presented by both Pt–Co/C and Pt/C catalysts, the improvement in the cell performance were ascribed to the higher ORR activity of the alloy catalyst. In 2009, Varela and Savadogo [98] obtained similar results with a commercial (ETek) Pt_1Co_1/C catalyst.

One year later, in 2008, Lopes et al. synthetized a series of Pt-Pd alloy catalysts supported on carbon black and investigated the ethanol tolerance of these materials onto gas diffusion electrodes, both in liquid (0.5 M H_2SO_4) and polymer electrolyte (Nafion™)

test systems. The ethanol tolerance of the PtPd alloy observed in ORR fundamental studies was translated in PEMDEFCs to produce higher performances in the practical device, Figure 9. These studies presented the basis and advantages of developing ethanol tolerant oxygen reduction reaction catalysts towards promoting DEFCs in practical applications. In Pt-M alloys, there is a tendency to occur a coupling between the d-orbitals of the alloy metals. Upon this interaction, if the d-orbital of the Pt atoms has a higher density of states (DOS) than the M atoms, there will be a tendency of decrease in the DOS at the Fermi level of Pt, which down shift its d-band center energy (more anti-bonding states below the Fermi level). A similar result is also explained in terms of the compression strain theory [99]. In this case, if the allowing metal M is smaller than Pt atom, this will lead to a compressive strain on the Pt surface atoms, which weakens the binding to oxygen-containing reaction intermediates. A mild weakening increases the catalytic activity of Pt for oxygen reduction [100].

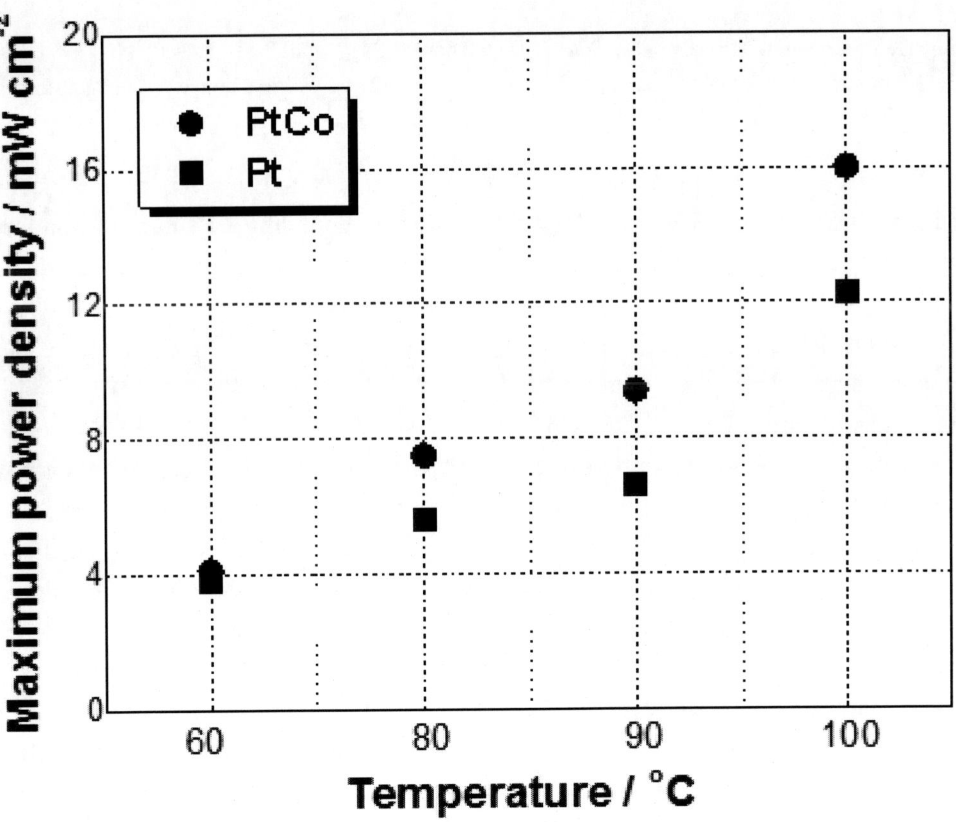

Figure 8. Histograms of the maximum power density of direct ethanol fuel cells operating at 60, 80, 90 and 100 °C with Pt/C and Pt–Co/C (3:1) as cathode electrocatalyst. Data from Ref. [85].

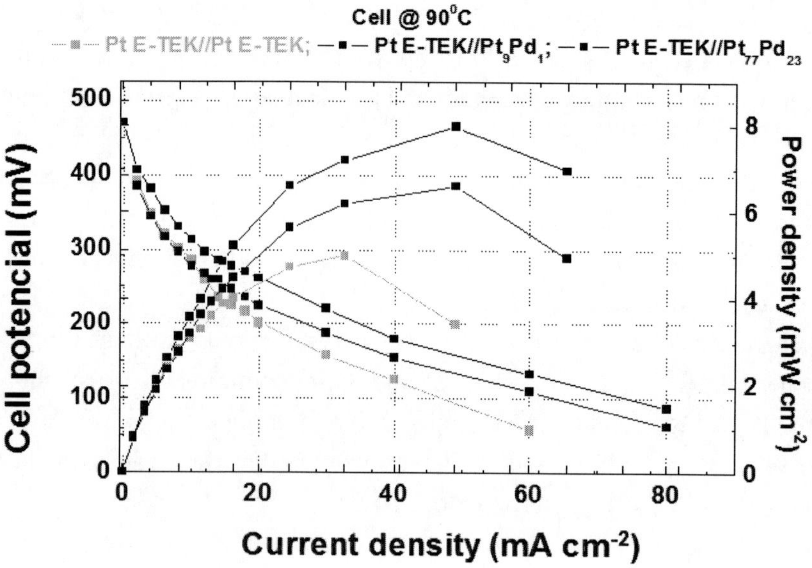

Figure 9. Polarization curves in a single DEFC with Pt–Pd/C and Pt/C (E-Tek) electrocatalysts as cathode materials for oxygen reduction at 90 °C/3 atm O_2 pressure using a 1 Methanol solution. Cathode Pt loading 1 mg cm^{-2}. Anode Pt loading 1 mg cm^{-2}. Data from Ref. [86].

Based on both theories, chemisorption of (hydr)oxy species on Pt sites occurs at potentials higher than 0.8 V vs RHE (weaker adsorbate bonding – Pauli repulsion lead by the higher population of antibonding states). This results in an enhancement of the ORR kinetics [99, 100] and on an inhibition of the oxidation of alcohols [101]. In 2009, Rao and Viswanathan [102] investigated a series of carbon supported Pt-M (M = Fe, Co and Cr) alloy catalysts as cathode materials in PEMDEFCs showing maximum power densities in the order Pt-Cr/C > Pt-Co/C > Pt-Fe/C > Pt/C. This tendency was in suitable agreement with the ORR activities and with the potential at which hydroxy species formed on the synthesized catalysts. One year later, Luna et al. [103] also reported on the higher performance of DEFCs employing PtFe/C as cathode catalyst compared to Pt/C. Also in 2010, Rao and Viswanathan [104] presented the suitable ethanol tolerance and ORR activity of a Pd-Co-Mo alloy supported on carbon black, which were explained by the suppression of (hydr)oxy species formation on Pd promoted by the alloying elements. In line with these studies, in 2013 Fu et al. [105] reported on a Polyallylamine functionalization of Pd icosahedra nanoparticles through a hydrothermal method to generate high ORR active materials with promising ethanol tolerance in alkaline media (0.1 M NaOH). It is interesting to note in the cyclic voltammograms (N_2-saturated) of the catalyst that the formation of (hydr)oxy species is shifted to more positive potentials in the functionalized material versus a standard Pt/C catalyst. Further to this fact, there is a work by Tanaskovic et al. [106] in the same year, which has shown the effect of dimethylsufoxide adsorption on platinum suppressing the formation of (hydr)oxy groups on the metal and suppressing the oxidation of ethanol across the potential window where

ORR takes place. Utilizing DFT calculations, these authors suggested that the electrochemical effects of adsorbed DMSO observed experimentally were primarily of geometrical nature. However, utilizing polyethyleneimine (PEI) layers on Pd nanowires (via a functionalization process), Xu et al. [107] proposed an explanation for the ethanol tolerance of this composite catalyst compared to a Pd/C due to a physical blockage of PEI ("molecular window gauze") to the passage of ethanol to access Pd active sites due to the larger size of the alcohol molecule compared to that of the oxygen molecule. In corroboration to this work, Li et al. [108] noted a high selectivity for the ORR in the presence of ethanol on polyallylamine-functionalized Pd-Pt nanodendrites. Nevertheless, the nitrogen atom on these molecules are prone to induce changes in the electronic structure of the metal, however, no clear changes can be noted in the cyclic voltammograms (N_2-saturated) of the catalysts in both works.

3.2. Non-PGM-Based Electrocatalysts

One of the first studies to report on the ethanol tolerance of platinum group metal (PGM) free oxygen reduction catalysts was that of Meng and Shen in 2006 [109]. In this study, tungsten carbide nanocrystals promoted Ag composite catalysts were studied in alkaline media (1 mol dm^{-3} KOH aqueous solution) showing similar activities to a Pt/C standard towards the ORR, though with a remarkable insensitiveness to the presence of ethanol in the electrolyte when compared to the referent catalyst (Pt/C). In 2008 Park et al. [110] presented a significantly different ethanol tolerant ORR catalyst in practical PEMDEFCs. A Ni-Co-Fe catalyst was synthesized by mixing the metal precursors in a complex form into a polymer matrix followed by heat treatment at elevated temperatures (800°C). The insensitivity of the Ni-Fe-Co catalyst towards the oxidation of ethanol rendered a higher open circuit potential on PEMDEFCs utilizing this material as cathode when compared to a standard Pt/C catalyst, however the later cathode material presented a higher maximum power density when compared with the Ni-Fe-Co catalyst in PEMDEFCs. In the same year, Saito et al. [111] investigated pyrochlore-type oxides as ORR in alkaline media. In 0.1 M KOH solutions, these authors found the onset potential for the ORR, in presence of 1 M ethanol, to be ~0.95 V vs RHE for the $NdRuMn_{0.25}$ catalyst and ~0.70 V for a standard Pt/C catalyst. However, the ORR catalytic activity of the materials was significantly lower when compared to a standard Pt/C catalyst. The first report of porphyrin-derived catalysts as ethanol tolerant ORR materials was perhaps that of Modestov et al. in 2009 [112]. In this work, a pyrolyzed cobalt porphyrin supported on carbon black presented a low ethanol oxidation rate and a relatively suitable ORR activity, however the performance of AAEMDEFC utilizing these catalysts as cathode was not higher than that of a standard Pt/C catalyst. In 2010, Tarasevich et al. [113] also reported on the ethanol tolerance of a Co porphyrin-derived catalyst, however a comparison to Pt/C was not given in AAEMDEFC tests.

One of the first suitable studies to present a comparative study of alkaline direct ethanol fuel cells with non-noble metal cathodic materials was perhaps that of Zhiani et al. [114] in 2011. In this study, a commercial HypermecTM K14 catalyst presented a suitably higher AAEMDEFC performance than a Pt/C catalyst at a lower metal loading, reaching high power densities (>10%) with the cathodic compartment of the cell at ambient pressures. As a general trend, non-PGMs tend to present a reasonable ORR activity in alkaline media when compared to Pt, however with a suitable higher tolerance to the presence of alcohols [115, 116].

One of the first studies to report on the ethanol tolerance of non-precious metal (FeCo) and metal-free carbon composite catalysts was that of Uhm et al. in 2011 [117]. However, the ORR activity of the catalyst was relatively far from a standard Pt/C and AAEMDEFC results were not given. One year later, Rao and Ishikawa [118] also reported on the high ethanol tolerance of virtually metal-free nitrogen-doped carbon nanotubes in alkaline media (0.5 M KOH), however, no AAEMDEFC results were presented, which seems to be a general trend in this class of ORR catalysts [119, 120].

4. Summary and Perspective

As it can be seen, there are different catalysts materials with diverse atomic compositions, morphology and also support materials proposed in the literature related to the ethanol electro-oxidation in alkaline media. Palladium-based materials are pointed out as promising electrocatalysts and graphene as support material for the nanoparticles. However, a lot of improvement is required for the commercialization of the fuel cell based in alkaline anion exchange membranes fed with ethanol.

Considering the oxygen reduction catalysts, even though palladium-based and silver-based materials hold great promise in terms of catalytic activity and tolerance to ethanol, there is still much to be explored in materials development and electrode optimization.

Certainly, there are excellent works that were not mentioned in this chapter. However, the intention was to present a focused overview of the recent literature on the development of electrocatalysts for ethanol electro-oxidation and oxygen reduction reaction in alkaline media related to direct ethanol fuel cells.

REFERENCES

[1] Nguyen, ST; Ling Tan, DS; Lee, JM; Chan, SH; Wang, JY; Wang, X. Tb promoted Pd/C catalysts for the electrooxidation of ethanol in alkaline media, *International Journal of Hydrogen Energy*, 36, (2011), 9645-9652.

[2] Antolini, E; Gonzalez, ER. Alkaline direct alcohol fuel cells, *Journal of Power Sources*, 195, (2010), 3431-3450.

[3] Wang, P; Lin, X; Yang, B; Jin, JM; Hardacre, C; Yu, NF; Sun, SG; Lin, WF. Activity Enhancement of Tetrahexahedral Pd Nanocrystals by Bi Decoration towards Ethanol Electrooxidation in Alkaline Media, *Electrochimica Acta*, 162, (2015), 290-299.

[4] Li, G; Jiang, L; Jiang, Q; Wang, S; Sun, G. Preparation and characterization of PdxAgy/C electrocatalysts for ethanol electro-oxidation reaction in alkaline media, *Electrochimica Acta*, 56, (2011), 7703-7711.

[5] Zhu, LD; Zhao, TS; Xu, JB; Liang, ZX. Preparation and characterization of carbon-supported sub-monolayer palladium decorated gold nanoparticles for the electro-oxidation of ethanol in alkaline media, *Journal of Power Sources*, 187, (2009), 80-84.

[6] Cai, J; Huang, Y; Guo, Y. Facile synthesis of PdSbx/C nanocatalyst with high performance for ethanol electro-oxidation in alkaline medium, *International Journal of Hydrogen Energy*, 39, (2014), 18256-18263.

[7] Geraldes, AN; da Silva, DF; Pino, ES; da Silva, JCM; de Souza, RFB; Hammer, P; Spinacé, EV; Neto, AO; Linardi, M; dos Santos, MC. Ethanol electro-oxidation in an alkaline medium using Pd/C, Au/C and PdAu/C electrocatalysts prepared by electron beam irradiation, *Electrochimica Acta*, 111, (2013), 455-465.

[8] Rizo, R; Sebastián, D; Lázaro, MJ; Pastor, E. On the design of Pt-Sn efficient catalyst for carbon monoxide and ethanol oxidation in acid and alkaline media, *Applied Catalysis B: Environmental*, 200, (2017), 246-254.

[9] Yu, EH; Scott, K. Development of direct methanol alkaline fuel cells using anion exchange membranes, *Journal of Power Sources*, 137, (2004), 248-256.

[10] Matsuoka, K; Iriyama, Y; Abe, T; Matsuoka, M; Ogumi, Z. Alkaline direct alcohol fuel cells using an anion exchange membrane, *Journal of Power Sources*, 150, (2005), 27-31.

[11] da Silva, SG; Assumpção, MHMT; de Souza, RFB; Buzzo, GS; Spinacé, EV; Neto, AO; Silva, JCM. Electrochemical and Fuel Cell Evaluation of PtIr/C Electrocatalysts for Ethanol Electrooxidation in Alkaline Medium, *Electrocatalysis*, 5, (2014), 438-444.

[12] Switzer, EE; Olson, TS; Datye, AK; Atanassov, P; Hibbs, MR; Cornelius, CJ. Templated Pt-Sn electrocatalysts for ethanol, methanol and CO oxidation in alkaline media, *Electrochimica Acta*, 54, (2009), 989-995.

[13] Gao, H; Liao, S; Liang, Z; Liang, H; Luo, F. Anodic oxidation of ethanol on core-shell structured Ru@PtPd/C catalyst in alkaline media, *Journal of Power Sources*, 196, (2011), 6138-6143.

[14] Neto, AO; da Silva, SG; Buzzo, GS; de Souza, RFB; Assumpção, MHMT; Spinacé, EV; Silva, JCM. *Ethanol electrooxidation on PdIr/C electrocatalysts in alkaline media: electrochemical and fuel cell studies*, *Ionics*, (2014), 1-9.

[15] Silva, JCM; Anea, B; De Souza, RFB; Assumpcao, MHMT; Calegaro, ML; Neto, AO; Santos, MC. Ethanol Oxidation Reaction on IrPtSn/C Electrocatalysts with low Pt Content, *Journal of the Brazilian Chemical Society*, 24, (2013), 1553-1560.

[16] Tayal, J; Rawat, B; Basu, S. Bi-metallic and tri-metallic Pt–Sn/C, Pt–Ir/C, Pt–Ir–Sn/C catalysts for electro-oxidation of ethanol in direct ethanol fuel cell, *International Journal of Hydrogen Energy*, 36, (2011), 14884-14897.

[17] Shen, SY; Zhao, TS; Xu, JB. Carbon supported PtRh catalysts for ethanol oxidation in alkaline direct ethanol fuel cell, *International Journal of Hydrogen Energy*, 35, (2010), 12911-12917.

[18] Ethanol-Water Adsorption on Commercial 3A Zeolites: Kinetic and Thermodynamic Data, *Brazilian Journal of Chemical Engineering*, 14, (1997), 00-00.

[19] Silva, JCM; Parreira, LS; De Souza, RFB; Calegaro, ML; Spinacé, EV; Neto, AO; Santos, MC. PtSn/C alloyed and non-alloyed materials: Differences in the ethanol electro-oxidation reaction pathways, *Applied Catalysis B: Environmental*, 110, (2011), 141-147.

[20] Ribeiro, J; dos Anjos, DM; Kokoh, KB; Coutanceau, C; Léger, JM; Olivi, P; de Andrade, AR; Tremiliosi-Filho, G. Carbon-supported ternary PtSnIr catalysts for direct ethanol fuel cell, *Electrochimica Acta*, 52, (2007), 6997-7006.

[21] An, L; Zhao, TS. Transport phenomena in alkaline direct ethanol fuel cells for sustainable energy production, *Journal of Power Sources*, 341, (2017), 199-211.

[22] Neto, AO; da Silva, SG; Buzzo, GS; de Souza, RFB; Assumpção, MHMT; Spinacé, EV; Silva, JCM. Ethanol electrooxidation on PdIr/C electrocatalysts in alkaline media: electrochemical and fuel cell studies, *Ionics*, 21, (2015), 487-495.

[23] Zignani, SC; Baglio, V; Linares, JJ; Monforte, G; Gonzalez, ER; Aricò, AS. Performance and selectivity of PtxSn/C electro-catalysts for ethanol oxidation prepared by reduction with different formic acid concentrations, *Electrochimica Acta*, 70, (2012), 255-265.

[24] Sun, CL; Tang, JS; Brazeau, N; Wu, JJ; Ntais, S; Yin, CW; Chou, HL; Baranova, EA. Particle size effects of sulfonated graphene supported Pt nanoparticles on ethanol electrooxidation, *Electro-chimica Acta*, 162, (2015), 282-289.

[25] Beyhan, S; Uosaki, K; Feliu, JM; Herrero, E. Electrochemical and *In situ* FTIR studies of ethanol adsorption and oxidation on gold single crystal electrodes in alkaline media, *Journal of Electroanalytical Chemistry*, 707, (2013), 89-94.

[26] Sieben, JM; Duarte, MME. Methanol, ethanol and ethylene glycol electro-oxidation at Pt and Pt–Ru catalysts electrodeposited over oxidized carbon nanotubes, *International Journal of Hydrogen Energy*, 37, (2012), 9941-9947.

[27] González-Quijano, D; Pech-Rodríguez, WJ; Escalante-García, JI; Vargas-Gutiérrez, G; Rodríguez-Varela, FJ. Electrocatalysts for ethanol and ethylene glycol oxidation reactions. Part I: Effects of the polyol synthesis conditions on the characteristics and catalytic activity of Pt–Sn/C anodes, *International Journal of Hydrogen Energy*, 39, (2014), 16676-16685.

[28] Jiang, L; Hsu, A; Chu, D; Chen, R. Ethanol electro-oxidation on Pt/C and PtSn/C catalysts in alkaline and acid solutions, *International Journal of Hydrogen Energy*, 35, (2010), 365-372.

[29] Han, SB; Song, YJ; Lee, JM; Kim, JY; Park, KW. Platinum nanocube catalysts for methanol and ethanol electrooxidation, *Electrochemistry Communications*, 10, (2008), 1044-1047.

[30] Burda, C; Chen, X; Narayanan, R; El-Sayed, MA. Chemistry and Properties of Nanocrystals of Different Shapes, *Chemical Reviews*, 105, (2005), 1025-1102.

[31] Maillard, F; Savinova, ER; Simonov, PA; Zaikovskii, VI; Stimming, U. Infrared Spectroscopic Study of CO Adsorption and Electro-oxidation on Carbon-Supported Pt Nanoparticles: Interparticle versus Intraparticle Heterogeneity, *The Journal of Physical Chemistry B*, 108, (2004), 17893-17904.

[32] Golikand, AN; Lohrasbi, E; Asgari, M. Enhancing the durability of multi-walled carbon nanotube supported by Pt and Pt–Pd nanoparticles in gas diffusion electrodes, *International Journal of Hydrogen Energy*, 35, (2010) 9233-9240.

[33] Brouzgou, A; Song, SQ; Tsiakaras, P. Low and non-platinum electrocatalysts for PEMFCs: Current status, challenges and prospects, *Applied Catalysis B: Environmental*, 127, (2012), 371-388.

[34] Antolini, E. Nitrogen-doped carbons by sustainable N- and C-containing natural resources as nonprecious catalysts and catalyst supports for low temperature fuel cells, *Renewable and Sustainable Energy Reviews*, 58, (2016), 34-51.

[35] Capeness, MJ; Edmundson, MC; Horsfall, LE. Nickel and platinum group metal nanoparticle production by Desulfovibrio alaskensis G20, *New Biotechnology*, 32, (2015), 727-731.

[36] da Silva, SG; Silva, JCM; Buzzo, GS; De Souza, RFB; Spinacé, EV; Neto, AO; Assumpção, MHMT. Electrochemical and fuel cell evaluation of PtAu/C electrocatalysts for ethanol electro-oxidation in alkaline media, *International Journal of Hydrogen Energy*, 39, (2014), 10121-10127.

[37] Baranova, EA; Padilla, MA; Halevi, B; Amir, T; Artyushkova, K; Atanassov, P. Electrooxidation of ethanol on PtSn nanoparticles in alkaline solution: Correlation between structure and catalytic properties, *Electrochimica Acta*, 80, (2012), 377-382.

[38] Tusi, MM; Polanco, NSO; da Silva, SG; Spinacé, EV; Neto, AO. The high activity of PtBi/C electrocatalysts for ethanol electro-oxidation in alkaline medium, *Electrochemistry Communications*, 13, (2011), 143-146.

[39] Fontes, EH; Piasentin, RM; Ayoub, JMS; da Silva, JCM; Assumpção, MHMT; Spinacé, EV; Neto, AO; de Souza, RFB. Electrochemical and *In situ* ATR-FTIR studies of ethanol electro-oxidation in alkaline medium using PtRh/C electrocatalysts, *Materials for Renewable and Sustainable Energy*, 4, (2015), 3.

[40] Antoniassi, RM; Oliveira Neto, A; Linardi, M; Spinacé, EV. The effect of acetaldehyde and acetic acid on the direct ethanol fuel cell performance using PtSnO2/C electrocatalysts, *International Journal of Hydrogen Energy*, 38, (2013), 12069-12077.

[41] Dutta, A; Mahapatra, SS; Datta, J. High performance PtPdAu nano-catalyst for ethanol oxidation in alkaline media for fuel cell applications, *International Journal of Hydrogen Energy*, 36, (2011), 14898-14906.

[42] Balasubramanian, S; Lakshmanan, B; Hetzke, CE; Sethuraman, VA; Weidner, JW. Quantifying oxidation rates of carbon monoxide on a Pt/C electrode, *Electrochimica Acta*, 58, (2011), 723-728.

[43] Farias, MJS; Cheuquepan, W; Camara, GA; Feliu, JM. Disentangling Catalytic Activity at Terrace and Step Sites on Selectively Ru-Modified Well-Ordered Pt Surfaces Probed by CO Electro-oxidation, *ACS Catalysis*, 6, (2016), 2997-3007.

[44] Mao, H; Wang, L; Zhu, P; Xu, Q; Li, Q. Carbon-supported PdSn–SnO2 catalyst for ethanol electro-oxidation in alkaline media, *International Journal of Hydrogen Energy*, 39, (2014), 17583-17588.

[45] Zhang, F; Zhou, D; Zhou, M. Ethanol electrooxidation on Pd/C nanoparticles in alkaline media, *Journal of Energy Chemistry*, 25, (2016), 71-76.

[46] Huang, Y; Guo, Y; Wang, Y; Yao, J. Synthesis and performance of a novel PdCuPb/C nanocatalyst for ethanol electrooxidation in alkaline medium, *International Journal of Hydrogen Energy*, 39, (2014), 4274-4281.

[47] Cai, J; Huang, Y; Guo, Y. PdTex/C nanocatalysts with high catalytic activity for ethanol electro-oxidation in alkaline medium, *Applied Catalysis B: Environmental*, 150–151, (2014), 230-237.

[48] Chen, H; Xing, Z; Zhu, S; Zhang, L; Chang, Q; Huang, J; Cai, WB; Kang, N; Zhong, CJ; Shao, M. Palladium modified gold nanoparticles as electrocatalysts for ethanol electrooxidation, *Journal of Power Sources*, 321, (2016), 264-269.

[49] Peng, C; Hu, Y; Liu, M; Zheng, Y. Hollow raspberry-like PdAg alloy nanospheres: High electrocatalytic activity for ethanol oxidation in alkaline media, *Journal of Power Sources*, 278, (2015), 69-75.

[50] Ma, L; Chu, D; Chen, R. Comparison of ethanol electro-oxidation on Pt/C and Pd/C catalysts in alkaline media, *International Journal of Hydrogen Energy*, 37, (2012), 11185-11194.

[51] Liang, ZX; Zhao, TS; Xu, JB; Zhu, LD. Mechanism study of the ethanol oxidation reaction on palladium in alkaline media, *Electrochimica Acta*, 54, (2009), 2203-2208.

[52] Liu, C; Cai, X; Wang, J; Liu, J; Riese, A; Chen, Z; Sun, X; Wang, SD. One-step synthesis of AuPd alloy nanoparticles on graphene as a stable catalyst for ethanol electro-oxidation, *International Journal of Hydrogen Energy*, 41, (2016), 13476-13484.

[53] Wang, Y; Nguyen, TS; Liu, X; Wang, X. Novel palladium–lead, (Pd–Pb/C) bimetallic catalysts for electrooxidation of ethanol in alkaline media, *Journal of Power Sources*, 195, (2010), 2619-2622.

[54] Nguyen, ST; Law, HM; Nguyen, HT; Kristian, N; Wang, S; Chan, SH; Wang, X. Enhancement effect of Ag for Pd/C towards the ethanol electro-oxidation in alkaline media, *Applied Catalysis B: Environmental*, 91, (2009), 507-515.

[55] Monyoncho, EA; Ntais, S; Soares, F; Woo, TK; Baranova, EA. Synergetic effect of palladium–ruthenium nanostructures for ethanol electrooxidation in alkaline media, *Journal of Power Sources*, 287, (2015), 139-149.

[56] Qin, YH; Li, Y; Lv, RL; Wang, TL; Wang, WG; Wang, CW. Pd-Au/C catalysts with different alloying degrees for ethanol oxidation in alkaline media, *Electrochimica Acta*, 144, (2014), 50-55.

[57] He, Q; Chen, W; Mukerjee, S; Chen, S; Laufek, F. Carbon-supported PdM (M = Au and Sn) nanocatalysts for the electrooxidation of ethanol in high pH media, *Journal of Power Sources*, 187, (2009), 298-304.

[58] Wei, YC; Liu, CW; Kang, WD; Lai, CM; Tsai, LD; Wang, KW. Electro-catalytic activity enhancement of Pd–Ni electrocatalysts for the ethanol electro-oxidation in alkaline medium: The promotional effect of CeO2 addition, *Journal of Electroanalytical Chemistry*, 660, (2011), 64-70.

[59] Xu, JB; Zhao, TS; Shen, SY; Li, YS. Stabilization of the palladium electrocatalyst with alloyed gold for ethanol oxidation, *International Journal of Hydrogen Energy*, 35, (2010), 6490-6500.

[60] Cai, J; Zeng, Y; Guo, Y. Copper@palladium–copper core–shell nanospheres as a highly effective electrocatalyst for ethanol electro-oxidation in alkaline media, *Journal of Power Sources*, 270, (2014), 257-261.

[61] Nguyen, ST; Yang, Y; Wang, X. Ethanol electro-oxidation activity of Nb-doped-TiO2 supported PdAg catalysts in alkaline media, *Applied Catalysis B: Environmental*, 113–114, (2012), 261-270.

[62] Silva, JCM; Piasentin, RM; Spinacé, EV; Neto, AO; Baranova, EA. The effect of antimony-tin and indium-tin oxide supports on the catalytic activity of Pt nanoparticles for ammonia electro-oxidation, *Materials Chemistry and Physics*, 180, (2016), 97-103.

[63] Wu, J; Yuan, XZ; Martin, JJ; Wang, H; Zhang, J; Shen, J; Wu, S; Merida, W. A review of PEM fuel cell durability: Degradation mechanisms and mitigation strategies, *Journal of Power Sources*, 184, (2008), 104-119.

[64] Zhang, S; Yuan, XZ; Hin, JNC; Wang, H; Friedrich, KA; Schulze, M. A review of platinum-based catalyst layer degradation in proton exchange membrane fuel cells, *Journal of Power Sources*, 194, (2009), 588-600.

[65] Hiltrop, D; Masa, J; Maljusch, A; Xia, W; Schuhmann, W; Muhler, M. Pd deposited on functionalized carbon nanotubes for the electrooxidation of ethanol in alkaline media, *Electrochemistry Communications*, 63, (2016), 30-33.

[66] Habibi, B; Mohammadyari, S. Facile synthesis of Pd nanoparticles on nano carbon supports and their application as an electrocatalyst for oxidation of ethanol in alkaline media: The effect of support, *International Journal of Hydrogen Energy*, 40, (2015), 10833-10846.

[67] Karousis, N; Tagmatarchis, N; Tasis, D. Current Progress on the Chemical Modification of Carbon Nanotubes, *Chemical Reviews*, 110, (2010), 5366-5397.

[68] Li, S; Ma, J; Huo, H; Jin, J; Ma, J; Yang, H. Ionic liquids-noncovalently functionalized multi-walled carbon nanotubes decorated with palladium nanoparticles: A promising electrocatalyst for ethanol electrooxidation, *International Journal of Hydrogen Energy*, 41, (2016), 12358-12368.

[69] Krishna, R; Fernandes, DM; Ventura, J; Freire, C; Titus, E. Facile synthesis of reduced graphene oxide supported Pd@NixB/RGO nanocomposite: Novel electrocatalyst for ethanol oxidation in alkaline media, *International Journal of Hydrogen Energy*, 41, (2016), 11811-11822.

[70] Dong, Q; Zhao, Y; Han, X; Wang, Y; Liu, M; Li, Y. Pd/Cu bimetallic nanoparticles supported on graphene nanosheets: Facile synthesis and application as novel electrocatalyst for ethanol oxidation in alkaline media, *International Journal of Hydrogen Energy*, 39, (2014), 14669-14679.

[71] Ghosh, S; Remita, H; Kar, P; Choudhury, S; Sardar, S; Beaunier, P; Roy, PS; Bhattacharya, SK; Pal, SK. Facile synthesis of Pd nanostructures in hexagonal mesophases as a promising electro-catalyst for ethanol oxidation, *Journal of Materials Chemistry A*, 3, (2015), 9517-9527.

[72] Wu, P; Huang, Y; Zhou, L; Wang, Y; Bu, Y; Yao, J. Nitrogen-doped graphene supported highly dispersed palladium-lead nanoparticles for synergetic enhancement of ethanol electrooxidation in alkaline medium, *Electrochimica Acta*, 152, (2015), 68-74.

[73] Liu, Q; Jiang, K; Fan, J; Lin, Y; Min, Y; Xu, Q; Cai, WB. Manganese Dioxide Coated Graphene Nanoribbons Supported Palladium Nanoparticles as an Efficient Catalyst for Ethanol Electrooxidation in Alkaline Media, *Electrochimica Acta*, 203, (2016), 91-98.

[74] Qin, YH; Zhuang, Y; Lv, RL; Wang, TL; Wang, WG; Wang, CW. Pd nanoparticles anchored on carbon-doped TiO2 nanocoating support for ethanol electrooxidation in alkaline media, *Electrochimica Acta*, 154, (2015), 77-82.

[75] Shao, Y; Liu, J; Wang, Y; Lin, Y. Novel catalyst support materials for PEM fuel cells: current status and future prospects, *Journal of Materials Chemistry*, 19, (2009), 46-59.

[76] Hasan, M; Newcomb, SB; Rohan, JF; Razeeb, KM. Ni nanowire supported 3D flower-like Pd nanostructures as an efficient electrocatalyst for electrooxidation of ethanol in alkaline media, *Journal of Power Sources*, 218, (2012), 148-156.

[77] Lamy, C; Coutanceau, C; Leger, JM. *The Direct Ethanol Fuel Cell: a Challenge to Convert Bioethanol Cleanly into Electric Energy, Catalysis for Sustainable Energy Production*, Wiley-VCH Verlag GmbH & Co. KGaA, 2009, pp. 1-46.

[78] Huang, CY; Lin, JS; Pan, WH; Shih, CM; Liu, YL; Lue, SJ. Alkaline direct ethanol fuel cell performance using alkali-impregnated polyvinyl alcohol/functionalized carbon nano-tube solid electrolytes, *Journal of Power Sources*, 303, (2016), 267-277.

[79] Matos, BR; Isidoro, RA; Santiago, EI; Fonseca, FC. Performance enhancement of direct ethanol fuel cell using Nafion composites with high volume fraction of titania, *Journal of Power Sources*, 268, (2014), 706-711.

[80] Matos, BR; Isidoro, RA; Santiago, EI; Linardi, M; Ferlauto, AS; Tavares, AC; Fonseca, FC. *In situ* Fabrication of Nafion–Titanate Hybrid Electrolytes for High-Temperature Direct Ethanol Fuel Cell, *The Journal of Physical Chemistry C*, 117, (2013), 16863-16870.

[81] Roelofs, KS; Hirth, T; Schiestel, T. Sulfonated poly(ether ether ketone)-based silica nanocomposite membranes for direct ethanol fuel cells, *Journal of Membrane Science*, 346, (2010), 215-226.

[82] Wu, XW; Wu, N; Shi, CQ; Zheng, ZY; Qi, HB; Wang, YF. Proton conductive montmorillonite-Nafion composite membranes for direct ethanol fuel cells, *Applied Surface Science*, 388, Part A,, (2016), 239-244.

[83] Gottesfeld, S. *Polymer Electrolyte and Direct Methanol Fuel Cells, Encyclopedia of Electrochemistry*, Wiley-VCH Verlag GmbH & Co. KGaA, 2007.

[84] Antolini, E. Catalysts for direct ethanol fuel cells, *Journal of Power Sources*, 170, (2007), 1-12.

[85] Lopes, T; Antolini, E; Colmati, F; Gonzalez, ER. Carbon supported Pt–Co, (3:1) alloy as improved cathode electrocatalyst for direct ethanol fuel cells, *Journal of Power Sources*, 164, (2007), 111-114.

[86] Lopes, T; Antolini, E; Gonzalez, ER. Carbon supported Pt–Pd alloy as an ethanol tolerant oxygen reduction electrocatalyst for direct ethanol fuel cells, *International Journal of Hydrogen Energy*, 33, (2008), 5563-5570.

[87] Antolini, E; Lopes, T; Gonzalez, ER. An overview of platinum-based catalysts as methanol-resistant oxygen reduction materials for direct methanol fuel cells, *Journal of Alloys and Compounds*, 461, (2008), 253-262.

[88] Gentil, R; Villullas, HM. Oxygen reduction activity and methanol tolerance of carbon-supported PtV nanoparticles and the effects of heat treatment at low temperatures, *Journal of Solid State Electrochemistry*, 20, (2016), 1119-1129.

[89] Malheiro, AR; Perez, J; Villullas, HM. Well-Alloyed PtFe/C Nanocatalysts of Controlled Composition and Same Particle Size: Oxygen Reduction and Methanol Tolerance, *Journal of The Electrochemical Society*, 156, (2009), B51-B58.

[90] Corradini, PG; Antolini, E; Perez, J. Activity, short-term stability (poisoning tolerance) and durability of carbon supported Pt–Pr catalysts for ethanol oxidation, *Journal of Power Sources*, 251, (2014), 402-410.

[91] Debe, MK. Electrocatalyst approaches and challenges for automotive fuel cells, *Nature*, 486, (2012), 43-51.

[92] Lopes, T; Kucernak, A; Malko, D; Ticianelli, EA. Mechanistic Insights into the Oxygen Reduction Reaction on Metal–N–C Electrocatalysts under Fuel Cell Conditions, *ChemElectroChem*, 3, (2016), 1580-1590.

[93] Yeager, E. Dioxygen electrocatalysis: mechanisms in relation to catalyst structure, *Journal of Molecular Catalysis*, 38, (1986), 5-25.

[94] Gasteiger, HA; Marković, NM. Just a Dream—or Future Reality?, *Science*, 324, (2009), 48-49.

[95] Israel, Y; Vromen, A; Paschkes, B. Polarographic study of inhibitors and catalysts for the reaction of dissolved oxygen with sulphite ion, *Talanta*, 14, (1967), 925-931.

[96] Savadogo, O; Varela, FJR. Ethanol-Tolerant Oxygen Reduction Reaction (ORR) Cathodes for Direct Ethanol Fuel Cell Applications, *ECS Transactions*, 1, (2006), 331-338.

[97] Savadogo, O; Valera, FJR. Palladium-Alloy Catalysts as Ethanol Tolerant Cathodes for Direct Alcohol Fuel Cell (DEFC) Applications, *ECS Transactions*, 1, (2006), 247-254.

[98] Rodríguez Varela, FJ; Savadogo, O. Ethanol-tolerant Pt-alloy cathodes for direct ethanol fuel cell (DEFC) applications, *Asia-Pacific Journal of Chemical Engineering*, 4, (2009), 17-24.

[99] Strasser, P; Koh, S; Anniyev, T; Greeley, J; More, K; Yu, C; Liu, Z; Kaya, S; Nordlund, D; Ogasawara, H; Toney, MF; Nilsson, A. Lattice-strain control of the activity in dealloyed core–shell fuel cell catalysts, *Nat Chem*, 2, (2010), 454-460.

[100] Stephens, IEL; Rossmeisl, J; Chorkendorff, I. Toward sustainable fuel cells, *Science*, 354, (2016), 1378-1379.

[101] Xu, Y; Shao, M; Mavrikakis, M; Adzic, RR. *Recent Developments in the Electrocatalysis of the O2 Reduction Reaction, Fuel Cell Catalysis*, John Wiley & Sons, Inc., 2008, pp. 271-315.

[102] Venkateswara Rao, C; Viswanathan, B. ORR Activity and Direct Ethanol Fuel Cell Performance of Carbon-Supported Pt–M (M = Fe, Co, and Cr) Alloys Prepared by Polyol Reduction Method, *The Journal of Physical Chemistry C*, 113, (2009), 18907-18913.

[103] Castro Luna, AM; Bonesi, A; Triaca, WE; Di Blasi, A; Stassi, A; Baglio, V; Antonucci, V; Aricò, AS. Investigation of a Pt–Fe/C catalyst for oxygen reduction reaction in direct ethanol fuel cells, *Journal of Nanoparticle Research*, 12, (2010), 357-365.

[104] Rao, CV; Viswanathan, B. Carbon supported Pd–Co–Mo alloy as an alternative to Pt for oxygen reduction in direct ethanol fuel cells, *Electrochimica Acta*, 55, (2010), 3002-3007.

[105] Fu, G; Jiang, X; Tao, L; Chen, Y; Lin, J; Zhou, Y; Tang, Y; Lu, T. Polyallylamine Functionalized Palladium Icosahedra: One-Pot Water-Based Synthesis and Their Superior Electrocatalytic Activity and Ethanol Tolerant Ability in Alkaline Media, *Langmuir*, 29, (2013) 4413-4420.

[106] Tanasković, V; Pašti, IA; Gavrilov, N; Mentus, SV. Dimethylsulfoxide as a modifier of platinum electrocatalytic activity toward oxygen reduction reaction in aqueous solutions: Combined theoretical and experimental study, *Journal of Electroanalytical Chemistry*, 714–715,, (2014), 11-18.

[107] Xu, GR; Liu, FY; Liu, ZH; Chen, Y. Ethanol-tolerant polyethyleneimine functionalized palladium nanowires in alkaline media: the "molecular window gauze" induced the selectivity for the oxygen reduction reaction, *Journal of Materials Chemistry A*, 3, (2015), 21083-21089.

[108] Li, FM; Gao, XQ; Li, SN; Chen, Y; Lee, JM. Thermal decomposition synthesis of functionalized PdPt alloy nanodendrites with high selectivity for oxygen reduction reaction, *NPG Asia Mater*, 7, (2015), e219.

[109] Meng, H; Shen, PK. Novel Pt-free catalyst for oxygen electro-reduction, *Electrochemistry Communications*, 8, (2006), 588-594.

[110] Park, N; Shiraishi, T; Kamisugi, K; Hara, Y; Iizuka, K; Kado, T; Hayase, S. NiCoFe/C cathode electrocatalysts for direct ethanol fuel cells, *Journal of Applied Electrochemistry*, 38, (2008), 371-375.

[111] Saito, M; Saito, Y; Konishi, T; Kawai, H; Kuwano, J; Shiroishi, H; Uchimoto, Y. Electrocatalytic O2 Reduction Properties of Pyrochlore-Type Oxides for Alkaline DAFCs, *ECS Transactions*, 16, (2008), 891-900.

[112] Modestov, AD; Tarasevich, MR; Leykin, AY; Filimonov, VY. MEA for alkaline direct ethanol fuel cell with alkali doped PBI membrane and non-platinum electrodes, *Journal of Power Sources*, 188, (2009), 502-506.

[113] Tarasevich, MR; Bogdanovskaya, VA; Mazin, PV. Electrocatalysts and membrane for direct ethanol-oxygen fuel cell with alkaline electrolyte, *Russian Journal of Electrochemistry*, 46, (2010), 542-551.

[114] Zhiani, M; Gasteiger, HA; Piana, M; Catanorchi, S. Comparative study between platinum supported on carbon and non-noble metal cathode catalyst in alkaline direct ethanol fuel cell (ADEFC), *International Journal of Hydrogen Energy*, 36, (2011), 5110-5116.

[115] Zhao, TS; Li, YS; Shen, SY. *Anion-exchange membrane direct ethanol fuel cells: Status and perspective*, *Frontiers of Energy and Power Engineering in China*, 4, (2010), 443-458.

[116] Jiang, R; Moton, E; McClure, JP; Bowers, Z. A Highly Active and Alcohol-Tolerant Cathode Electrocatalyst Containing Ag Nanoparticles Supported on Graphene, *Electrochimica Acta*, 127, (2014), 146-152.

[117] Uhm, S; Jeong, B; Lee, J. A facile route for preparation of non-noble CNF cathode catalysts in alkaline ethanol fuel cells, *Electrochimica Acta*, 56, (2011), 9186-9190.

[118] Venkateswara Rao, C; Ishikawa, Y. Activity, Selectivity, and Anion-Exchange Membrane Fuel Cell Performance of Virtually Metal-Free Nitrogen-Doped Carbon Nanotube Electrodes for Oxygen Reduction Reaction, *The Journal of Physical Chemistry C*, 116, (2012), 4340-4346.

[119] Liu, Y; Li, J; Li, W; Li, Y; Chen, Q; Zhan, F. Nitrogen-doped graphene aerogel-supported spinel CoMn2O4 nanoparticles as an efficient catalyst for oxygen reduction reaction, *Journal of Power Sources*, 299, (2015), 492-500.

[120] Guo, C; Liao, W; Li, Z; Sun, L; Ruan, H; Wu, Q; Luo, Q; Huang, J; Chen, C. Coprinus comatus-derived nitrogen-containing biocarbon electrocatalyst with the addition of self-generating graphene-like support for superior oxygen reduction reaction, *Science Bulletin*, 61, (2016) 948-958.

In: Advanced Materials and Systems …
Editor: Mauro Coelho dos Santos

ISBN: 978-1-53613-852-8

Chapter 3

NANOSTRUCTURED MATERIALS FOR BIOLOGICAL FUEL CELLS

Sidney Aquino Neto and Adalgisa R. De Andrade *

Departamento de Química, Faculdade de Filosofia Ciências e Letras de Ribeirão Preto, Universidade de São Paulo, Ribeirão Preto, São Paulo, Brasil

ABSTRACT

In contrast to conventional fuel cells that rely on noble metal catalysts to ensure that the electrode is active, biofuel cells employ biocatalysts to provide clean, renewable energy and have potential application as energy source for low-power electronic devices. Enzymatic biofuel cells have promising characteristics, but their future use depends on obtaining longer enzyme lifetime, higher enzyme stability, higher power density, and more facile electron transfer between the enzyme and the electrode surface. In this context, designing improved techniques to immobilize enzymes is an important research topic. Some new trends in biocatalysts for biofuel cells include incorporation of nanoscaled materials such as carbon nanotubes, nanofibers and nanocomposites as well as introduction of metal nanoparticles into the structure of the bioelectrode. This chapter reviews the most recent trends in the development of enzymatic biofuel cells containing nanoscaled materials.

Keywords: biological fuel cells, hybrid bioelectrodes, nanobioelectrocatalyst

* Corresponding Author Email: ardandra@ffclrp.usp.br.

INTRODUCTION

Biofuel Cells: Concepts and Technology

Overview

Among the various alternative processes for energy conversion, fuel cells seek efficient and sustainable production of electricity from electrochemical oxidation and reduction reactions [1]. This green system converts chemical energy into electricity through oxidation of a fuel catalyzed by either metal catalysts (traditional fuel cells) or biomolecules (biological fuel cells) [2]. Biological fuel cells have attracted the attention of several research groups because they display technological potential for many applications. For example, biological fuel cells can be used as an energy source in biomedical devices and microchips or as battery in small electronic devices [3-9]. Three main types of biological fuel cells exist: enzymatic, microbial, and organelle-based fuel cells [9]. Compared to traditional metal-based fuel cells, these devices present interesting features such as the ability to operate at lower temperatures and under physiological conditions [10]. Although research in the area of biological fuel cells has advanced in the past decades, the practical application of these cells requires that several key issues, like enzyme lifetime, electrode power density, and electrical connection between the active site of the enzyme and the surface of the electrode, be improved [11]. In the context of enzyme-based fuel cells, most literature papers have focused on the anodic side. Few investigations have dealt with the cathodic counterpart, which displays more complex reaction kinetics and has challenged many researchers [12, 13].

Electron Transfer

Electron transfer between the active site of the enzyme and the surface of the electrode can occur in two different ways (Figure 1): directly, known as direct electron transfer (DET), or via a mediator molecule, known as mediated electron transfer (MET). About 1400 oxidoreductase enzymes (enzymes that catalyze reactions involving electron transfer) are known to date (www.enzymedatabase.org). These enzymes could be used as biocatalysts for biological fuel cells. Less than a hundred of the known oxidoreductases can communicate with the surface of the electrode by DET, which makes the use of mediator molecules necessary for electrical connection between the enzyme and the electrode [14]. Although MET requires that a mediator molecule (usually organic dyes or organometallic complexes that remain in solution or in the form of a polymeric film) be added to the system, this process is often preferred over DET: MET systems generally afford higher power density and can employ commercially available enzymes, including alcohol dehydrogenase and glucose oxidase [15].

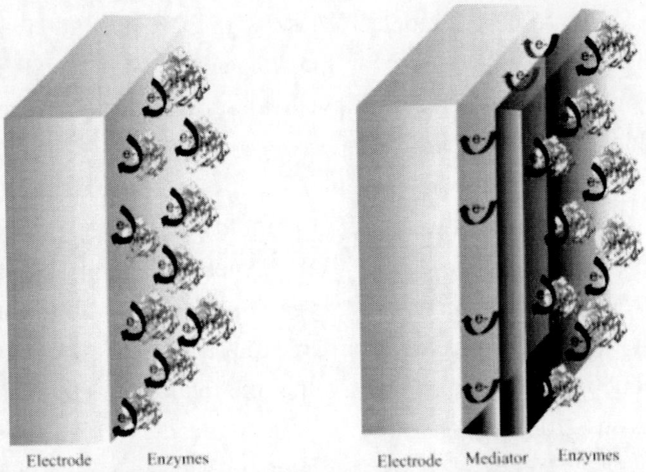

Figure 1. Types of electron transfer processes between enzymes and electrode surfaces in enzymatic biofuel cells, *Journal of the Brazilian Chemical Society*, *24*, p. 1895, Copyright 2013 by Sociedade Brasileira de Química.

Considering the biological fuel cells that operate through DET, they offer advantages such as the simpler architecture of the bioelectrode (which favors miniaturization of the process) and smaller performance losses due to differences between the potentials of the enzyme and the mediator species, which are the reason why researchers have increasingly targeted enzymes that can conduct DET. However, the investigated systems commonly yield lower power density values because it is difficult to connect enough molecules of the enzyme to the surface of the electrode to obtain satisfactory power values. Moreover, most of the enzymes that can perform this type of electron transfer (such as the pyrroloquinoline quinone-PQQ-dependent enzymes) are not commercially available, which demands their extraction and purification in the laboratory [16]. DET usually involves metals and metal complexes that are present in the structure of the protein (e.g., copper, iron, heme-c, etc.), whereas MET focuses on the use of ferrocene, quinones, and phenothiazines derivatives, among others [7].

Bioanodes and Biocathodes

Owing to their unique specificity, various oxidoreductase enzymes have been used to construct enzyme-based electrodes. The choice of enzyme frequently depends on the reaction to be catalyzed. Dehydrogenase enzymes (like hydrogenase and alcohol, aldehyde, and glucose dehydro-genase) and oxidases (such as glucose oxidase, peroxidase, catalase, xanthine, choline, urease, and lactate) are usually employed to prepare bioanodes [9]. Nicotinamide adenine dinucleotide (NAD)-dependent and flavin adenine dinucleotide (FAD)-dependent enzymes are the most often used. These co-factors, which may weakly or strongly bind to the structure of the protein, act as electron carriers in many enzyme-based bioelectrodes [15]. PQQ-dependent enzymes are also worthy of mention: these enzymes bear the cofactor (PQQ) covalently linked to the structure of the protein along with multiple heme-c moieties that can directly transfer electrons to conductive surfaces.

Concerning the cathodic side, the complex kinetics of the reaction taking place in this compartment has led numerous researchers to employ laccase or bilirubin oxidase in an attempt to achieve high catalytic activity for oxygen reduction [12, 13]. These multi-copper oxidases catalyze the reduction of oxygen to water via a four-electron mechanism with concomitant oxidation of aromatic organic substrates. The tri-nuclear metal cluster underlies the reduction of oxygen whilst the mononuclear site acts as the primary acceptor of electrons from a variety of organic molecules [17]. Among the different organic substrates that can act as redox mediators in laccase and bilirubin oxidase-based biocathodes, 2,2′-azino-bis(3-ethylbenzothiazoline-6-sulfonic acid) (ABTS) stands out—it is relatively inexpensive, has low toxicity, and affords high turnover rate. However, literature has shown that both laccase and bilirubin can be employed to prepare biocathodes that act via a DET mechanism. In this situation, the enzymes can directly transfer electrons to the surface of the electrode without the assistance of a mediator [16, 18]. Among the different methods reported in the specialized literature, modification of multi-walled carbon nanotubes (MWCNTs) with hydrophobic groups, like anthracene, promote the appropriate orientation of both laccase and bilirubin oxidase for the efficient bioelectrocatalytic reduction of oxygen in quiescent buffered solutions [19].

Immobilization

The immobilization of enzymes on solid surfaces has various biotechnological applications, including the use of immobilized enzymes in the industrial and analytical areas [15]. In the context of biological fuel cells, immobilization techniques largely influence the overall efficiency of the bioelectrode, which has increased researcher's interest in developing and improving the lifetime of these electrodes. Moving the enzyme from its natural solution environment to a solid surface requires specific methods that will ensure the conformation and catalytic properties of the protein will be retained while a mechanically and chemically stable layer for electron transfer will be achieved [20]. Because various functional groups are present in the structures of enzymes, different methods of immobilization can be used to anchor enzymes on solid surfaces. Chemical immobilization methods, during which the enzymes are directly bound to a solid support via covalent binding, sol-gel synthesis, or crosslinking, are commonly employed [21, 22]. When formation of a chemical bond is not necessary, simple and effective physical processes are more advantageous. Adsorption processes, membranes, entrapment into polymeric gels, and microcapsules are generally used to prepare bioelectrodes by physical processes [21, 22].

Electrode Materials

Besides the novel and improved immobilization methodologies presented in the specialized literature, advances in nanoscience and nanotechnology have provided a new direction for the development of micro- and nanomaterials to fabricate superior electrodes in the field of biological fuel cell technology. Some of the new trends in the design of catalysts for biological fuel cells include incorporation of materials at the nanoscale; e.g., carbon nanotubes, nanofibers, nanocomposites, and metallic nanoparticles, into the structure of the bioelectrode [16, 23]. Nanostructured materials are optimal environment to immobilize macromolecules: they provide enhanced surface area to anchor enzymes (which enables incorporation of a higher load of the desired enzyme), and they accelerate the kinetics through the surface of the bioelectrode. Moreover, the presence of nanomaterials increases the electroactivity of the bioelectrode markedly and improves the electrical contact between the active site of the enzyme and the surface of the electrode [24, 25]. Furthermore, incorporation of metal nanoparticles and enzymes into bioelectrodes can generate hybrid materials with synergistic properties, thus improving the overall performance [26, 27].

Carbon nanotubes (CNTs) have attracted considerable attention from the scientific community since their discovery in 1991. Various studies have shown that incorporation of these nanotubes into materials improves performance significantly and gives rise to extraordinary electronic, thermal, and mechanical properties [28, 29]. The high electrical conductivity of CNTs along with their biocompatibility makes them interesting components for the fabrication of bioelectrodes—the nanotubes favor both DET, by decreasing the electron transfer distance, and MET, through introduction of redox molecules into CNTs by functionalization or electropolymerization [29]. Both non-covalent and covalent approaches can be employed to immobilize enzymes on CNTs. Adsorption techniques and chemical modifications of their sidewalls are great options to prepare both bioanodes and biocathodes.

Metallic nanoparticles have gained researchers' attention as because they have potential application in numerous fields. The most important characteristics of these nanometric materials are their increased surface area volume, excellent electron transfer characteristics, and catalytic power. In the case of biological fuel cells, the ability of metallic nanoparticles to act as electron carriers between the enzyme and the surface of the electrode (thus favoring the bioelectrocatalytic process) underlies the great interest in incorporating them into the structure of the bioelectrode. The fact that enzymes and metallic nanoparticles have similar size reduces the distance of the electron transfer without harming the structure of the biomolecule [24]. Therefore, the integration of nanoparticles that exhibit unique electronic and catalytic properties with biomaterials that have particular specificity and catalytic efficiency can afford hybrid nanobiomaterials with synergistic properties and functions, to culminate in higher catalytic power and power densities [26].

More recently, graphene has also been considered for the design of nanobioelectrodes [30-32]. Many carbon allotropes including graphite, charcoals, and CNTs, among others, display excellent thermal conductivity, electrical conductivity, and mechanical strength, high surface-to-volume ratio, and great biocompatibility, which makes them potentially applicable in bioelectronic devices such as biosensors and biological fuel cells. It is possible to functionalize graphene with different compounds and to use the final material to construct either DET-based bioelectrodes or biomaterials based on electron transfer mediated by a third entity.

Nanostructured Materials for the Assembly of Bioelectrodes

The benchmark report about CNTs was authored by S. Iijuma [33]. Since then, countless investigations into these materials have been published in journals of different research areas, and these papers have also included biomacromolecules like enzymes or even microorganisms [34]. Carbon nanotubes are allotropic forms of carbon built in the sp_2 form; they comprise closed structures with hexagonal honeycomb lattices that present well-defined regions [35]. Multi-walled CNTs (MWCNTs) are the most common and the least expensive of CNTs. MWCNTs consist of multiple closed tubular layers of graphite sheets with sizes that typically range from 2 to 50 nm; they are separated by 0.34 nm [35].

The electroactivity of CNTs is attributed to the presence of reactive groups such as carboxylic and oxygenated species on their surface. These species emerge after pre-treatment of CNTs to eliminate metallic species. These surface groups aid immobilization of biomolecules on top of CNTs. Compton's group proposed that the activity of CNTs is related to the edge-plane located in the area of defects of the tubes [36]. As a general rule, electrodes modified with CNTs always display decreased overvoltage as well as increased power density or peak currents, which indicates better electroactivity of the 3D composite. Wang and co-workers confirmed this good performance when they reported a 490-mV shift in the NADH peak for a glassy-carbon electrode coated with MWCNTs [37]. Another research group has corroborated this behavior [35]. Several good reviews covering the attachment of enzymes to CNTs are available in the literature for further reading. See [29, 38-41], for instance.

Successful attachment of enzymes to CNTs and graphene may be obtained without the native structure of the enzyme being altered or the enzyme being leached to the solution during operation. In this context, covalent attachment provides more structurally stable and durable devices. Functionalization of CNTs with organic, polymeric, and biological molecules provides a more biocompatible environment and favors the attachment of enzymes while maintaining their activity [37]. Table 1 summarizes the most common protocols to immobilize enzymes on CNTs [29]. The great advantage of CNTs is that they provide a large three-dimensional surface area to immobilize the enzyme and its mediator, to enhance the current density. Also, CNTs allow close interaction with the active sites, to favor direct electron transfer (DET) from the deeply buried active site of the enzyme. These two factors increase the open circuit potential (which decreases the overpotential-thermodynamic gain) and the current density (which improves kinetics).

According to Cosnier and co-workers, the number of publications investigating CNTs in biological biofuel cells increased exponentially between 2004 and 2010 [49]. Figure 2 depicts a literature search profile employing the Web of Science database showing that the use of CNTs in biological fuel cells has increased systematically increasing over the last seven years. The use of graphene in biofuel cells is also gaining importance.

Table 1. Protocols to immobilize enzymes on CNT materials

Protocol	Immobilization of Enzyme	Reference
Non-covalent attachment	Enzyme is physically adsorbed onto the CNTs	[42]
Direct physical adsorption	Hydrophobic and π–π interaction between the sidewalls of CNTs and enzymes	[43], [44]
Enzymes adsorbed onto CNTs functionalized with polymers	Functionalization of CNTs favors aqueous dispersion and specific binding of enzymes	[45]
Covalent attachment of enzymes to CNTs with linking molecules	Linking molecules bind to CNTs through hydrophobic and π–π interactions. Then, a covalent binding ensues in specific sites of the enzyme	[46], [47]
Layer-by-layer technique to immobilize enzymes	Electrostatic adsorption assembly of oppositely charged layers	[48]

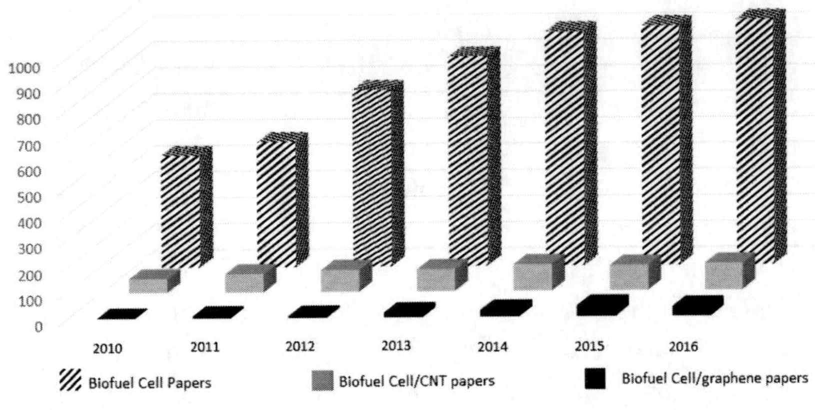

Figure 2. Literature search carried out in the Web of Science database covering biological fuel cells, biofuel cell/CNT, and biofuel cell/graphene.

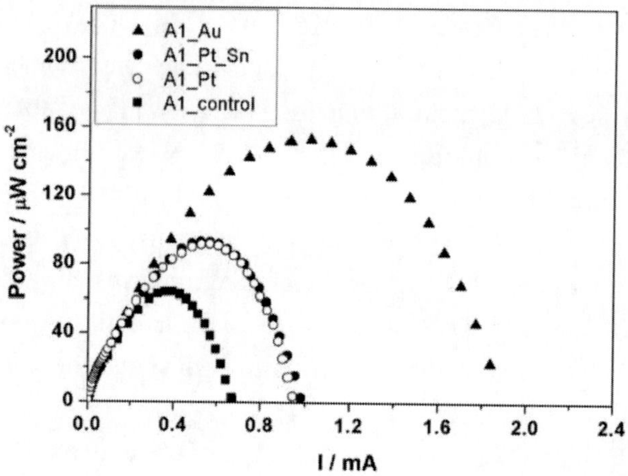

Figure 3. Power density curves obtained in an ethanol/O_2 biofuel cell by using the hybrid nanostructured bioanodes containing the commercially available enzyme alcohol dehydrogenase and either gold, platinum, or $Pt_{0.75}Sn_{0.25}$ nanoparticles. Reprinted from "Hybrid nanocatalysts containing enzymes and metallic nanoparticles for ethanol/O_2 biofuel" by Aquino Neto S. et al. 2014, *Journal of Power Sources*, *259*, p. 29, Copyright 2014 by Elsevier.

Our group has used a carbon cloth platform containing immobilized MWCNTs and the enzyme alcohol dehydrogenase (ADH) to improve electron transfer between the electrode and the enzyme. Systematic wiring of MWCNTs with a biological system in the best possible configuration and proper choice of wiring (Figure 3) generate a sixteen-fold higher power density as compared to bioanodes without CNTs [50]. At the cathode side, direct electron transfer of BOD depends on the amount of π-π conjugated system and on whether the carboxy groups at the surface of the electrode can be modulated to control the length of MWCNTs [51].

Metallic Nanoparticles

According to the literature, there are different ways to incorporate metallic nanoparticles with biomolecules into bioelectrodes. Most literature papers on this topic have investigated hybrid bioelectrodes containing metallic gold or platinum nanoparticles along with molecules of enzymes [26]. Gold nanoparticles have great biocompatibility and conductivity; compared to gold nanoparticles, platinum nanoparticles display similar properties as well as excellent electrocatalytic power [52]. The specialized literature brings publications on hybrid biosensors and enzymatic biofuel cells based on the oxidation of glucose and fructose containing gold nanoparticles [53, 54]. Likewise, incorporation of platinum nanoparticles with biomolecules into bioelectrodes has proven efficient in the cases of amperometric glucose sensor and enzymatic biofuel cells [26].

A good strategy to employ metallic nanoparticles when assembling bioelectrodes is to take advantage of the capacity of these nanoparticles to act as an electronic bridge for direct electron transfer. Gutiérrez-Sánchez et al. prepared a hybrid material that enhanced DET between laccase and an electroactive surface modified with functionalized gold nanoparticles for the bioelectrocatalytic reduction of oxygen at low overpotentials [55]. On the basis of the diazonium salt reduction strategy, these authors obtained a stable nanostructure through the step-by-step covalent attachment of gold nanoparticles along with laccase molecules to a porous graphite electrode. Through their Cu T1 site, the oriented enzymes performed fast DET with the surface of the electrode ($k_0 \gg 400\ s^{-1}$).

Iron can provide biocompatible metallic nanoparticles with interesting properties. By combining graphene with magnetic nanoparticles, Teymourian et al. developed a glassy carbon electrode modified with Fe_3O_4 magnetic nanoparticles/reduced graphene oxide nanosheets as a platform to catalyze lactate with lactate dehydrogenase [56]. These authors synthesized Fe_3O_4 nanoparticles by Massart's method [57] and obtained reduced graphene oxide nanosheets by reducing the oxide with hydrazine. The resulting hybrid nanocomposite showed excellent electrocatalytic activity toward different analytes due to the synergistic integration of the two nanomaterials, including low oxidation potential for NADH (0.05 V vs. Ag/AgCl) and reduced overpotentials for the reduction of both oxygen and peroxide [56].

Palladium is another noble metal that can be efficiently employed to manufacture bioelectrodes. Wen et al. prepared hybrid bioelectrodes containing palladium aerogels with high surface area along with glucose oxidase [58]. The high degree of porosity provided by the metallic matrix and its excellent electrical conductivity made it a promising material for bioelectronic purposes. The developed method was based on the use of calcium ions as destabilizing agent. In this method, growth was simply controlled by addition of different concentrations of Ca^{2+} to obtain the three-dimensional network of the composite. This was one of the first works to describe the application of aerogels for bioelectrocatalysis. The resulting hybrid material exhibited high activity toward the ferrocene-mediated oxidation of glucose and became a promising alternative to fabricate high-performance glucose biosensors and glucose/O_2 biofuel cells [58].

In 2015, Gai et al. prepared a ternary hybrid composite based on carbon nanotubes, graphitic carbon nitride nanosheets, and gold nanoparticles to obtain a membraneless glucose/oxygen biological fuel cell [59]. At the anodic side, the idea was to decorate a CNT/nitride nanocomposite with carboxyl-functionalized gold nanoparticles through electrostatic interaction, followed by immobilization of pyrroloquinoline quinone-dependent glucose dehydrogenase via a condensation reaction between the amino groups of the enzyme and the carboxyl groups of the gold nanoparticles. At the cathodic side, the same architecture was proposed for the structure of the electrode, and bilirubin oxidase afforded the bioelectrocatalytic reduction of oxygen. Concerning the assays conducted with the membraneless biofuel cells; the maximum power density was about 249 μW cm^{-}

2, which was 3.3 times higher as compared to the control bioelectrodes. In addition, evaluation of the stability of the electrode after 30 days revealed that only 27% of the original power density had been lost.

Still in 2015, gold nanoparticle-based electrodes were proposed as a platform to immobilize a hydrogenase enzyme for the bioelectrooxidation of hydrogen [60]. The anodic side consisted of a matrix of gold nanoparticles functionalized with thiol and a hydrogenase from the hyperthermophilic bacterium *Aquifex aeolicus*; the cathodic side was based on a DET cathode composed of bilirubin oxidase immobilized on unmodified gold nanoparticles or on a carboxylate-terminated self-assembled monolayer. The prepared hydrogen/oxygen biofuel cell allowed direct oxidation of hydrogen over a large range of temperatures and generated current density as high as 1.85 mA cm^{-2}, which was 170 times higher as compared to the control electrode based on a bare gold electrode. In addition, the authors described an impressive open circuit voltage of about 1.08 V, but the prepared biocathode lost about 50% of its initial current value. The current in the bioanode remained stable after the polarization experiment [60].

Regarding the oxidation of ethanol, Aquino Neto et al. presented different strategies to synthesize distinct bioelectrodes containing metallic nanoparticles and immobilized dehydrogenase enzymes [61-63]. The authors first evaluated hybrid nanostructured bioanodes containing the commercially available enzyme alcohol dehydrogenase and either gold, platinum, or $Pt_{0.75}Sn_{0.25}$ nanoparticles in an ethanol/O_2 biofuel cell [61]. The bioanodes consisted of a Nafion matrix modified with MWCNT-decorated metallic nanoparticles/tetrabutylammonium bromide (TBAB) or of a matrix of dendrimer-encapsulated nanoparticles to anchor the enzyme. A 20% platinum gas diffusion membrane cathode reduced oxygen. Addition of metallic nanoparticles to the structure of the bioanode enhanced the overall performance of the biofuel cell. The system containing gold nanoparticles performed the best: an open circuit potential of 0.61 V and a maximum power density of 155 μW cm^{-2} were recorded (Figure 3). On the basis of the voltammetric and chronoamperometric results, the authors claimed that the hybrid material containing gold nanoparticles performed better as a result of the catalytic behavior of the metallic nanoparticles toward the oxidation of NADH. In fact, among the different materials tested by the authors, the gold nanoparticles supported on MWCNTs provided a shift of 200 mV vs SCE toward negative potential for the desired reaction as compared to the most commonly employed NADH electrocatalyst, methylene green.

After Aquino Neto et al. verified that incorporation of metallic nanoparticles into the structure of bioanodes was beneficial, and that the gold-based material performed the best [42-44], the same research group investigated alternatives to synthesize nanoparticles and to improve the functioning of biofuel cells [62]. This was the first description of the synthesis of gold nanoparticles by the dendrimer-encapsulated methodology with effective anchoring onto the surface of CNTs. The authors observed that the hybrid bioanodes containing smaller and better-distributed gold nanoparticles on the surface of the CNT led to higher catalytic currents. Also, the type of functional group present on the surface of the CNTs directly influenced the final size of the crystallite and the performance of the nanotubes. The hybrid bioelectrode containing gold nanoparticles supported on NH_2-functionalized MWCNTs afforded the best results. In addition, this material shifted the potential of NADH oxidation to lower values, to generate higher power output values. This confirmed that the electrochemical regeneration of NAD^+ plays an important role in the performance of ethanol/O_2 biofuel cells based on the bioelectrocatalysis of dehydrogenase enzymes [62]. The same group also investigated the first two steps of the bioelectrooxidation of ethanol using the PQQ-dependent enzymes alcohol and aldehyde dehydrogenases [63]. The architectures of the bioanodes were designed to work through either mechanism (DET or MET); MWCNT-decorated gold nanoparticles along with different immobilization matrixes were employed. For MET, the highly efficient redox polymer 1′-dimethylferrocene-modified linear poly-ethyleneimine ($FcMe_2$-C_3-LPEI) was used. The results showed that the hybrid material performing DET still lacked high current density whereas the hybrid bioanode containing the ferrocene-based redox polymer generated J_{max} as high as 925 μA cm^{-2} in ethanol/air biofuel cell tests (Figure 4) [63].

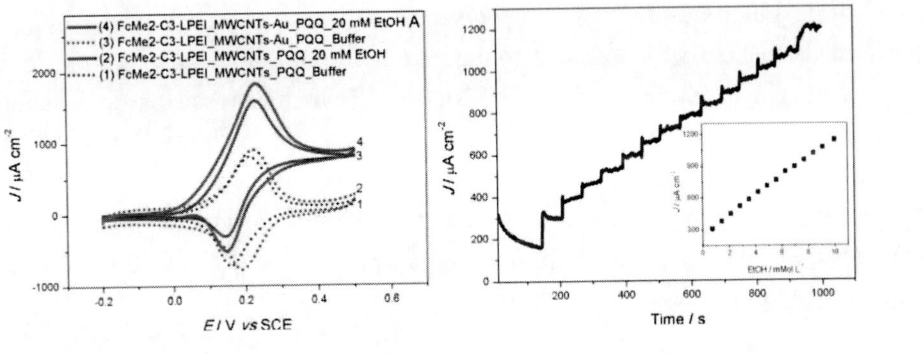

Figure 4. Representative cyclic voltammograms of the bioanodes based on $FcMe_2$-C_3-LPEI prepared with MWCNTs or MWCNTs-Au in the absence and in the presence of 20 mM ethanol. (B) Chronoamperometric assay performed with a bioanode prepared with the $FcMe_2$-C_3-LPEI redox polymer in the presence of MWCNTs-Au at a fixed oxidation potential (0.05 V larger than the oxidation peak potential) by gradually increasing the concentration of substrate in the electrolyte solution. 0.2 mol L^{-1} phosphate buffer pH 7.5, at 0.010 V s^{-1}. Reprinted from “High current density PQQ-dependent alcohol and aldehyde dehydrogenase bioanodes” by Aquino Neto S. et al. 2015, *Biosensors and Bioelectronics*, *72*, p. 253, Copyright 2014 by Elsevier.

Rojas-Carbonell and co-workers successfully integrated enzymes and non-platinum based catalysts into a single hybrid material targeting the oxygen reduction reaction [64]. When operating as an air breathing gas diffusional electrode, the prepared catalytic hybrid bioelectrode presented lower overpotential (as compared to a purely non-platinum catalyst) and generated higher current densities as compared to an enzyme-based catalyst.

In addition, the hybrid catalyst reduced oxygen through the desired four-electron transfer process, which afforded water as a final product. In such test, the authors achieved maximum current density of 1.2 mA cm^{-2}, which is one of the highest values reported for bilirubin oxidase-based gas-diffusion cathode to date [64].

PERSPECTIVES ON BIOMATERIALS FOR BIOFUEL CELL

Since the first description of an electrical signal obtained from a biological species in 1911, the goals of immobilizing enzymes and microorganisms onto the surface of an electrode have remained the same: to increase the current generated from a bioelectrocatalytic process and to improve the stability of the prepared biomaterial. For a long time, glucose-based biofuel cells were the standard material scientists desired to develop because they thought it might be possible to apply these cells as battery for pacemakers *in vivo*. However, after decades of research, it is still hard to predict when a biofuel cell will be commercially available for that purpose. After the great research conducted in the past years, all we can say is that a lot of improvement has been achieved. Considering the two main goals of biological fuel cells, the first one is much more within our reach because the power densities published in the past years have increased dramatically—they have moved from few microwatts to a few milliwatts in different enzymatic systems. What is also clear is that nanoscience and nanomaterials underlie most of this progress. On the other hand, the stability of the bioelectrodes reported in the past years is still far from the values needed for long-term practical applications. It is quite clear that research teams worldwide focus on stability, nanostructured materials will help them to achieve this goal and will definitely become an important part of the history of biofuel cells.

REFERENCES

[1] Lamy, C; et al., Recent advances in the development of direct alcohol fuel cells (DAFC). *Journal of Power Sources*, 2002, 105(2), p. 283-296.

[2] GTR, P; GM, W. Microbial and enzymatic biofuel cell, in Enzymatic Conversion of Biomass for Fuels Production, E. Himmel, Editor. 1994, *American Chemical Society.*, p. 271-290.

[3] Osman, MH; Shah, AA; Walsh, FC. Recent progress and continuing challenges in bio-fuel cells. Part I: Enzymatic cells. *Biosensors and Bioelectronics*, 2011, 26(7), p. 3087-3102.

[4] Zebda, A; et al., Mediatorless high-power glucose biofuel cells based on compressed carbon nanotube-enzyme electrodes. *Nat Commun*, 2011, 2, p. 370.

[5] Willner, I; Katz, E. Integration of Layered Redox Proteins and Conductive Supports for Bioelectronic Applications. *Angewandte Chemie International Edition*, 2000, 39(7), p. 1180-1218.

[6] Mao, F; Mano, N; Heller, A. Long Tethers Binding Redox Centers to Polymer Backbones Enhance Electron Transport in Enzyme "Wiring" Hydrogels. *Journal of the American Chemical Society*, 2003, 125(16), p. 4951-4957.

[7] Kavanagh, P. & Leech, D. Mediated electron transfer in glucose oxidising enzyme electrodes for application to biofuel cells: recent progress and perspectives. *Physical Chemistry Chemical Physics*, 2013, 15(14), p. 4859-4869.

[8] Shao, M; et al., Optimization of a Membraneless Glucose/Oxygen Enzymatic Fuel Cell Based on a Bioanode with High Coulombic Efficiency and Current Density. *Chem Phys Chem*, 2013, 14(10), p. 2260-2269.

[9] Cooney, MJ; et al., Enzyme catalysed biofuel cells. *Energy & Environmental Science*, 2008, 1(3), p. 320-337.

[10] Aquino Neto, S; De Andrade, AR. New energy sources: the enzymatic biofuel cell. *Journal of the Brazilian Chemical Society*, 2013, 24, p. 1891-1912.

[11] Leech, D; Kavanagh, P; Schuhmann, W. Enzymatic fuel cells: Recent progress. *Electrochimica Acta*, 2012, 84(0), p. 223-234.

[12] Omar Garcia, S; et al., Gas-Diffusion Cathodes Integrating Carbon Nanotube Modified-Toray Paper and Bilirubin Oxidase. *Journal of The Electrochemical Society*, 2014, 161(9), p. H523-H528.

[13] Le Goff, A; Holzinger, M; Cosnier, S. Recent progress in oxygen-reducing laccase biocathodes for enzymatic biofuel cells. *Cellular and Molecular Life Sciences*, 2015, 72(5), p. 941-952.

[14] Ramanavicius, A; Ramanaviciene, A. Hemoproteins in Design of Biofuel Cells. *Fuel Cells*, 2009, 9(1), p. 25-36.

[15] Bullen, RA; et al., Biofuel cells and their development. *Biosensors & Bioelectronics*, 2006, 21(11), p. 2015-2045.

[16] Falk, M; Blum, Z; Shleev, S. Direct electron transfer based enzymatic fuel cells. *Electrochimica Acta*, 2012, 82(0), p. 191-202.

[17] Solomon, EI; Sundaram, UM; Machonkin, TE. Multicopper Oxidases and Oxygenases. *Chemical Reviews*, 1996, 96(7), p. 2563-2606.

[18] Blanford, CF; Heath, RS; Armstrong, FA. A stable electrode for high-potential, electrocatalytic O2 reduction based on rational attachment of a blue copper oxidase to a graphite surface. *Chemical Communications*, 2007, (17), p. 1710-1712.

[19] Meredith, MT; et al., Anthracene-Modified Multi-Walled Carbon Nanotubes as Direct Electron Transfer Scaffolds for Enzymatic Oxygen Reduction. *ACS Catalysis*, 2011, 1(12), p. 1683-1690.

[20] Aquino Neto, S; Forti, JC; De Andrade, AR. An Overview of Enzymatic Biofuel Cells. *Electrocatalysis*, 2010, 1(1), p. 87-94.

[21] Adlercreutz, P. *Immobilized Enzymes*, in *Enzymes in Food Processing*, N. T. and R. G., Editors. 1993, Academic Press: New York. p. 103-119.

[22] Minteer, SD. *Enzyme Stabilization and Immobilization: Methods and Protocols.* 2011, New York: Humana Press.

[23] Campbell, AS; et al., Membrane/Mediator-Free Rechargeable Enzymatic Biofuel Cell Utilizing Graphene/Single-Wall Carbon Nanotube Cogel Electrodes. *ACS Applied Materials & Interfaces*, 2015, 7(7), p. 4056-4065.

[24] Vincent, KA; et al., Enzymatic catalysis on conducting graphite particles. *Nat Chem Biol*, 2007, 3(12), p. 761-762.

[25] Lalaoui, N; et al., A membraneless air-breathing hydrogen biofuel cell based on direct wiring of thermostable enzymes on carbon nanotube electrodes. *Chemical Communications*, 2015, 51(35), p. 7447-7450.

[26] Yang, XY; et al., Immobilization technology: a sustainable solution for biofuel cell design. *Energy & Environmental Science*, 2012, 5(2), p. 5540-5563.

[27] Trifonov, A; et al., Electrically Contacted Bienzyme-Functionalized Mesoporous Carbon Nanoparticle Electrodes: Applications for the Development of Dual Amperometric Biosensors and Multifuel-Driven Biofuel Cells. *Advanced Energy Materials*, 2015, 5(8), p. 1853-1853.

[28] Wang, Y; et al., Multifunctional carbon nanotubes for direct electrochemistry of glucose oxidase and glucose bioassay. *Biosensors and Bioelectronics*, 2011, 30(1), p. 107-111.

[29] Feng, W; Ji, P. Enzymes immobilized on carbon nanotubes. *Biotechnology Advances*, 2011, 29(6), p. 889-895.

[30] Karimi, A; et al., Graphene based enzymatic bioelectrodes and biofuel cells. *Nanoscale*, 2015, 7(16), p. 6909-6923.

[31] Wen, D; Eychmuller, A. Enzymatic Biofuel Cells on Porous Nanostructures. *Small*, 2016, 12(34), p. 4649-4661.

[32] Koushanpour, A; et al., Graphene-Functionalized 3D-Carbon Fiber Electrodes - Preparation and Electrochemical Characterization. *Electroanalysis*, 2016, 28(9), p. 1943-1946.

[33] Iijima, S. Helical microtubules of graphitic carbon. *Nature*, 1991, 354(6348), p. 56-58.

[34] Ajayan, PM; Zhou, OZ. *Applications of Carbon Nanotubes, in Carbon Nanotubes: Synthesis, Structure, Properties, and Applications*, M.S. Dresselhaus, G. Dresselhaus, and P. Avouris, Editors. 2001, Springer Berlin Heidelberg: Berlin, Heidelberg., p. 391-425.

[35] Rivas, GA; et al., Carbon nanotubes for electrochemical biosensing. *Talanta*, 2007, 74(3), p. 291-307.

[36] Banks, CE; et al. Electrocatalysis at graphite and carbon nanotube modified electrodes: edge-plane sites and tube ends are the reactive sites. *Chemical Communications*, 2005(7), p. 829-841.

[37] Musameh, M; et al., Low-potential stable NADH detection at carbon-nanotube-modified glassy carbon electrodes. *Electrochemistry Communications*, 2002, 4(10), p. 743-746.

[38] de Poulpiquet, A; Ciaccafava, A; Lojou, E. New trends in enzyme immobilization at nanostructured interfaces for efficient electro-catalysis in biofuel cells. *Electrochimica Acta*, 2014, 126, p. 104-114.

[39] Minteer, SD; et al., New materials for biological fuel cells. *Materials Today*, 2012, 15(4), p. 166-173.

[40] Cosnier, S; et al., Recent advances on enzymatic glucose/oxygen and hydrogen/oxygen biofuel cells: Achievements and limitations. *Journal of Power Sources*, 2016, 325, p. 252-263.

[41] Zhou, M; Dong, S. Bioelectrochemical Interface Engineering: Toward the Fabrication of Electrochemical Biosensors, Biofuel Cells, and Self-Powered Logic Biosensors. *Accounts of Chemical Research*, 2011, 44(11), p. 1232-1243.

[42] Lee, CA; Tsai, YC. Preparation of multiwalled carbon nanotube-chitosan-alcohol dehydrogenase nanobiocomposite for amperometric detection of ethanol. *Sensors and Actuators B: Chemical*, 2009, 138(2), p. 518-523.

[43] Pedrosa, VA; et al., Enhanced stability of enzyme organophosphate hydrolase interfaced on the carbon nanotubes. *Colloids and Surfaces B: Biointerfaces*, 2010, 77(1), p. 69-74.

[44] Gao, Y; Kyratzis, I. Covalent Immobilization of Proteins on Carbon Nanotubes Using the Cross-Linker 1-Ethyl-3-(3-dimethylamino-propyl)carbodiimide—a Critical Assessment. *Bioconjugate Chemistry*, 2008, 19(10), p. 1945-1950.

[45] Wu, X; et al., Effects of Ionic Liquids on Enzymatic Catalysis of the Glucose Oxidase toward the Oxidation of Glucose. *The Journal of Physical Chemistry B*, 2009, 113(40), p. 13365-13373.

[46] Pang, HL; et al., Immobilization of laccase onto 1-aminopyrene functionalized carbon nanotubes and their electrocatalytic activity for oxygen reduction. *Electrochimica Acta*, 2010, 55(22), p. 6611-6616.

[47] Kim, BJ; et al., Immobilization of horseradish peroxidase on multi-walled carbon nanotubes and its enzymatic stability. *Current Applied Physics*, 2009, 9(4, Supplement), p. e263-e265.

[48] Arugula, M; Simonian, A. Review—Nanocarbon-Based Multi-Functional Biointerfaces: Design and Applications. *ECS Journal of Solid State Science and Technology*, 2016, 5(8), p. M3045-M3053.

[49] Holzinger, M; Le Goff, A; Cosnier, S. Carbon nanotube/enzyme biofuel cells. *Electrochimica Acta*, 2012, 82, p. 179-190.

[50] Fenga, PG; et al., Multiwalled carbon nanotubes to improve ethanol/air biofuel cells. *Electrochimica Acta*, 2013, 106(0), p. 109-113.

[51] So, K; et al., Significance of the Length of Carbon Nanotubes on the Bioelectrocatalytic Activity of Bilirubin Oxidase for Dioxygen Reduction. *Electrochimica Acta*, 2016, 192, p. 133-138.

[52] Li, Y; et al., Platinum Nanoparticles (PtNPs) - Laccase Assisted Biocathode Reduction of Oxygen for Biofuel Cells. *International Journal of Electrochemical Science*, 2011, 6(12), p. 6398-6409.

[53] Murata, K; et al., High performance bioanode based on direct electron transfer of fructose dehydrogenase at gold nanoparticle-modified electrodes. *Electrochemistry Communications*, 2009, 11(3), p. 668-671.

[54] Wang, X; et al., Mediatorless sugar/oxygen enzymatic fuel cells based on gold nanoparticle-modified electrodes. *Biosensors and Bioelectronics*, 2012, 31(1), p. 219-225.

[55] Gutiérrez-Sánchez, C; et al., Gold Nanoparticles as Electronic Bridges for Laccase-Based Biocathodes. *Journal of the American Chemical Society*, 2012, 134(41), p. 17212-17220.

[56] Teymourian, H; Salimi, A; Khezrian, S. Fe3O4 magnetic nanoparticles/reduced graphene oxide nanosheets as a novel electrochemical and bioeletrochemical sensing platform. *Biosensors & Bioelectronics*, 2013, 49, p. 1-8.

[57] Massart, R. Preparation of aqueous magnetic liquids in alkaline and acidic media. *IEEE Transactions on Magnetics*, 1981, 17(2), p. 1247-1248.

[58] Wen, D; et al., Controlling the Growth of Palladium Aerogels with High-Performance toward Bioelectrocatalytic Oxidation of Glucose. *Journal of the American Chemical Society*, 2014, 136(7), p. 2727-2730.

[59] Gai, PP; et al., A ternary hybrid of carbon nanotubes/graphitic carbon nitride nanosheets/gold nanoparticles used as robust substrate electrodes in enzyme biofuel cells. *Chemical Communications*, 2015, 51(79), p. 14735-14738.

[60] Monsalve, K; et al; Hydrogen bioelectrooxidation on gold nanoparticle-based electrodes modified by Aquifex aeolicus hydrogenase: Application to hydrogen/oxygen enzymatic biofuel cells. *Bioelectrochemistry*, 2015, 106, p. 47-55.

[61] Aquino Neto, S; et al., Hybrid nanocatalysts containing enzymes and metallic nanoparticles for ethanol/O2 biofuel cell. *Journal of Power Sources*, 2014, 259(0), p. 25-32.

[62] Aquino Neto, S; et al., Enhanced Reduced Nicotinamide Adenine Dinucleotide electrocatalysis onto multi-walled carbon nanotubes-decorated gold nanoparticles and their use in hybrid biofuel cell. *Journal of Power Sources*, 2015, 273(0), p. 1065-1072.

[63] Aquino Neto, S; et al., High current density PQQ-dependent alcohol and aldehyde dehydrogenase bioanodes. *Biosensors and Bio-electronics*, 2015, 72(0), p. 247-254.

[64] Rojas-Carbonell, S; et al., Hybrid electrocatalysts for oxygen reduction reaction: Integrating enzymatic and non-platinum group metal catalysis. *Electrochimica Acta*, 2016, 190, p. 504-510.

In: Advanced Materials and Systems … ISBN: 978-1-53613-852-8
Editor: Mauro Coelho dos Santos

Chapter 4

ACTIVATED NITROAROMATICS ARE USEFUL SURFACE REDOX MODIFIERS IN ELECTROANALYSIS

Cleylton B. Lopes[1,2], Erivaldo O. Costa[1], Leonardo V. da Silva[1,2], Jadriane A. Xavier[1], Andresa K. A. de Almeida[1], Francisco A. S. Silva[1,3], Phabyanno R. Lima[1,2], Lauro T. Kubota[4] and Marília O. F. Goulart[1,*]

[1]Institute of Chemistry and Biotechnology, Federal University of Alagoas, Maceió, Alagoas, Brazil

[2]Federal Institute of Education, Science and Technology of Alagoas, Maceió, Alagoas, Brazil

[3]Federal Institute of Education, Science and Technology Baiano, Catu, Bahia, Brazil

[4]Department of Analytical Chemistry, Brazil Institute of Chemistry, University of Campinas (UNICAMP), Campinas, São Paulo, Brazil

ABSTRACT

Chemically modified electrodes (CME) are electrode surfaces that have been modified by adequate molecules or polymers through different techniques to render an electrochemical function that is either not possible or difficult to achieve and suited for a particular function, such as sensing, protection against corrosion, electrocatalysis, and electro-chromic devices, among others. This area has blossomed following the

[*] Corresponding Author Email: mofg@qui.ufal.br.

development of a molecular-level understanding of the synthesis and characterization of interfacial architectures using modern superficial analysis techniques. The construction of chemically modified electrodes has attracted increasing interest over the past few decades, with a large number of publications describing different applications, including food quality control, environmental monitoring, clinical diagnostics, safety/ security, and industrial applications, among others. Several authoritative books and reviews in the area have been published. In the present chapter, we emphasize a special class of surface modifiers based on the reductive activation of nitroaromatic ($ActArNO_2$) redox mediators that form the reversible couple ArNHOH/ArNO in the presence or absence of multi-walled carbon nanotubes (MWCNT), as well as some of the most recent applications in electroanalysis. Here, we present a detailed discussion of their preparation, properties and applications in the electrochemical (bio)sensing field. The combination $ActArNO_2$ + MWCNT leads to a better performance due to synergistic effects and/or complementary electrical, electrochemical and mechanical properties. Finally, we provide an overview of the current state and future trends in this area.

Keywords: chemically modified electrodes, redox mediators, activated nitroaromatics, electroanalysis

INTRODUCTION

The science of electrochemistry is concerned with electron transfer, mainly at the solution/electrode interface. Electrochemistry plays key roles in a broad range of fundamental and applied areas involving many biological relevant processes that exploit the synthesis and reactivity of inorganic, organic and organometallic compounds, including nanoscale investigations, corrosion, current industrial technologies for energy conversion and storage, and importantly, (bio)analytical chemistry [1]. Electrochemical methods have found important applications in sample analysis, as the electrode surface itself has been used as a variable free energy source (or sink) of electrons.

Chemical sensors, consisting of a transduction element covered by a recognition layer (Figure 1), hold a prime position in quantitative/ qualitative (bio)analysis [2, 3].

The analytical output of optical sensors results from the interaction of the analyte with the transducer in terms of optical properties such as absorbance, reflectance, luminescence, fluorescence, refractive index, surface plasmon resonance (SPR), optothermal effect and light scattering [4]. The piezoelectric effect arises when pressure applied to a dielectric material deforms its crystal lattice, causing the separation of centers of opposite charges and producing dipole moments in molecules [5]. Variable values, which are measured with piezoelectric sensing devices, are usually digital quantities. The resonant frequency (fundamental, $f0$, or harmonic, fn), change in the resonant frequency (Δf), phase angle (ϕ), acoustic wave speed (ν), change in the acoustic wave speed ($\Delta\nu$), or relative changes in these quantities with respect to some reference values ($\Delta f/f$ or $\Delta\nu/\nu$) are measured, depending on the type of a piezoelectric sensing device used. A frequency counter or a phase matching circuit is often used for that purpose. Piezoelectric chemical sensors are generally selective surface film-coated piezoelectric crystal oscillators. Two of the most commonly explored techniques for electrochemical mass sensing are the quartz crystal microbalance (QCM) and cantilevers, in which the resonant frequency of the crystal depends on its chemical nature, size, shape and mass [4]. Thermal sensors rely on measurements of the heat effects of a specific chemical reaction or adsorption event that involves the analyte. For example, in catalytic sensors, the heat of a combustion reaction or an enzymatic reaction is measured by a thermistor [4].

As shown in Figure 1, the nature of the electrochemical changes being detected allows their classification into the categories listed below [2].

1. Amperometric sensors are based on the current generated as a function of the redox behavior of the target analyte on the surface of a working electrode. The magnitude of this current is proportional to the concentration of electroactive species present in solution.
2. Potentiometric sensors convert analytical information into a potential signal that is proportional (on a logarithmic scale) to the concentration (activity) of the species involved in the recognition event. Normally, potentiometric sensors measure the potential across an interface, often a membrane.
3. Conductimetric sensors measure electrolyte conductivity, which varies with changes in the concentration of ionic species; electrical conductance/resistance is the measured parameter.
4. Impidimetric sensors measure impedance, both resistance and reactance.

Electrochemical sensors show advantages for monitoring biomolecules due to several important characteristics (Figure 2) [3, 6]. The capacity to operate in complex matrixes, which is central to a clinical diagnosis and for monitoring the treatment of diseases, demands qualities such as stability, biocompatibility and miniaturization [7]. The field of point-of-care (POC) diagnostics offers the possibility of providing rapid diagnostic results in nonlaboratory settings [8]. Bioanalytical sensors that are capable of accurately measuring the chemical dynamics of important physiological analytes *in vivo* have been transformative on several fronts, including disease/health management and improving our understanding of biological processes [9, 10].

The quantifiable terms, which are essential for ensuring the quality of the analysis, are validated by assessing analytical parameters (Figure 2) [6, 11]. Sensitivity is the slope of the analytical calibration curve and it refers to a small change in analyte concentration that causes a change in the response. Selectivity refers to the extent to which a method can determine the concentrations of particular analytes in mixtures or matrices without interference from other components [12]. A method is selective when the response of the analyte is differentiated from every other response. Repeatability and reproducibility are different ways to express the precision. Repeatability applies to replicate analyses performed by one person or using one method, whereas reproducibility applies to replicate analyses performed by different analysts, using different methods, etc. LOD (limit of detection) is the lowest detectable level with acceptable degree of certainty for a given analytical procedure [6, 11]. The signal to background noise ratio is a measure of signal strength relative to background noise, which is *represented* as the *mean* value of a *signal* (S) divided by the standard deviation of the *background* (N). All analytical parameters are used to compare the performance of sensors and biosensors, as well as their reliability, capacity and variability.

Despite the great versatility and applications of the electrochemical sensors, they have some limitations. At least one heterogeneous step is always involved. The gradual surface passivation and/or deactivation is one of the most important drawbacks [13] in addition to the large background current, which results in inferior analytical performance (Figure 2). Passivation occurs through the adsorption of the products of the redox reaction used for detection or through by-products of these reactions that polymerize and accumulate on the surface of the electrodes.

The direct electrochemical detection of many biologically and environmentally important compounds usually necessitates large overpotentials, which exceed the potential window for solid electrodes, to obtain reasonable current densities.

Moreover, surface modifications of the electrodes are required and, depending on the results, have been widely applied in electroanalysis. Surface modifications directly influence the ability to recognize or amplify the current signal either through catalytic effects (decrease in the overpotential of the electrodic processes), redox catalysis (use of a catalyst with homogeneous electron transfer) or increasing selectivity by restricting the passage of interfering species using appropriate membranes [13].

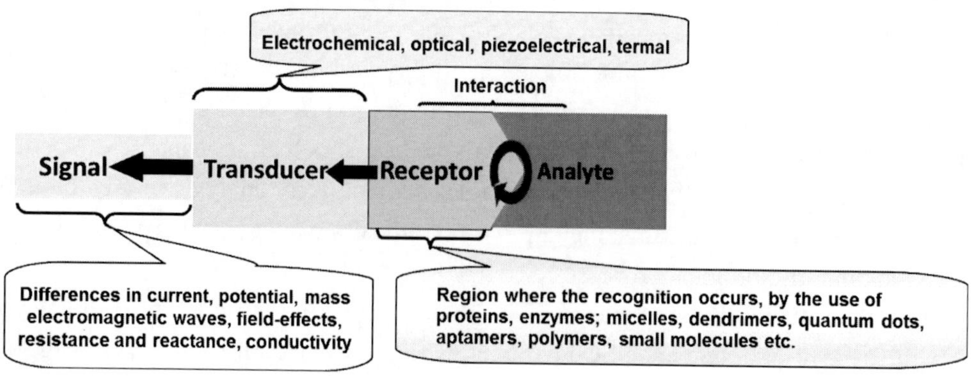

Figure 1. Simplified scheme of a chemical sensor.

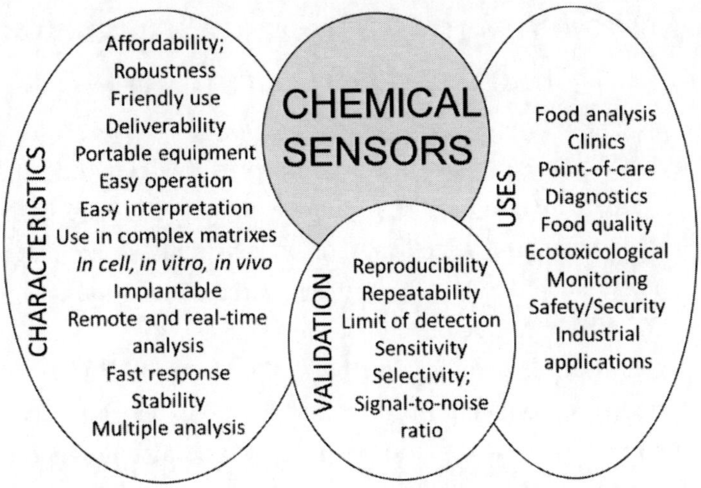

Figure 2. Uses and main characteristics of chemical sensors, including analytical parameters used to validate the process.

The present chapter focuses on a special class of surface modifiers based on activated nitroaromatic redox mediators, and describes some of the most recent materials applied in electroanalysis by providing several examples of findings from our and other research groups and their applications in the electrochemical (bio)sensing field.

Chemically Modified Electrodes: General Comments

Electrode surfaces are modified to render an electrochemical function that is impossible or difficult to achieve using conventional electrodes.

The main interests include increased selectivity, sensitivity, chemical and electrochemical stability (Figure 2), as well as a larger usable potential window and improved resistance to fouling. These electrodes are characterized in fundamental studies aimed at understanding interfacial processes [2, 14-17]. Therefore, the need for improved electrode performance and logically designed interfaces is rapidly growing in many areas of science.

In particular, a chemically modified electrode (CME) is a conducting or semiconducting material coated with a monomolecular, multimolecular, ionic, or polymeric film that modifies the electrochemical properties of the interface. The conductive and semiconductive substrates are derived from conventional electrode materials, whereas the adlayers have a wide range of origins and properties [14]. The first reports on electrodes with a modified surface appeared in the 1970s in studies using olefins on Pt [18, 19].

APPROACHES TO GENERATING CHEMICALLY MODIFIED ELECTRODES (CME)

First, we need to consider the substrate, the platform electrode that supports the adlayer. Commercially available or custom fabricated substrates are used. Although any conventional electrode material can serve as a substrate, some materials are more convenient to modify, such as Au, Pt, Ag and carbon electrodes, as well as carbon nanofibers, which display reproducible and clean surfaces [14].

In most schemes used to prepare electrodes, the electrode surface is treated before modification to provide reproducible surface attributes, e.g., crystallinity, roughness, and chemical functionality. Specific functional groups may also be necessary for some modification strategies. In these cases, pretreatment methodologies can include steps designed to create a high population of active sites on the substrate surface. Likewise, the requirements of adlayer characterization methods may require specific substrate properties. For example, atomically smooth surfaces are preferable for high-resolution scanning tunneling microscopic (STM) characterization of adlayer structures. Thus, the intended application of the modified electrode, requirements of surface characterization techniques, and the modification chemistry often direct the choice of substrates and pretreatment procedures [14].

Important questions concerning CME that must be considered have been posited.

1. What is the surface coverage?
2. What is the orientation of the admolecule or molecular layer on the surface?
3. What is the spatial uniformity of the admolecule or molecular layer over the surface?
4. What type of electrical connection exists between the admolecule or molecular layer and the electrode surface?

After the choice and preparation of the substrate, the surface can be modified using one of several methods (substrate or platform).

Figure 3 displays some of the most important approaches [20, 21].

Adsorption (Chemisorption)

The procedure is simple and useful. The film is strongly adsorbed, or chemisorbed, onto the surface of the electrode. This process is performed by dipping the electrode into a mediator solution or by preparing supporting materials prior to mixing with the electrode material. This approach involves substrate-coupled self-assembled monolayers (SAMs), where molecules are spontaneously chemisorbed to the surface of the electrode because the substrate environment is energetically more favorable than the solution environment. In cases of limited stability of the layer, improvements are possible by introducing functional groups into the molecule.

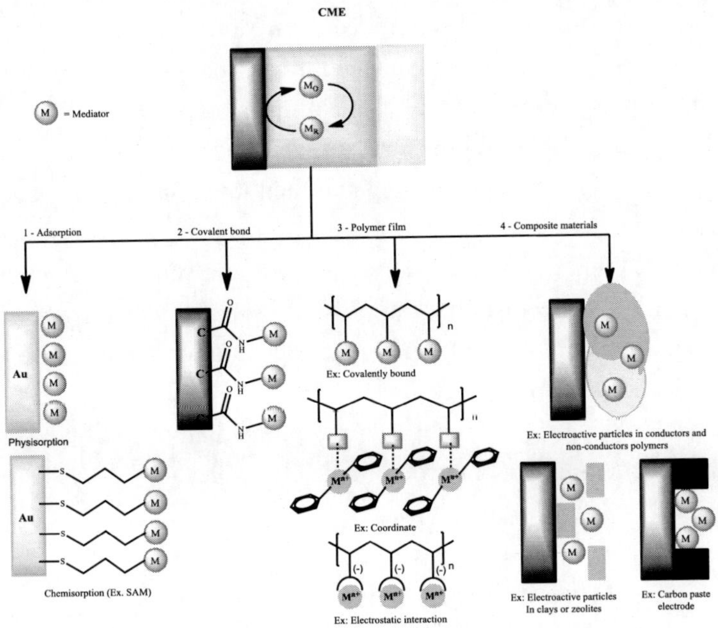

Figure 3. Scheme showing some of the approaches employed for the chemical modification of electrodes. CME: chemically modified electrodes.
Adapted from Zen et al., 2003; Kochius et al., 2012 [20, 21].

Covalent Bonding

A method that uses chemical agents to create a covalent bond between one or more monomolecular layers of the chemical modifier and the electrode surface. Examples include the bonds between a disulfide or a protein on Hg, an olefin on Pt, etc.

Polymer Coating

A method that introduces electron-conductive and nonconductive polymer films on the electrode surface through physical or chemical adsorption. The polymer films are organic, organometallic or inorganic, and either contain the chemical modifier or have the chemical added to the polymer in a subsequent process. Polymers are interesting materials as a support matrix for the immobilization of biomolecules and for electrocatalysis. They are polymerized on an electrode via cyclic voltammetry or chronoamperometry [21].

Entrapping/Incorporation

A method in which the chemical modifier is mixed with an electrode matrix material. An example of this method is to mix an electron-transfer mediator (the chemical modifier) with carbon nanoparticles/graphenes and other modifiers in a carbon paste electrode (the electrode matrix). Encapsulation of the mediator is also possible by trapping it within pores on the surface of a tridimensional multiwalled carbon nanotube (MWCNT) array [22]. If insoluble, an organic solvent is used to dissolve the electroactive species in the entrapment step and then it is transferred to an aqueous buffer for the detection step.

Nanostructures offer a large surface area-to-volume ratio to immobilize labels and biological recognition elements, thus amplifying the analytical signal and improving sensitivity [21, 23]. The use of CMEs is a developing area and new methods of preparation are described almost daily [24].

In the present chapter, we focus on redox mediator-modified electrodes, more specifically on reduction-activated nitroaromatic electrodes.

Redox Mediators: Importance

The efficiency of electroanalysis is increased by the use of indirect electrochemistry through redox mediators, which, ideally, engage in reversible electron transfer initiated at the electrode (heterogeneous) and followed by a chemical reaction (homogeneous), thus utilizing a electrochemical catalytic (E_rC_{cat}) mechanism (Figure 4) [2, 15, 20].

Indirect electrochemical conversions are hybrids between direct electrochemical conversions (Figure 4A) and homogeneous redox processes (both reagents in solution). The heterogeneous electron transfer between the electrode and the substrate is replaced by a homogeneous redox reaction in solution, which occurs between an electrochemically activated species (mediator) and the substrate [16, 17] (Figure 4B). The conditions are chosen to ensure that only the mediator undergoes oxidation or reduction at the electrode, and the substrate, intermediate(s), and the product(s) do not interfere with the electrochemical regeneration of the catalyst. In some cases, the mediator is not directly regenerated to its redox partner, but suffers structural changes; however, redox catalysis is still observed (Figure 4C).

Mechanistically, two types of redox electron transfer processes have been described (Figure 5).

In the first process, a non-bonded or outer-sphere, electron is transferred between the starting material (Red or Oxi) and the mediator (M_R or M_O). In the second so-called inner-sphere pathway, a homogeneous chemical reaction involving bond formation between the active form of the mediator and the substrate occurs and is followed by bond cleavage to regenerate the mediator (M_R or M_O) along with the oxidized (reduced) form of the substrate (Oxi or Red). Alternatively, inner-sphere electron transfer may occur within a charge transfer complex [16, 17], either after the incorporation of the substrate or an intermediate into the inner-coordination sphere of a metal ion or in the form of a hydrogen or hydride atom abstraction, among other possibilities [25]. In the case of the outer-sphere mechanism, the mediator catalyzes electron exchange between the substrate and the electrode.

In indirect anodic oxidation reactions, the electron transfer equilibrium resides on the side of the substrate if the redox potential of the mediator is lower than the oxidation potential of the starting material (conversely for indirect cathodic reductions). However, the position of this equilibrium opposite to the standard potential gradient occurs through two different mechanisms. In the first mechanism, a thermodynamically unfavorable electron transfer equilibrium may be shifted toward the product when one or more of the subsequent steps consist of an irreversible chemical reaction [16, 17, 25] (Figure 5). In this case, the reaction rate is not only controlled by the equilibrium constant K but also by the rate constant k of the subsequent chemical reaction.

Surface-confined redox mediators display several advantages in electroanalysis, as listed below.

1. The kinetic inhibition of the heterogeneous electron transfer reaction between the electrode and substrate (Figure 4A) can be eliminated, enabling overpotentials to be avoided and reactions to be accelerated. These properties are particularly significant for large biomolecules that exhibit strong kinetic inhibition due to steric shielding of their redox centers.
2. Electron transfer mediators may exhibit higher or completely different selectivity.
3. When the direct electrochemical conversion (Figure 4A) causes passivation of the electrode, the employment of a mediator can be helpful, since the direct interaction of the substrate with the electrode surface is avoided.

4. Since the electroanalysis is conducted at potentials lower than the redox potential of the starting material, the reaction is performed under milder conditions and side reactions are avoided. These features are particularly significant when sensitive functional groups, which are not intended to react, are present.

Despite their utmost important, enzymes will not be analyzed in the present chapter. Several reviews are available in this area [2].

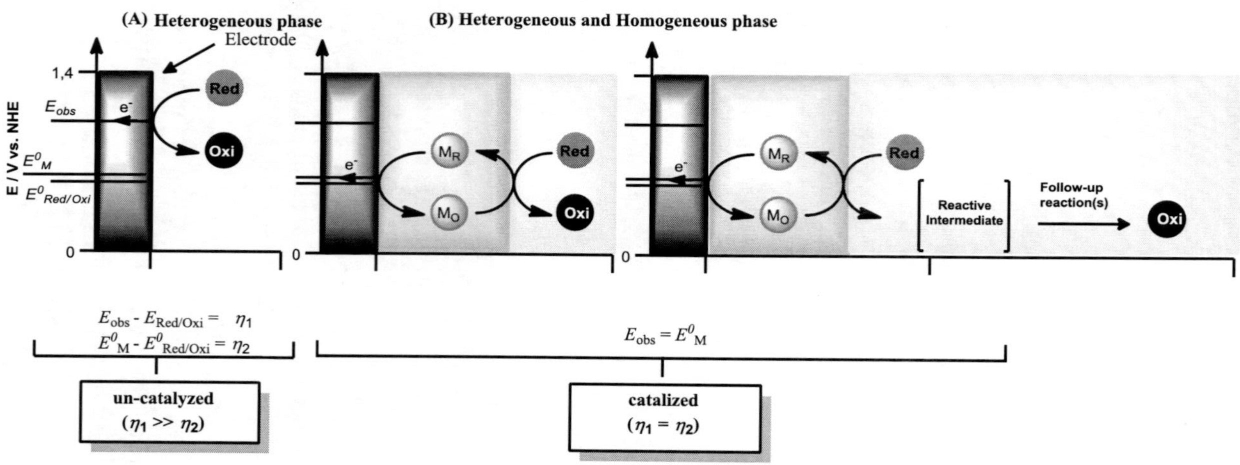

Figure 4. Comparison between oxidation processes: A) direct and B) and C) indirect or mediated processes. (Red) and (Oxi) represent the analyte in its reduced and oxidized states, respectively. M_R and M_O represent the mediator in the reduced and oxidized states, both of which are confined to the electrode surface. Adapted from Zen et al., 2003 [20].

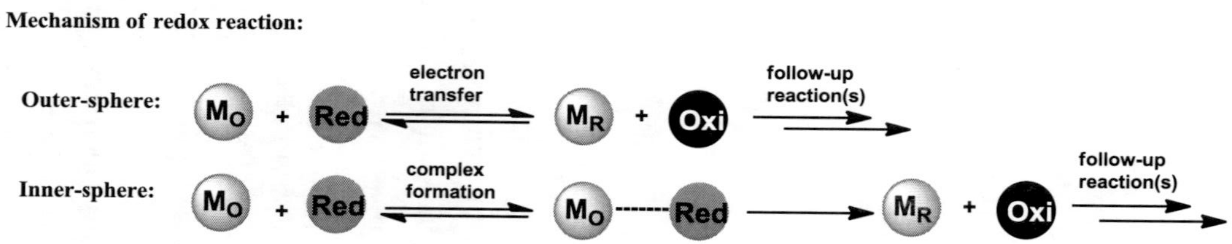

Figure 5. Different mechanisms for mediator-induced electron transfer. Adapted from Francke and Little, 2014 [16, 17].

The mediator is generally a reversible redox system that catalyzes redox reactions. Low reversibility usually means that the mediator is prone to side reactions, leading to lower catalytic turnover numbers [16, 17, 25].

The main desired characteristics of a mediator include:

1. a low formal potential (less negative or less positive, depending on whether the process is reduction or oxidation, respectively), to avoid interferences as much as possible;
2. a lower dependence or complete independence of the oxygen concentration;
3. stability in both oxidized and reduced forms (toward hydrolysis, light decomposition, and chemical oxidation), as even slight side reactions that produce unregenerable compounds compromise their catalytic activity;
4. a fast and reversible electron transfer reaction between the electrode and the immobilized mediator; a second-order rate constant of at least 10^6-10^7 M^{-1} s^{-1}, preferably higher, approaching a diffusion-controlled process [26];
5. a fast charge transfer rate within the film (for polymeric and multilayer coatings);
6. a fast electron transfer rate between the confined mediator and the analyte;
7. inertness toward other substances present in the samples, specifically solvents and products, nucleophiles or electrophiles;
8. non toxic;
9. cheap;
10. long-term stability of the mediator-electrode assembly;
11. a well-defined stoichiometry toward the analyte; and
12. the catalyst turnover number must be high since the amount of material that can be attached to the electrode surface is limited.

With an electrocatalytic CME, the oxidation (or reduction) of the analyte is designed to occur at the redox potential of the mediator catalyst couple unless a catalyst-substrate adduct is formed, in which case the reaction occurs at the potential for the adduct [2].

Low molecular weight (LMW) redox mediators include:

1. organic species or their redox products (for instance, nitroaromatic compounds, quinones, aromatic hydrocarbons, phenols, *para*-substituted triarylamines, triarylimidazoles, fullerenes, N-oxyl radicals, iodobenzene derivatives, phenoxazines, phenothiazines, phenazines, etc.) [27],
2. organometallic compounds (ferrocenes, phthalocyanines, metalloporphyrins, Ni- and Co-salen complexes, etc.), and
3. low-valent metal salts and highly dispersed base metal particles, among others [16, 17].

The search for new molecules with strategic functional groups has been a way to connect organic chemistry and electrochemistry.

Direct immobilization of redox mediators on the carbon surface is an elegant way to tailor the electrode surface.

Regarding the construction of CMEs, mediators have been immobilized by direct application or by chemical bonding on the electrode surface using different methods, as shown above (Figure 3). Several materials are being explored for the development of new nanostructured interfaces.

Nanomaterials have a significant role in new developments in this area, as nanomaterials can help researchers address some of the key issues observed in the development of all sensors [28]. These issues include the characteristics illustrated in Figure 2 and mentioned above. More specific challenges include compatibility with biological matrices to ensure that they can be used in complex biological samples or even *in vivo* [29, 30], the fabrication of viable sensors that operate within confined environments such as inside cells [30] and multiplexing sensors that detect multiple analytes with a single device [28, 31].

Examples of these materials include semiconductor quantum dots, metallic nanoparticles, metallic or semiconductor nanowires, CNTs, nanostructured conductive polymers or nanocomposites and various other nanomaterials [28], as analyzed in another chapter of this book.

We focus on carbon nanotubes (CNTs). CNTs have attracted increasing attention since their discovery due to their good electrical conductivity, their nanometer size, lack of toxicity, high surface area, significant mechanical strength, good chemical stability, possibility of functionalization with carboxylic acids and excellent adsorption [32, 33]. The electrocatalytic response observed with CNTs was mainly attributed to the presence of edge plane graphite sites within the walls and at the ends of CNTs. The use of the redox mediators and CNTs in a matrix may provide a synergistic increase in sensor performance [33, 34]. CNTs also offer more efficient ways of communicating between sensor electrodes and the redox-active sites of biological molecules, which are frequently embedded deep inside surrounding peptides [28].

In theory, CNTs are pure carbon; in reality, they always contain some impurities, such as:

1. metallic compounds or nanoparticles derived from the catalysts used in nanotube growth and
2. oxygen-containing moieties created during the washing steps.

These impurities, particularly the metallic compounds, are probably responsible for the “electrocatalysis” observed at some nanotube-modified electrodes. Careful removal of these metallic nanoparticle impurities is necessary. The other major source of discrepancy in reports of the electrochemical behavior of CNTs is compositional heterogeneity. The poor dispersion and weak interfacial interactions of CNTs with other materials are the major limitations to their dispersion and the fabrication of composites with enhanced properties [35, 36].

Among the different approaches used for surface modification, acid treatments are the most frequently reported in literature, generally using nitric acid (HNO_3) or a mixture of HNO_3 and sulfuric acid (H_2SO_4) [35]. Acid treatments introduce oxygenated functional groups on the surface of the CNTs, particularly carboxyl, phenolic hydroxyl and lactones/lactol groups, which may lead to subsequent functionalization, such as amidation, alkylation, esterification and thiolation. The development of methodologies that combine a high degree of functionalization with proper preservation of CNTs is of great importance [37]. The acid treatment method using ultrasound, moderate heating and mechanical stirring has been successfully applied to the effective functionalization of MWCNTs with high yield [35]. Nevertheless, the resulting CNTs must be characterized. Some of the tools cited in literature that have been used to characterize carbon nanotubes are [34]:

1. high-resolution transmission electron microscopy (HRTEM)
2. scanning tunneling microscopy (STM)
3. atomic force microscopy (AFM)
4. nuclear magnetic resonance (NMR)
5. X-ray photoemission spectroscopy (XPS)
6. grazing incidence small angle X-ray scattering (GISAXS)
7. X-ray absorption near-edge structure spectroscopy (XANES)
8. Raman spectroscopy

Nuclear magnetic resonance (NMR) [38] and X-ray absorption near-edge structure spectroscopy (XANES) have proven to be the most powerful tools in detecting and resolving the various bonding environments. The main drawback of NMR is that it requires the detachment and powderization of the coating to produce a relatively large amount of sample. XANES, instead, can be performed on a single coating without any sample preparation [39].

Several studies on surface modifications of CNTs, which have a highly hydrophobic surface, designed to improve their dispersion, solubilization, and chemical compatibility for further processing and applications have been published [15, 40].

Surface functionalization of carbon nanotubes may be classified into:

1. covalent functionalization and
2. non-covalent functionalization

In covalent functionalization, chemical reactions that result in bond formation through addition reactions with the surface of nanotube sidewalls are performed. The covalent modification involves the formation of covalent bonds in the wall of the CNTs, changing the hybridization of the sp^2 carbon atoms to sp^3. This change affects the conductivity of the CNTs due to the loss of π electrons from the nanotube structure [41]. The loss of conductivity decreases the electrochemical response or feature, which is a disadvantage for an electrode material. Other chemical methods that are also being investigated include the grafting of multiwalled CNTs (MWCNTs) with monomers or polymers. CNT functionalization by polymers occurs through covalent reactions between the long polymer chains and CNTs, known as polymer grafting. Polymers are either "grafted to" or "grafted from" CNT. In the first case, monomers are initially immobilized onto the CNTs, followed by *In situ* polymerization, and in the second case, functionalized polymer molecules are attached to the functionalized CNTs via chemical reactions [15].

In contrast, non-covalent functionalization presents the advantage of altering the structure of the CNTs to a lesser extent and relies on physical adsorption based on weak interactions (e.g., hydrogen bonding, π–π stacking, van der Waals forces and hydrophobic interactions). This method is particularly attractive because it offers the possibility of attaching a chemical functional group while preserving the sp^2 structure of the nanotubes, leaving the electronic structure unchanged [22]. Electrostatic interactions have also been observed following CNT functionalization, which can be implemented in such a way as to produce a surface, either negative, as is the case for the frequently used acidic functionalization that provides —COO^- groups, or positive, as is the case for amino-functionalized CNTs [34, 42].

Surfactants or polymers have also been used for surface modification (Figure 6).

Surface modification with redox mediators may surpass the drawback of the superficial heterogeneity of CNTs. Surface characterization uses the methods listed above [43, 44], together with electrochemical quartz crystal microbalance (EQCM), electrochemical impedance and ATR spectroscopy [45].

NITROAROMATIC-BASED MEDIATORS

The electrochemical reduction of nitrocompounds has been studied since approximately 1900, and has played an impressive role in the development of organic chemistry.

The cathodic reduction of nitro compounds is generally very well understood [46, 47].

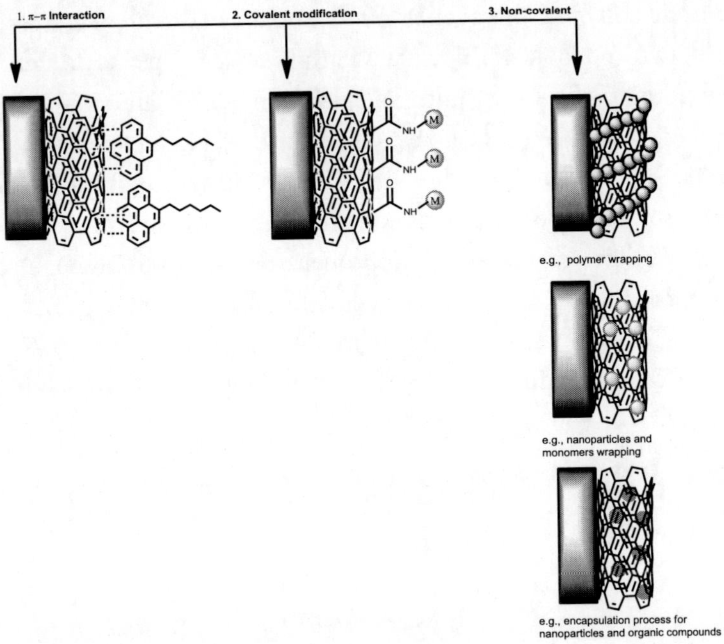

Figure 6. Surface modifications of CNTs introduced using covalent and non-covalent approaches.

The interest in this class of compounds has grown, due to their medicinal uses [48]. Their electrochemical behaviors depend strongly on their structures and experimental conditions: the solvent, supporting electrolyte, pH values and nature of the electrode. In the present chapter, a brief overview of the electroreduction mechanisms of some nitrocompounds in different media is presented. Complete monographs on this topic have appeared in chapters of several books [46, 49].

The reduction of nitroaromatics in aqueous solutions occurs through a single irreversible step of four electrons and four protons, leading to the production of a hydroxylamine (Eqs. 1 and 2) (Figure 7). The cyclic voltammogram for the *meta*-nitro-dihydropyridine derivative (Figure 7A) displays the standard profile for nitro reduction in neutral protic medium, as represented by the first irreversible wave (Ic), followed by the redox pair (IIa/IIc). In the case of 3,5-dinitrobenzamide glycoside, two irreversible cathodic waves (Ic and IIc) appear. In the anodic part of the voltammogram, two oxidation waves (Ia, which is related to Ic, and IIa, which is related to IIc) are characteristics of the oxidation of both electrogenerated hydroxylamines on the Hg electrode, with similar behaviors observed on solid electrodes (Figure 7B) [47].

In contrast, in mixed media (aqueous + aprotic medium in different proportions), reduction occurs in separate steps: a first quasi-reversible monoelectronic wave to the nitro radical anion, followed by a further reduction at more negative potential, reflecting a kinetic limitation imposed by protonation of the radical anion that is thermodynamically unfavorable [50].

In pure aprotic solvents, the cyclic voltammograms (Figure 8A, Eqs. 3 and 4) show the presence of two monoelectronic reduction processes – peaks Ic and IIc, which are related to the formation of the nitroradical anion and dianion and associated with oxidation peaks Ia and IIa, respectively. The third reduction peak corresponds to an irreversible process that leads to the formation of further reduced products, without evidence of the formation of the hydroxylamine [51] (Figure 8A). Depending on the presence of an acidic proton, as in 3,5-dinitrobenzoic acid (3,5-DNB), a self-protonation mechanism occurs (Figure 8B). This reaction leads to a first wave, at more positive potential (E_{pc} = - 0.75 V) that is related to the full reduction of 1/5 of the original compound to the corresponding hydroxylamine, followed by the reduction of 4/5 of the 3,5-dinitrobenzoate, the conjugated base, at E_{pcI} = -1.02 V and E_{paI} = -0.92 V (Figure 10, eq. 5) [52].

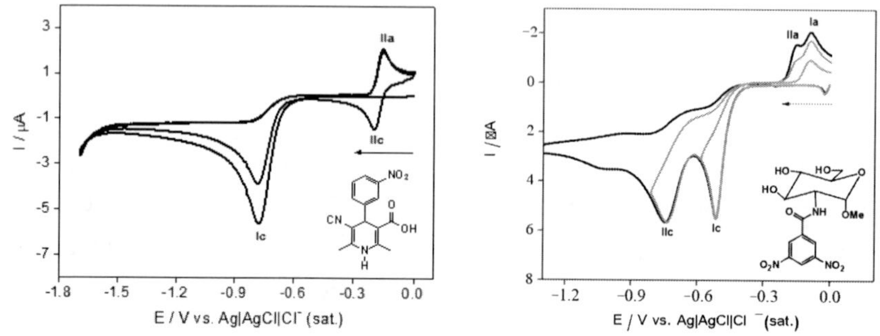

A B

Figure 7. Typical cyclic voltammograms (CV) of nitroaromatics in protic medium with an Hg electrode: phosphate buffer (0.1 mol L^{-1}), pH 6.9. ν= 0.100 V s^{-1}. c_{ArNO2} = 2 mmol L^{-1}. A) Successive cycles showing the reversible electrogenerated redox system. B) A dinitrocompound (1 mmol L^{-1}) presents two irreversible reduction peaks (Ic and IIc) and two oxidation peaks related to ArNHOH (Ia and IIa).

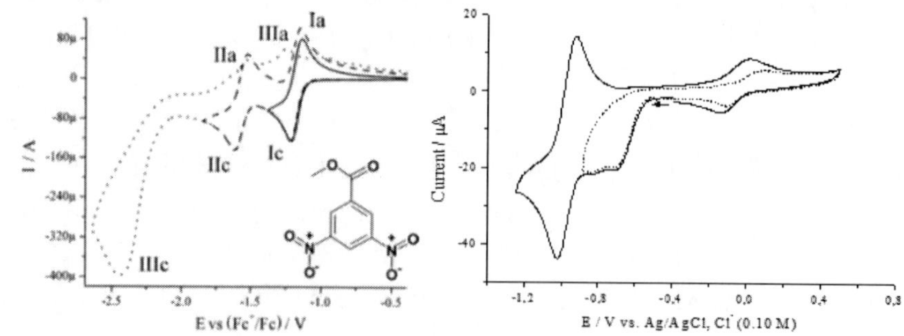

A B

Figure 8. A) CV of 3,5-dinitrobenzoic methyl ester (1 mmol L^{-1}) in 0.1 mol L^{-1} n-Bu_4NPF_6/CH_3CN with different inversion potentials, v: 0.100 V s^{-1}. GCE. Reprinted with permission from Hernandez-Munoz et al., 2010, copyright 2010, Elsevier. B. CV of 3,5-dinitrobenzoic acid in n-Bu_4NClO_4/DMF (0.1 mol L^{-1}), potential range: 0.45 up to -1.2 V, ν: 0.100 V s^{-1}.

In their oxidized state, aromatic nitrocompounds do not exhibit electrocatalytic activity. However, when electrochemically reduced to a hydroxylamine functionality in protic medium (Eq. 1), a stable reversible system is obtained, depending on the structure. The resulting hydroxylamine is reversibly oxidized to the corresponding nitroso compound (ArNO/ArNHOH couple) by a $2e^-/2H^+$ redox process (Eq. 2) that may be useful for electroanalysis.

Some earlier articles [53, 54] investigated surface electrode modifications with long chain dialkyl disulfide containing terminal dinitrophenyl groups and used them as impedimetric sensors [54].

Approximately 20 years ago, Casero and coworkers (1999) [45] used 5,5'-dithiobis(2-nitrobenzoic acid) (DTNB) and 2,2'-dithio(5-nitropyridine) (DTNPy), whereas Mano and Kuhn (1999) [55, 56] employed 2,4,7-trinitro-9-fluorenone (TNF) for NAD^+ regeneration. Since those pioneer works, which deliberately trigger a surface reaction through an electrochemical method, several papers have appeared reporting the use of activated nitroaromatics as redox mediators (Figure 10 and Tables 1 and 2) in electroanalysis.

The very first step in all cases is generally the electrochemical characterization of the nitroaromatic compound by cyclic voltammetry or any other technique using an adequate electrolyte on a glassy carbon electrode, with a platinum wire as the counter and a reference electrode. Notably, the nitroreduction based-redox pair (ArNO/ArNHOH) must be stable in the chosen medium, insoluble in the supporting electrolyte and/or surface-confined to avoid lixiviation.

The nitroaromatic mediator is also chosen by examining different chemical classes of a family of nitroaromatic derivatives and identifying the best experimental stability/formal potential of the $ArNO_2$-electrogenerated redox couple ArNHOH/ArNO using conventional electrochemical techniques (cyclic voltammetry, differential pulse voltammetry, etc.), in the appropriate cathodic region [55, 56].

The number and positions of the nitro groups in several types of aromatic rings, the presence of other electroactive substituents in multi-redox compounds, their acid-base properties, interferences among the redox groups, the presence of several reversible redox systems, and other factors should be preliminarily examined and experimentally investigated.

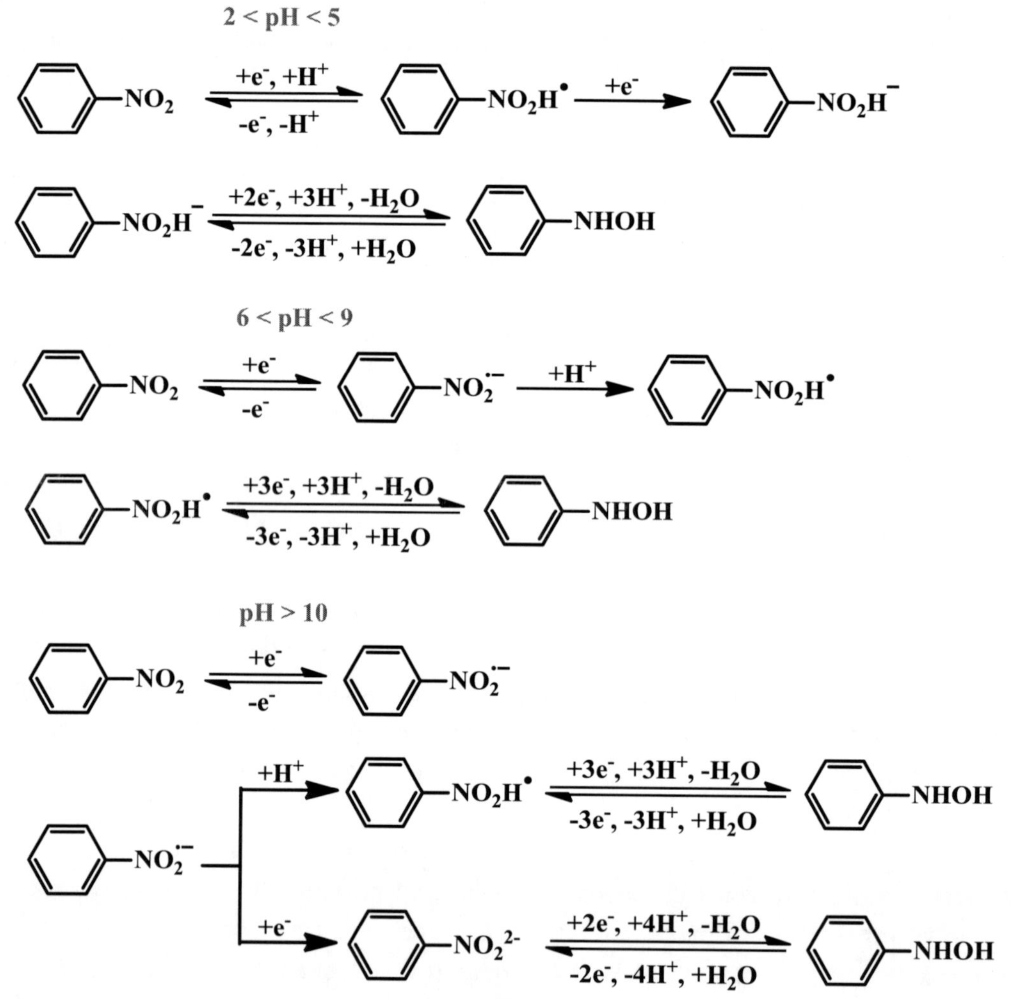

A

$$ArNO_2 + 4e^- + 4H^+ \longrightarrow ArNHOH + H_2O \qquad \text{Eq. 1}$$

$$ArNHOH \rightleftharpoons ArNO + 2e^- + 2H^+ \qquad \text{Eq. 2}$$

B

$$ArNO_2 + e^- \rightleftharpoons ArNO_2^{\bullet -} \qquad \text{Eq. 3}$$

$$ArNO_2^{\bullet -} + 3e^- + 4H^+ \longrightarrow ArNHOH + H_2O \qquad \text{Eq. 4}$$

C

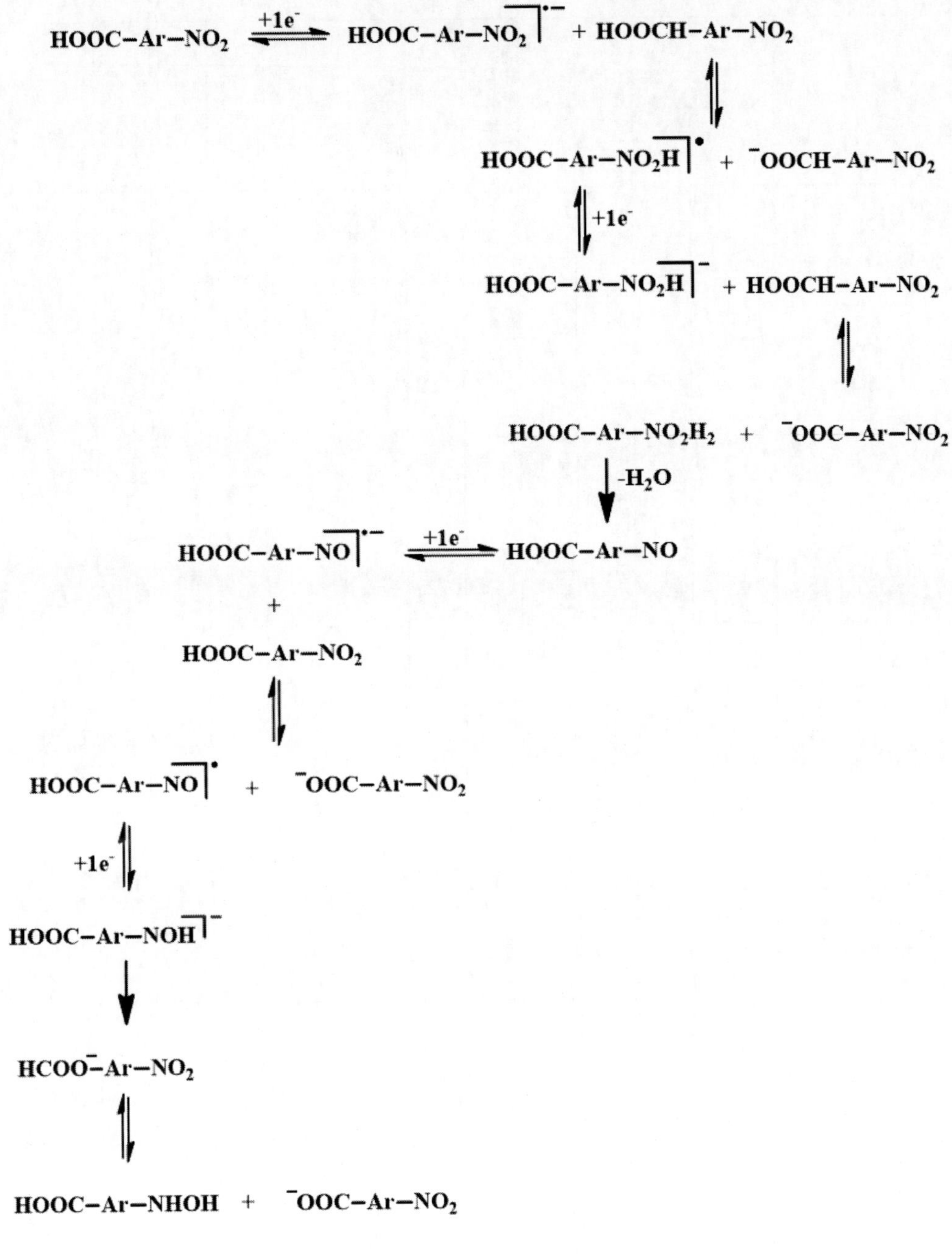

$$5\ HOOC{-}Ar{-}NO_2 + 4e^- \rightleftharpoons HOOC{-}Ar{-}NHOH + 4\ {}^-OOC{-}Ar{-}NO_2 \qquad \text{Eq. 5}$$

D

Figure 9. Proposed mechanisms for the electrochemical reduction of nitrobenzene, a representative of simple nitroaromatics. A: Reduction in protic medium at different pH values, B: general scheme in protic medium, C: general scheme in mixed medium or in aprotic medium, and D: self-protonation mechanism in aprotic medium. The general equation (eq. 5) shows the mechanism of reduction in aprotic medium at the first wave. Adapted from Hammerich, 2016 [49].

Table 1. Nitroaromatics used as redox mediators, thermodynamic data and the immobilization technique employed

Mediator	Nitroaromatic Reduction Potentials (V)	Redox pair E°' *vs.* NHE (V) (*vs.* other references)	Sensor Immobilization technique	References
		0.245 (40 mV *vs.* SCE)	GCE/MWCNT/ CHIT/act3,5-DNBA. **EI**	[692]
3,5-DNBA	*E*pIc ~ -0.49	0.255 (50 mV *vs.* SCE)	GCE/MWCNT/ PEI/act3,5-DNBA. **CL**	[57]
	*E*pIc = -0.68 *E*pIIc = -0.97	0.169 (-28 mV *vs.* Ag/AgCl/NaCl)	GCE/MWCNT/act3,5-DNBA **EI**	[58]
BisDNBA		0.237 (40 mV *vs.* Ag/AgCl/NaCl)	GCE/actBisDNBA	[59]
	*E*pIc ~-0.36	(-50 mV *vs.* Ag/AgCl/NaCl)	GCE/act**2,4,7**-TNF **AD**	[55]
	*E*pIc ~-0.36	(-46 mV *vs.* Ag/AgCl/NaCl)	GCE/act**2,4,7**-TNF + Ca^{2+} **AD**	[56]
2,4,7-TNF		(250 mV *vs.* Ag/AgCl)	CPE-ZP/act**2,4,7**-TNF **AD**	[60]
	*E*pIc ~0.00 *E*pIIc = -0.50	0.449 (250 mV *vs.* Ag/AgCl/KCl)	CPE-ZP/act**2,4,7**-TNF + Ca^{2+} **AD**	[61]
	*E*pIc ~ -0.35 *E*pIIc ~ -0.50 *E*pIIIc ~ -0.60	(~ -80 mV *vs.* Ag/AgCl)	GCE/act2,4,7-TNF **AD**	[62-63]
2,5,7-TNFM		0.105 (-90 mV *vs.* Ag/AgCl)	Au/act2,5,7-TNFM **AD**	
2,5,7-TNFAA		(~ -100 mV *vs.* Ag/AgCl)	Pt/act2,5,7-TNFAA N	
2,5,7-TNFCA	*E*pIc = -0.37 *E*pIIc = -0.52	0.160 (-30 mV *vs.* Ag/AgCl)	GCE/act2,5,7-TNFCA **AD**	[64]
		0.429 (230 mV *vs.* Ag/AgCl/KCl)	CPE/act2,5,7-TNFCA **AD**	[61]
2,7-DNF	*E*pIc = -0.40 *E*pIIc = -0.54	Ag/AgCl (1 M KCl)	GCE/MWCNT/act2,7-DNF **AD**	[44]
DTNB	*E*pIc ~ -0.64	0.209 (-35 mV *vs.* SCE)	GCE/MWCNT/PEI-GNP/actDTNB **EI + AD**	[65]

	*E*pIc = -0.55	(-40 mV *vs.* SSCE)	Au **AD**	[45]
DTNPy	*E*pIc = -0.55	(-40 mV *vs.* SSCE)	Au **AD**	[45]
DTNBP	*E*pIc ~ -0.44 *E*pIIc -0.50	0.284 (40 mV *vs.* Ag/AgCl/Cl^-sat)	GCE/MWCN/actDTNB-pyr **CL**	[66]
4-NPN	Ep ~ -0.4 *vs.* Ag/AgCl	0.300 (95 mV *vs.* Ag/AgCl/Cl^- sat)	CPE/act4-NPN **EI**	[67-70]
NIC	*E*pIc = -0.52	0.110 (-95 mV *vs.* Ag/AgCl)	GCE/MWCNT/actNIC **EI**	[71]
4-NBA	*E*pIc = -0.6 V (*vs.* SCE) pH 6.8	-0.1 V *vs.* SCE	CPE/MWCNT/act4-NBA **AD**	[43]

AD: adsorption; EP: electropolymerization; EI: entrapment/incorporation; CL: covalent linkage. GCE: glassy carbon electrode. CPE: carbon paste electrode. MWCNT: multi-walled carbon nanotubes. CHIT: chitosan. 3,5-DNBA: 3,5-dinitrobenzoic acid. 2,4,7-TNF: 2,4,7-trinitro-9-fluorenone. BDNBA: 3,5-bis(3,5-dinitrobenzoyl amino) benzoic acid. 2,4,7-TNFCA: 2,5,7-trinitro-9-fluorenone-4-carboxylic acid. 4NPN: 4-nitrophthalonitrile.DTNB: 5,5'-dithiobis-2-nitrobenzoic acid. DTNBP: 5,5'-dithiobis(2-nitrobenzoic acid) pyrene. DTNPy: 2,2'-dithiobis(5-nitropyridine). Activated NIC: activated niclosamide. PEI: polyethyleneimine. ZP: zirconium phosphate. GNP: gold nanoparticles.

Table 2. Nitroaromatics used as redox mediators in electroanalysis, together with the analytes, techniques used and main analytical parameters

Mediator	Analytes	Technique Applied potential (V)	Medium pH	Linear range ($\mu mol\ L^{-1}$)	LOD ($\mu mol\ L^{-1}$)	Sensitivity	Second order rate constant ($M^{-1}\ s^{-1}$)	References
3,5-DNBA	GSSG	Amp. 0.100 *vs.* Ag/AgCl	7.0	2.0 - 35	0.6	6.24 mA L mol^{-1}	5.05 x 10^4	[269]
	NADH	Amp. 0.020 and – 0.130 *vs.* Ag/AgCl	8.0	4.0 - 42.0	1.2	71 nA L μmol^{-1}	3.3 x 10^4	[65]
	NADH	CV	7.0					[58]
BisDNBA	NADH	CV 0.120 *vs.* Ag/AgCl	7.0					[59]
2,4,7-TNF	NADH	CV -0.050 *vs.* SCE	8.0	5.0 - 200			10^4	[55]
	NADH	CV 0.260 *vs.* Ag/AgCl	7.0				10^6	[60]

	NADH	CV 0.260 *vs.* Ag/AgCl	7.0	5.0 - 260	0.4	3.30 nA μM^{-1}	10^6 - 10^7	[61]
2,4,7-TNF	NADH	CV -0.050 *vs.* Ag/AgCl	8.0			184 μA cm^{-2} mM^{-1}	2.5 x 10^4	[62, 63]
2,5,7-TNFM	NADH	CV 0.025 *vs.* Ag/AgCl	8.0			90 μA cm^{-2} mM^{-1}	4.2 x 10^3	
2,5,7-TNFCA	NADH	CV 0.100 *vs.* Ag/AgCl	8.0			120 μA cm^{-2} mM^{-1}	6.8 x 10^3	
2,5,7-TNFCA	NADH	CV -0.050 *vs.* Ag/AgCl	8.0			0.2 μA cm^{-2} mM^{-1}		[64]
2,5,7-TNFCA	NADH	CV	7.0	5.0 - 120	0.6	3.04 nA μM^{-1}	10^6 - 10^7	[61]
	Ethanol	CV -0.070 *vs.* Ag/AgCl	8.0			0.3 μA cm^{-2} L mmol^{-1}		[62, 63]
DTNB	*L*-Cysteine NADH	Amp. 0 *vs.* SCE	7.0	9.0 - 250	2.7	21 nA L μmol^{-1}	8.60×10^3	[45, 57]
DTNPy	NADH	CV -0.100 *vs.* SSCE	PBS pH 7.0					[45]
DTNBP	NADH	CV -0.09 *vs.* Ag/AgCl/KCl	PBS pH 7.0	0 - 400				[66]
4-NPN	NADH	Amp. 0.100 *vs.* Ag/AgCl	TRIS buffer 7.0	0.8 - 8.5	0.25	0.5 μA L μmol^{-1}	1 x 10^4 (NADH)	[67]
	L-Glutathione	Amp. 0.400 *vs.* Ag/AgCl	PBS 7.0	0.8 - 83.0	2.7	54 nA L μmol^{-1}		[68]
	L-Cysteine	Amp.	7.0	0.8 - 13.2	0.25	37 nA L μmol^{-1}		[69]
	Uric acid (UA) Ascorbic acid (AA)	Amp. 0.150 *vs.* Ag/AgCl (UA) and 0.400 *vs.* Ag/AgCl (AA)	7.0	5.0 - 120.0	1.3 (UA), 1.6 (AA)	0,680 μA L μmol^{-1} 0,54 μA L μmol^{-1}	5.5 x10^6 (AA)	[70]
NIC	NADH	Amp. 0.065 *vs.* Ag/AgCl	7.0	10 - 280	3.0	2.4 nA μM^{-1}		[71]

Amp.: amperometry; CV: cyclic voltammetry. PBS: phosphate buffer system. AA: ascorbic acid. UA: uric acid.

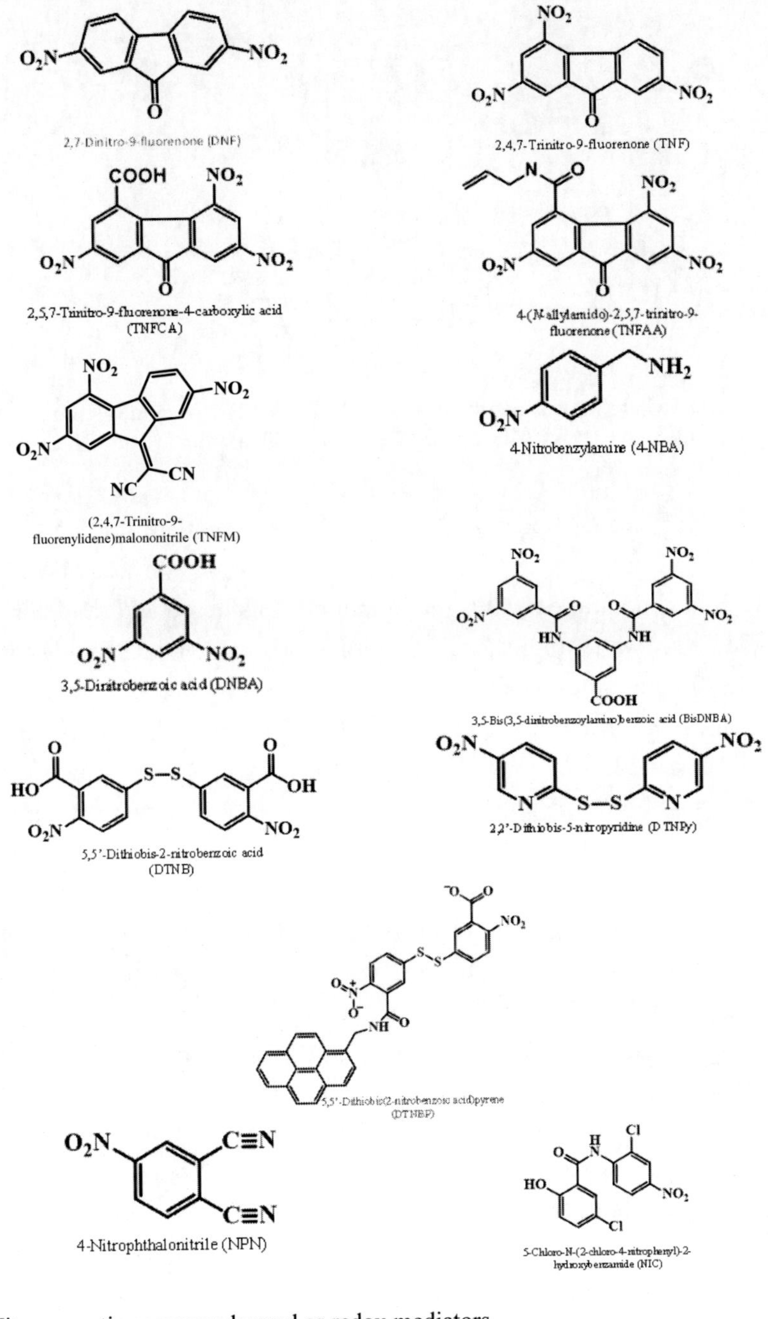

Figure 10. Nitroaromatic compounds used as redox mediators.

Parameters such as the possibility of π-π stacking, hydrophobic and electrostatic interactions in relation to surface modifiers, molar mass of the nitrocompound, affinity between electrode/supporting electrolyte, log P, log D, partition coefficients of the analyte between the aqueous and polymer phase, and solubility in different solvents are important to consider when choosing the best activated nitroaromatic mediator to avoid

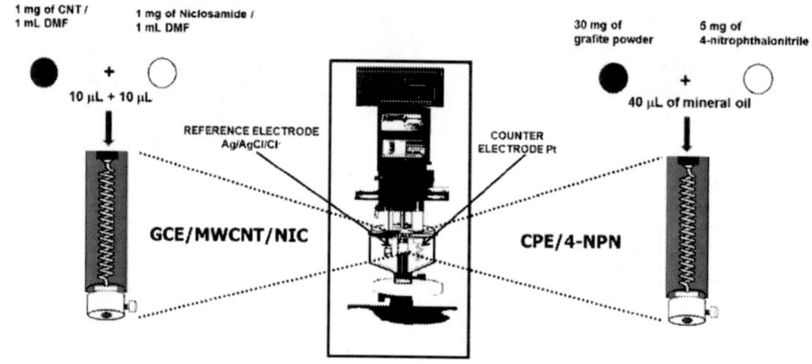

A B

Figure 11. Scheme showing the construction of the platforms. A) Niclosamide (NIC) + MWCNT on a glassy carbon electrode (GCE/MWCNT/NIC) and B) 4-nitrophthalonitrile (4-NPN) on a carbon paste electrode (CPE/4-NPN).

lixiviation. The presence of functional groups that are adequate for functionalization or covalent bonding formation or grafting on the electrode surface/MWCNT-modified electrodes are also structural demands that might be considered and rationalized in search for an ideal mediator. As examples, we cite thiols, carboxylic acids, amines, alcohols, and diazonium salts, among others. Questions related to toxicity are also relevant, as some of the nitroaromatics are hazardous materials [72]. Searches for safety and hazardous profiles are recommended.

Despite the possibility of homogeneously dissolved redox catalysts, in this chapter, we will focus on heterogeneous and surface-confined catalysts. The applicability of dissolved redox mediators has been consistently reported. Nevertheless, the use of mediators in solutions for sensor applications is not a very convenient strategy since the analysis is not reagentless and long stabilization times are necessary. The most important drawbacks of dissolved mediators in bioanalytical applications are interference produced during product purification, limited reusability of the mediators, and their cost-intensive elimination from wastewater. Furthermore, some of the substances, as well as their precursors and degradation products, are carcinogenic and/or mutagenic [73]. Therefore, retention of the mediators in the reaction system is strongly desired. Moreover, the use of immobilized mediators has both economic and ecological advantages [21].

Nitroaromatics, such as 4-nitrophthalonitrile (4-NPN), 4-nitrobenzyl-amine (4-NBA), and 2,4,7-trinitro-9-fluorenone (TNF) and its derivatives (2,4,7-trinitro-9-fluorenylidene)malononitrile (TNFM), 2,5,7-trinitro-9-fluorenone-4-carboxylic acid (TNFCA), 4-(*N*-allylamido)-2,5,7-trinitro-9-fluorenone (TNFAA), as well as 2,7-dinitrofluorenone (DNF), 3,5-dinitrobenzoic acid (DNBA) and its derivative 3,5-bis(3,5-dinitrobenzoyl amino) benzoic acid (BisDNBA), 5,5'-dithiobis(2-nitrobenzoic acid) (DTNB) and its pyrene derivative (DTNBP), along with niclosamide (NIC), have been applied as redox mediators in analyses of compounds with biological interest.

Figure 11 shows the structures and abbreviations of the principal nitroaromatics that are currently used.

Table 1 describes the main mediators used and presents data on thermodynamic and kinetic parameters, when reported, and the immobilization technique used for platform preparation.

Table 2 lists the uses of these mediators in electroanalysis, together with the analytes, techniques employed and main analytical parameters. These mediators were mainly used for the electroanalysis of NADH and, to a lesser extent, thiols.

Studies reported by our group are emphasized, and a possible rationale for the different mechanisms observed during the redox catalysis is provided. After the choice of the nitrocompound and the best material for the electrode and cleaning steps, the optimal conditions for redox-mediator surface preparation and application in analysis must be assessed in relation to [26, 74]:

1. the influence of pH,
2. the nature and concentration of the buffers,
3. the influence of oxygen,
4. the effect of potential range,
5. the number of cycles in cyclic voltammetry,
6. the surface coverage measurement,
7. the concentration of the redox mediator,
8. the type of modifier and its pre-treatment,
9. the immobilization technique and
10. surface characterization using different techniques.
11. When characterizing the reaction between the surface-confined mediator and the analyte, the following data must be obtained:
12. the number of transferred electrons,
13. thermodynamic parameters,
14. charge transfer kinetics and apparent kinetic constants,
15. analytical parameters, interference data, and real sample applications,
16. and last but not least, the definition of the electrochemical mechanisms.

Main Examples in Electroanalysis

Toward NAD(P)H

The electrooxidation of α-nicotinamide adenine dinucleotide (NADH) or its phosphate derivative (NADPH) is still of great interest in the field of bioelectrocatalysis, as this molecule is the cofactor for more than 350 dehydrogenases [66]. NADPH-dependent enzymes have attracted increasing interest for the synthesis of fine and bulk chemicals, as well as for sensor applications [21]. The NAD^+/NADH redox couple plays a key role in biological electron transfer as a two single electron (SET), one proton redox reagent [75]. It is the initial electron donor in the mitochondrial electron transport chain [26]. Thus, the development of modified electrodes that are able to oxidize NADH to the enzymatically active form NAD^+ has attracted substantial attention because of their potential applications in a wide range of biosensors, food processing, environmental analysis, and clinical diagnosis [76-79].

In vivo, reduced and oxidized cofactors are supplied by cellular metabolism. *In vitro*, the regeneration of nicotinamide cofactors is a crucial step for an economic use of the enzymes [80]. The cofactors cannot be used in stoichiometric amounts in the process [21].

Nitro-fluorenone derivatives have been used to modify the surface of glassy carbon electrodes (Table 1). The stable, adherent thin layer was electrochemically transformed into the corresponding hydroxylamine. The completely reversible two-electron oxidation of the hydroxylamine group leads to the formation of a nitroso compound that exhibits high catalytic activity in the electrooxidation of reduced nicotinamide coenzyme (NADH). A gradual shift in the redox potential toward a more positive value was observed when the number of $-NO_2$ substituents in the catalyst molecule increased.

Oxidation of NADH at potentials as low as 50 mV *versus* Ag/AgCl was obtained using catalytic currents proportional to the concentration of NADH over the range 5 μM – 2 mM. The rate constant for the catalytic process was evaluated by cyclic voltammetry and rotating disc electrode measurements, and typical values on the order of 10^4 were obtained [55, 56] (Tables 1 and 2).

The effects of the number and positions of the nitro groups, the charges of the compounds, cations, mainly Ca^{2+}, on NADH electrooxidation were investigated when it was possible to show the dependence of the catalytic activity upon these factors [61-64] (Table 1).

The addition of zirconium phosphate and Ca^{2+} enhanced the electroactivity, and the second rate constant for NADH (Eq. 7) was among the highest reported in the literature to date (Table 1) [61], almost reaching a diffusion-controlled reaction. The addition of Ca^{2+} did not affect the thermodynamic driving force for NADH oxidation. Its addition to 2,5,7-TNFCA might have caused the complexation of NADH with the cation, which played the role of a bridging unity between the carboxyl group of the mediator and the phosphates of the analyte.

$$\mathbf{Med_{red}} \xrightarrow{k_s} \mathbf{Med_{ox}} + \mathbf{2e^-} + \mathbf{2H^+} \qquad \text{(Eq. 7)}$$

$$\mathbf{NAD(P)H} + \mathbf{Med_{ox}} \xrightarrow{k_{obs}} \mathbf{NAD(P)^+} + \mathbf{Med_{red}} \qquad \text{(Eq. 8)}$$

Recently, a new type of mediator based on 4-nitrophthalonitrile (4-NPN)-modified carbon paste electrode (CPE/4-NPN) was studied by our laboratory [67-70] (Tables 1 and 2).

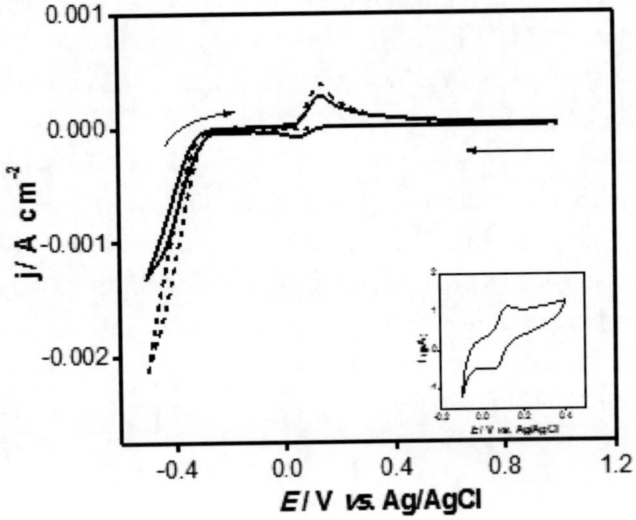

A

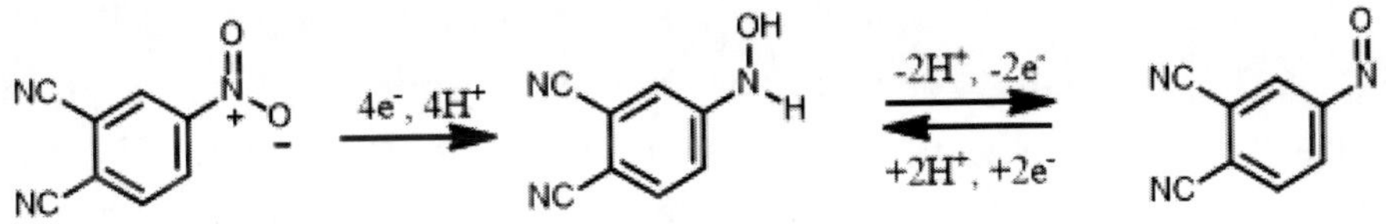

B

Figure 12. A) Cyclic voltammogram obtained during the first scan (–) and second scan (–) of 4-NPN-modified carbon paste electrode (CPE) in 0.1 mol L^{-1} phosphate buffer solution, pH 7.0, at a scan rate of 0.02 V s^{-1}. $E_{initial}$ = 1.0 V *vs.* Ag/AgCl and E_λ = 0.5 V. B) Proposed mechanism for the activation of 4-NPN-modified electrode. Reprinted with permission from Lima et al., 2008 [68], copyright 2008, Elsevier.

The modified carbon paste was prepared by mixing 35 mg of graphite powder with 5 mg of 4-NPN and 20 μL of mineral oil. The carbon paste was placed in the cavity of a homemade Teflon holder using graphite at the bottom for electric contact and then attached to the RDE device (Figure 11) [67-69].

Thus, the geometric area of the working electrode (sensor) was determined as ($A = \pi r^2$) and it presented the value of 0.2 cm^2. The reasons we study 4-NPN are related to its versatility, stability and maintenance on the CPE, without lixiviation. This mediator was applied to the analysis of NADH, *L*-glutathione, *L*-cysteine, uric and ascorbic acid.

The redox-active couple ArNHOH/ArNO was electrogenerated *In situ* from the mediator (4-NPN), as shown in Figure 12. The act4-NPN-modified electrode showed efficient electrocatalytic activity toward the oxidation of NADH, with an activation overpotential of 0.12 V *vs*. Ag/AgCl (Figure 13A).

As mentioned above, the nitro substituent in 4-NPN can be completely transformed into the hydroxylamine (4-NHOHPN) that in turn, is oxidized to the corresponding nitroso (4-NOPN) through a $2e^{-}/2\ H^{+}$ process (Figures 12A, insert, 13B), which is currently named act4-NPN [67].

For the electrogenerated reversible redox system (4-NHOHPN/4-NOPN) known as act4-NPN, the hydride route might be preferred due to the presence of a partial positive charge on the nitrogen of the ArNO group, due to electron delocalization. The formation of an intermediate charge transfer complex was proposed for the charge transfer reaction between NADH and the resulting act4-NPN system (Figure 13B).

For a mediator-catalyzed reaction, the kinetically limited current (I_{kin}) is generally described by the following equation:

$$I_{kin} = nFA\kappa_{obs}\,\Gamma^{*}\left[NADH\right]^{*} \tag{Eq. 9}$$

where Γ* is the total coverage of the mediator system ArNO/ArNHOH. A linear dependence of the catalytic current on ω1/2 is expected from the Levich equation. The observed deviation from the Levich equation results from the chemical-limiting step. Thus, the general expression for the overall current I is:

$$I^{-1} = I^{-1}_{d(L)} + I^{-1}_{Kin} \tag{Eq. 10}$$

The Koutecky-Levich equation for the catalytic reaction occurring at the mediator-modified electrodes for the special case of a rotating disk electrode with the indicated restrictions is [81]:

$$\frac{1}{I} = \frac{1}{nFA\kappa_{obs}\Gamma^{*}[NADH]^{*}} + \frac{1}{0.620nFA\nu^{-1/6}D^{2/3}[NADH]^{*}}\frac{1}{\omega^{1/2}} \tag{Eq. 11}$$

The number of electrons (n) participating in the reaction is calculated from the slope. From the slope and intercept, κ_{obs} (once the Γ^{*} is known and evaluated by integrating the area under the CV waves) is also calculated.

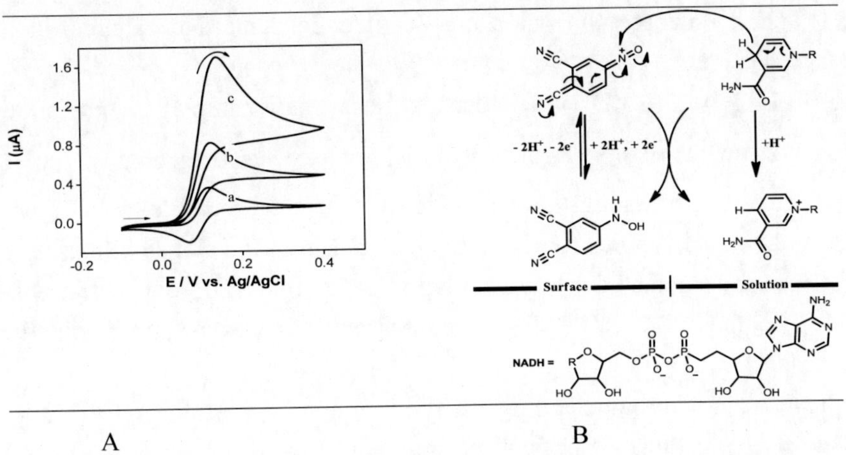

Figure 13. A) Cyclic voltammograms of CPE/act4-NPN in the absence (a) and presence of increasing concentrations of NADH/mmol L^{-1}: (b) 0.1 and (c) 0.3. Cyclic voltammograms of the unmodified carbon paste electrode in absence (d) and presence (e) of 0.3 mmol L^{-1} NADH. Scan rate: 2.0 mV s^{-1}. All studies were performed in 0.1 mol L^{-1} Tris buffer at pH 7.0. B) Proposed mechanism for the reaction between act4-NPN-modified electrode and NADH. Reprinted with permission from Lima et al., 2008 [67], copyright 2008, Elsevier.

In this case, the calculated value of the number of electrons was approximately 2. A second way to calculate *n* is to plot the slope of the Koutecky-Levich plots *vs.* (NADH)*; when a straight line is obtained, *n* can be calculate. In the present case, *n* is approximately 2.

However, the values of κ_{obs} strongly depended on (NADH)* for act4-NPN on CPE, consistent with the previous results reported for virtually all mediator-modified electrodes used for catalytic NADH oxidation [82-84] and has been explained by the actual reaction between NADH and the mediator that occurs through formation of a charge transfer (CT) complex before yielding the reaction products.

Starting from the deductions of the respective kinetic process as reported by Gorton in 1986, the heterogeneous second order reaction rate κ_{obs} for any concentration of NADH is calculated using the following equation:

$$\kappa_{obs} = \frac{\kappa_{+2}}{K_M + [NADH]^*} \quad \text{(Eq. 12)}$$

Inversion of Eq. (12) yields:

$$\frac{1}{\kappa_{obs}} = \frac{K_M}{\kappa_{+2}} + \frac{[NADH]^*}{\kappa_{+2}} \quad \text{(Eq. 13)}$$

where K_M and κ_{+2} denote the Michaelis-Menten constant and rate constant for the processes, respectively.

Plots of $1/\kappa_{obs}$ *vs.* (NADH)* yielded a linear relationship (data not shown), with correlation coefficient of 0.999 for $n = 5$, where κ_{obs}^{-1} is presented in mol L^{-1} s and (NADH) in mol L^{-1}, providing strong evidence that a CT complex is formed between act4-NPN and NADH. Values for κ_{+2} are calculated from the slopes of these graphs, and values for K_M are calculated by extrapolating to zero NADH concentrations (κ_{obs} for (NADH) = 0* equals κ_{+2}/K_M), i.e., from the intercept, as 1.0×10^4 mol L s^{-1}and 2.7×10^{-5} mol L^{-1}, respectively.

The present sensor provided a linear response range ranging from 0.8 to 8.5 μmol L^{-1}, with sensitivity, detection, quantification limits and time response of 0.50 nA L μmol^{-1}, 0.25 μmol L^{-1}, 0.82 μmol L^{-1} and 0.1 s, respectively. The repeatability of the measurements with the same sensor and different sensors, as evaluated in terms of relative standard deviation, were 4.1 and 5.0%, respectively, for $n = 10$.

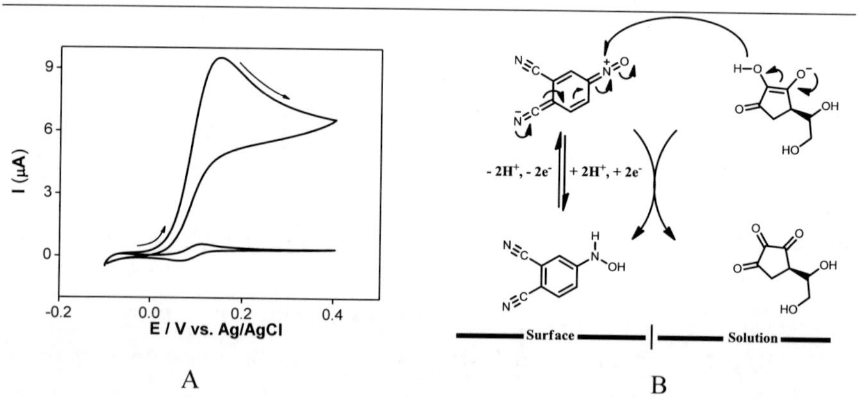

Figure 14. A) Cyclic voltammograms for the act4-NPN-modified electrode in the presence (a) and absence (b) of 0.2 mmol L^{-1} AA in phosphate buffer solution at pH 7. Scan rate: 0.002 V s^{-1}. B) Proposed mechanism for the reaction between the act 4-NPN-modified electrode and AA. Reprinted with permission from Lima et al., 2009 [70], copyright 2009, WILEY-VCH Verlag GmbH & Co.

A similar approach was used for ascorbic acid (AA) and uric acid (UA) (Figure 14). Electrocatalytical activity was observed for AA, but not for uric acid (Figure 15).

The kinetic constant κ for the catalytic electrocatalytic oxidation of AA evaluated by cyclic voltammetry, chronoamperometry and RDE voltammetry provided values of approximately 10^6 L mol^{-1} s^{-1}.

Double potential step chronoamperometry was employed to investigate electrochemical processes. The current–time curves of act4-NPN were obtained by setting the working electrode potential to 100 mV (first step) and 50 mV (second step) *versus* Ag/AgCl for various concentration of AA. No net cathodic current corresponding to the oxidation of mediator was observed in the presence of AA, whereas the forward and backward potential step chronoamperograms for the mediator in the absence of AA showed very symmetrical chronoamperometry with an equal charge consumed for the reduction and oxidation of the Ar-NO/Ar-NHOH redox system at the surface of carbon paste electrode. However, at the different concentrations of AA employed, the anodic current value associated with forward chronoamperogram was significantly greater than the value observed for backward chronoamperogram (data not shown).

Chronoamperometry was used to estimate the diffusion coefficient of AA in solution. For an electroactive material with diffusion coefficient D, the current corresponding to the electrochemical reaction (under diffusion control) is described by Cottrell's law [74]:

$$I = nFAD^{1/2}c_s\pi^{-1/2}t^{-1/2} \qquad \text{(Eq. 14)}$$

where D and c_s are the diffusion coefficient ($cm^2\ s^{-1}$) and the bulk concentration ($mol\ cm^{-3}$), respectively. The experimental plots of I *versus* $t^{-1/2}$ with the best fits for different concentrations of AA employed indicates that the observed current must be controlled by AA diffusion in solution. The slopes of the resulting straight lines were then plotted *vs.* the AA concentration (data not shown). From that slope and using the Cottrell equation (Eq. 14), we calculated a diffusion coefficient of $9.01 \times 10^{-6}\ cm^2\ s^{-1}$ for AA.

Furthermore, the rate constant κ for the chemical reaction between AA and redox sites confined on the surface of act4-NPN-modified electrode was evaluated using chronoamperometry according to the method described in the literature [85]:

$$I_{cat}/I_L = \gamma^{1/2}\left[\pi^{1/2}erf\left(\gamma^{1/2}\right) + \exp(-\gamma)/\gamma^{1/2}\right] \qquad \text{(Eq. 15)}$$

where I_{cat} is the catalytic current of the modified electrode in the presence of AA, I_L is the limiting current in the absence of AA and $\gamma = \kappa c_s t$ (c_s is the bulk concentration of AA) is the argument of the error function. In the cases where γ exceeds 2, the error function is approximately equal to 1 and the above equation can be reduced to:

$$I_{cat}/I_L = \gamma^{1/2}\pi^{1/2} = \pi^{1/2}\left(\kappa c_s t\right)^{1/2} \qquad \text{(Eq. 16)}$$

From the slope of the I_{cat}/I_L *versus* $t^{-1/2}$ plot, we can simply calculate the value of k for a given concentration of substrate. The kinetic constant κ for the catalytic reaction for the electrocatalytic oxidation of AA evaluated using cyclic voltammetry, chronoamperometry and RDE voltammetry provided values of approximately $10^6\ L\ mol^{-1}\ s^{-1}$. This value is consistent with the value of $1.0 \times 10^6\ L\ mol^{-1} s^{-1}$ for k (or $k\Gamma = 0.65\ cm\ s^{-1}$) obtained by cyclic voltammetry using extensive computations from the theoretical study by Andrieux and Savéant.

The sensor provided a linear response range for AA and UA from 5.0 to 120.0 μmol L^{-1}, with detection limits of 1.6 μmol L^{-1} and 1.3 μmol L^{-1}, respectively.

The sensor was applied to the simultaneous determination of AA and UA concentrations in urine samples, and the average recoveries for these samples were 99.8 (3.1)% and 99.9 (2.1)%, respectively (Figure 15).

The present platform was also used to analyze the thiols glutathione and cysteine (Figures 16 and 17). In this case, cylic voltammograms showed a completely different profile than NADH and AA (Figures 13, 14 and 15). After adding GSH to the solution, wave Ia (oxidation peak of the 4-NHOHPN) is still present, with a slight increase, and a second wave (IIa) appears, with a comparatively higher peak current at a potential of 0.350V *versus* Ag/AgCl. On the reverse scan, wave Ic is no longer observed. Based on these results, GSH/GS^- (pKa_{NH3+} = 9.65) reacts with the electrogenerated 4-NOPN and forms adducts, which are oxidizable species, as shown in biological media in a previous study [86]. A catalyst-substrate adduct is formed, and the reaction occurs at the potential for the adduct (Figure 4C).

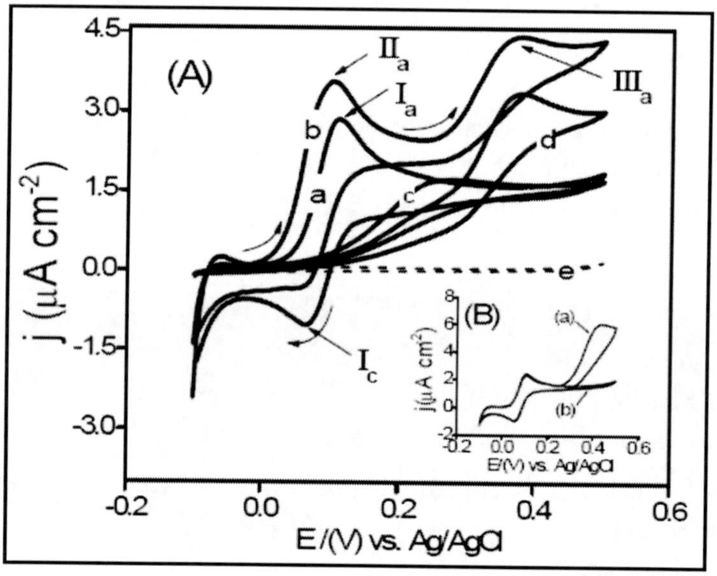

Figure 15. Cyclic voltammograms for the act4-NPN-modified electrode in the absence (a) and presence (b) of 20 μmol L^{-1} AA and UA; the unmodified carbon paste electrode in the presence of 20 μmol L^{-1}AA (c) or AA and UA (d) and the absence (e) of AA and/or AA and UA in phosphate buffer solution at pH 7. Scan rate: 0.002 V s^{-1}. Reprinted with permission from Lima et al., 2009 [70], copyright 2009, WILEY-VCH Verlag GmbH & Co.

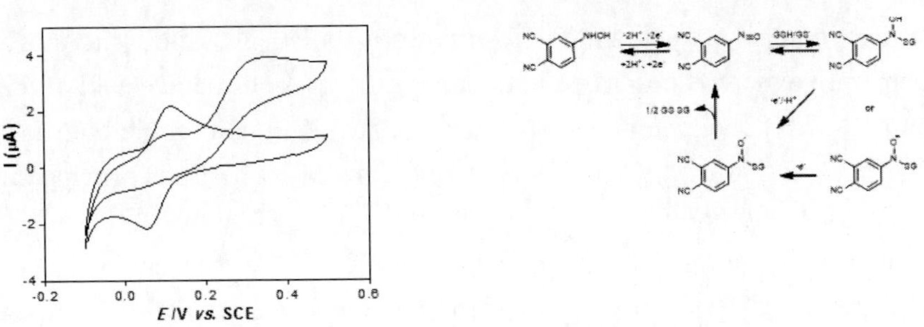

A B

Figure 16. A) Cyclic voltammograms for an act4-NPN-modified electrode in the presence (a) and absence (b) of 0.2 mmol L^{-1} GSH in phosphate buffer solution at pH 7.0. Scan rate: 0.05 V s^{-1}. $E_{initial}$ = 0.1 V and E_{λ} = 0.5 V, anodic scan. B) Proposed mechanism for the reaction between the activated 4-nitrophthalonitrile-modified electrode and GSH. Reprinted with permission from Lima et al., 2008 [69], copyright 2008, Elsevier.

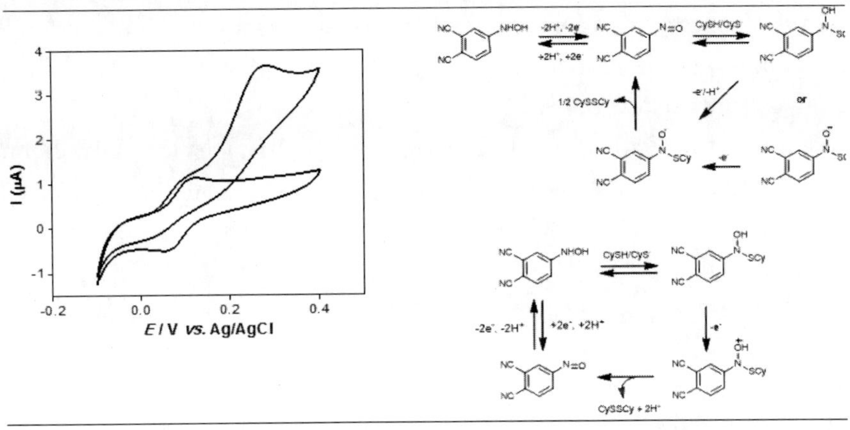

A B

Figure 17. A) CV for the act4-NPN-modified electrode in the presence (a) and absence (b) of 0.2 mmol L^{-1} CySH in phosphate buffer solution at pH 7.0. Scan rate: 0.05 V s^{-1}. $E_{initial}$ =0.1 V and E_{λ} = 0.4 V, anodic scan. B) Proposed mechanism for the reaction between the act4-NPN-modified electrode and CySH. Reprinted with permission from Lima et al., 2008 [68], copyright 2008, Elsevier.

L-Glutatione (GSH) levels were quantified in neutral medium at an applied potential of 0.4 V *versus* Ag/AgCl. After the optimization of operational conditions, the sensor provided a linear response range for GSH concentrations from 8.0 to 83.0 μmol L^{-1}, with sensitivity, detection and quantification limits of 54 nA L μmol^{-1}, 2.7 μmol L^{-1} and 8.0 μmol L^{-1}, respectively [69].

The proposed sensor presents higher sensitivity than other modified electrodes described in the literature and showed a stable response fot at least 100 successive determinations. The repeatability of the measurements with the same sensor and different sensors, as evaluated in terms of relative standard deviations, were 4.1 and 5%, respectively, for n = 10. The developed sensor was applied to determine GSH concentrations in yeast extract, and the results were statistically similar to those obtained by the comparative method described in the literature at a confidence level of 95% [69].

Cysteine (CySH) concentrations were also quantified in neutral medium at an applied potential of 0.33 V *versus* Ag/AgCl. The sensor showed its best performance in 0.1 mol L^{-1} phosphate buffer solution, pH 7.0 (Figure 17).

After the optimization of operational conditions, the sensor provided a linear response range for CySH concentrations from 0.8 to 13.2 μmol L^{-1}, with sensitivity, detection and quantification limits of 37 nA L μmol^{-1}, 0.25 μmol L^{-1} and 0.38 μmol L^{-1}, respectively. The sensitivity of the proposed sensor was much higher than other sensors described in the literature and showed a stable response for at least 80 successive determinations. The repeatability of the measurements with the sensor and the preparation of a series of electrodes, as evaluated in terms of relative standard deviations, were 4.1 and 5%, respectively, for n = 10. The developed sensor was applied to determine cysteine concentrations in food supplement samples and the results were statistically similar to those obtained using a comparative method described in the literature at a confidence level of 95% [68].

Niclosamide (2′,5-dichloro-4′-nitrosalicylanilide) (**NIC**) is the only commercially available molluscicide recommended by the WHO for large scale use in schistosomiasis control programs [87-89]. Niclosamide (**NIC**) was initially used as a redox mediator in the electroanalysis of NADH [71] Tables 1 and 2).

NIC contains the following strategic functional groups:

1. a nitro group that can be reduced electrochemically *In situ* to the hydroxylamine (ArNHOH), furnishing an important and reversible couple (ArNO/ArNHOH) through a redox process involving $2e^-/2H^+$ and resulting in a stable and confined system at the electrode surface;
2. an aromatic group that can strongly interact with CNTs, which have a large surface area, via van der Waals and/or hydrophobic interactions, or π-π stacking;
3. a phenol group that is susceptible to oxidation following the formation of dimers and/or polymers that strongly adhere to the surface of CNTs, increasing the stability and fixation of the mediator on the platform; and
4. an amide group in the *para* position, which itself is a reasonable deactivating group that yields a more electrophilic, electrogenerated NO that might undergo nucleophilic attack from NADH more efficiently, thus increasing the efficiency of the catalytic system.
5. Moreover, it is a compound with limited soluble in water, an adequate pKa, and thus a lower chance of leaching in aqueous solutions.

A simple and sensitive method for the electrocatalytic detection of dihydronicotinamide adenine dinucleotide (NADH) on an activated niclosamide (activated NIC)-based modified electrode prepared on a glassy carbon electrode modified with multi-walled carbon nanotubes (GCE/MWCNT/actNIC) was developed.

This modified electrode shows efficient electrocatalytic oxidation activity toward NADH at an applied potential of 0.065 V *vs.* Ag/AgCl, with a kinetic constant, k_{kin}, evaluated by chronoamperometry of 1.15 x 10^4 L mol^{-1} s^{-1}. The sensor presents a linear response ranging from 10.0 to 280.0 μM, with limits of detection and quantification and sensitivity of 3 μM, 10 μM and 2.4 nA μM^{-1}, respectively (Table 2).

As shown in the present study, the GCE/MWCNT/actNIC electrode exhibits good electrocatalytic activity toward NADH oxidation. The redox response of the modified electrode is the same response as anticipated for a surface-immobilized redox couple. The catalytic rate constant, 1.15 x 10^4 L mol^{-1} s^{-1}, was calculated from cyclic voltammetric and chronoamperometric responses, suggesting that **NIC** is a very interesting transducer candidate for use in dehydrogenase-based biosensors.

Conclusion and Perspectives

Electrochemical sensors represent a crucial analytical tool as the demand for the sensitive, rapid, and selective determination of analytes increases. Unlike spectroscopic and chromatographic instruments, electrochemical sensors are easily adapted to detect a wide range of analytes at a low cost. Additionally, these sensors are capable of being incorporated into robust, portable, or miniaturized devices, enabling tailoring to particular applications [7]. In general, the field of electrochemical sensors continues to grow and find new areas for application. The field will focus on the incorporation and interaction of unique materials, both nano and biological, in the future. These materials should be robust enough to be used in an emergency and in remote areas to rapidly detect pollutants, biological and chemical warfare agents, and components in natural and essential environments [28].

Two branches of electrochemical sensors are developing: sensors with increased specificity and sensors capable of simultaneous/multiplex determination. In both of these branches, the ability of the sensors to operate in complex biological matrixes will remain critical, forcing researchers to solve problems of biocompatibility and stability. However, further efforts are needed to identify the accurate theoretical description of electrochemistry at the nanoscale and satisfy the demand for more powerful characterization methods to eliminate experimental uncertainties.

The tailoring of the electrochemical interface with redox molecular assemblies is one of the fascinating approaches used to develop versatile platforms for catalytic and sensing applications. The surface-adsorbed redox molecules have important roles in controlling the electron transfer reaction at the interface. Various groups have attempted to design these interfaces; however, the search for new molecules with strategic functional groups continues and has been a way to connect organic chemistry and electrochemistry. The electrocatalysis of molecules continues to be a hot research topic in biological and energy-related chemistry. Electrochemical screening is a welcome choice [90, 91]. Typical parameters would be the indication of the formation of the product and the efficiency of the electrocatalytical conversion. The most prominent approach is based on the combined measurement using cyclic voltammetry (CV) to identify promising electrocatalysts [92].

In summary, the choice of the best redox mediator and the best method for incorporating it into the electrode configuration for each application is critical to ensure the electrochemical behavior of the resulting sensor.

The demand for the perfect mediator is very high [26, 93], and the rational choice of the mediator is a challenge, similar to the computational design of nitro-substituted phenanthro-imidazoles [16], which may be used in electroanalysis. Additionally, trial and error should not be used to search for electrocatalysis mediators. A rational approach would be time-saving and relevant to the topic.

The joint efforts of organic electrochemists, theoreticians and electroanalysts are needed to accelerate development in this area.

The steadily increasing number of examples of redox catalysis indicates that the principle has achieved wide application in electroanalysis and related fields, mainly because mediated processes frequently afford higher and/or different selectivity than a direct process. In addition, electrode reactions are accelerated and the overall energy consumption is reduced; moreover, reagent waste and difficult separation procedures are avoided. Redox catalysis provides a useful alternative when electrode passivation or over-reaction poses a problem in reaction control.

Concerning applications in electrocatalysis, only a few studies using nanoelectrodes to investigate the catalytic activities of single nanoparticles have been reported. The method could improve our understanding of the complex relationships between the size, geometry and activity of nanostructured materials used for energy conversion and storage. An increasing number of studies have examined the use of nanoelectrodes as probes to target single living cells. The first examples revealed the promising applications of these electrodes to the detection of analytes in the intracellular space. To date, these reports have been proof-of-concept studies to validate the new analytical methods, but need to be developed toward more robust applications. Establishing these methods might help to solve real biologically relevant questions that have been left unanswered due to the lack of suitable analytical tools. In this context, more selective nanosensors must be developed through specific design and modification of the nanoelectrode surface. Rapid progress in high-resolution electrochemical imaging was achieved recently. Nanoelectrodes have helped researchers visualize the electrochemical activity of individual objects in the nanometer range. Nevertheless, if the analyzed object is in the size range of the detecting tip, the influence of the electrochemical reaction at the tip must be considered. The ultimate limitation for amperometric detection of small analyte concentrations using nanoelectrodes is currently dictated by the capabilities of the electrochemical instrumentation. Completely new detection strategies must be developed to overcome this barrier.

Compounds that were originally designed as mediators of electron transfer have attracted the attention of researchers in other fields. For example, mediators used in electrosynthesis have been investigated as excited state electron transfer agents in photoinitiated processes [94] in electrosynthesis by simply adjusting the scale h [17]. Other applications include redox shuttles for overcharge protection [95] and in the design of biofuel cells [15]. Although direct electron transfer became a goal for enzymatic biofuel cells, most biofuel cells reported in the literature have continued to utilize mediated electron transfer, because these systems typically resulted in higher current densities. Nevertheless, they exhibit higher potential losses and issues with mediator stability [96].

Current developments are focused on the possibilities of tailoring thc individual and combined 3D structures of mediators and electrodes to achieve new improved modified electrodes. Future research may facilitate the development of advanced sensors with applications in the food industry, clinic and point-of-care diagnostics, among others.

Clearly, the field of mediated electron transfer constitutes a fertile arena in which to conduct research. We have no doubt that many important and exciting developments will be forthcoming in the future. As our ability to precisely and imaginatively manipulate the chemical nature of the electrode/solution interface continues to increase, we expect that applications in the areas listed above will expand and completely new applications be developed.

ACKNOWLEDGMENTS

The authors wish to thank the Brazilian financial agencies: CNPq, CAPES, FAPEAL, FAPESP and INCT – Bioanalítica for providing financial support.

REFERENCES

[1] Ciobanu, M., Wilburn, J. P., Krim, M. L., Cliffel, D. E. in Handbook of Electrochemistry, Edited by: Cynthia G. Zoski. ISBN: 978-0-444-51958-0. 2007, p. 3-29.

[2] Ciucu, A. A., in Journal of Biosensing Bioelectronics 2014, Vol. 5, p. 1.

[3] Labib, M., Sargent, E. H., Kelley, S. O., in Chemical Reviews 2016, Vol. 116, p. 9001.

[4] Farré, M., Kantiani, L., Pérez, S., Barceló, D., in Trends in Analytical Chemistry 2009, Vol. 28, p. 170.

[5] Buck, R. P., Lindner, E., Kutner W., Inzelt, G., in Pure and Applied Chemistry 2004, Vol. 76, p. 1139.

[6] Justino, C. I. L., Duarte, A. C., Rocha-Santos. T. A. P., in Trends in Analytical Chemistry 2016, Vol. 85, p. 36.

[7] Kimmel, D. W., LeBlanc, G., Meschievitz, M. E., Cliffel, D. E., in Analytical Chemistry 2012, Vol. 84, p. 685.

[8] Nayak, S., Blumenfeld, N. R., Laksanasopin, Sia, S. K., in Analytical Chemistry 2017, Vol. 89, p. 102.

[9] Xiao, T., Wu, F., Hao, J., Zhang, M., Yu, P., Mao, L., in Analytical Chemistry 2017, Vol. 89, p. 300.

[10] Soto, R. J., Hall, J. R., Brown, M. D., Taylor, J. B., Schoenfisch, M. H., in *Analytical Chemistry* 2017, *Vol. 89*, p. 276.

[11] Justino, C. I. L., Rocha-Santos, T. A. P., Duarte, A. C., in *Trends in Analytical Chemistry* 2010, *Vol. 29*, p. 1172.

[12] Dorkó, Z., Verbić, T., Horvai, G., in *Talanta* 2015, *Vol. 132*, p. 680.

[13] Lowinshon, D., Bertotti, M., in *Quimica Nova* 2006, *Vol. 29*, p. 1318.

[14] Edwards, G. A., Bergren, A. J., Porter, M. D. Chemically Modified Electrodes. In Elsevier B.V. *Handbook of Electrochemistry* Edited by C. G. Zoski. ISBN: 978-0-444-51958-0, 2007, p. 295.

[15] Barsan, M. M., Ghica, M. E., Brett. C. M. A., in *Analytica Chimica Acta* 2014, *Vol. 881*, p. 1.

[16] Francke R. and Little, R. D., in *Chemical Society Reviews* 2014, *Vol. 43*, p. 2492.

[17] Francke, R., Little, R. D., in *Journal of the American Chemical Society* 2014, *Vol. 136*, p. 427.

[18] Lane, R. F., Hubbard, A. T., in *The Journal of Physical Chemistry* 1973, *Vol. 77*, p. 1401.

[19] Murray, R. W., Ewing, A. G., Durst, R. A., in *Analytical Chemistry* 1987, *Vol. 59*, p. 379A.

[20] Zen, J. M., Kumar, S. A., Tsai, D. M., in *Electroanalysis* 2003, *Vol. 15*, p. 1073.

[21] Kochius, S., Anders O., in *Applied Microbiology and Biotechnology* 2012, *Vol. 93*, p. 2251.

[22] Moscoso, R., Inostroza, E., Squella. J. A., in *Electrochemistry Communications* 2017, *Vol. 81*, p. 61.

[23] D'Orazio, P., in *Clinica Chimica Acta* 2011, *Vol. 412*, p. 1749.

[24] Siqueira Jr., J. R., Luciano Caseli, L., Crespilho, F. N., Zucolotto, V., Oliveira Jr., O. N., in *Biosensors and Bioelectronics* 2010, *Vol. 25*, p. 1254.

[25] Steckhan, E., in *Topics in Current Chemistry* 1987, *Vol. 142*, p. 1.

[26] Gorton, L., Dominguez, E., in *Reviews in Molecular Biotechnology* 2002, *Vol. 82*, p. 371.

[27] Malinauskas, A., Ruzgas, T., Gorton, L., in *Journal of Colloid and Interface Science* 2000, *Vol. 224*, p. 325.

[28] Yang, W., Kyle, R., Ringer, R. S. P., Thordarson, P., Gooding, J. J., Braet, F., in *Angewandte Chemie, International Edition* 2010, *Vol. 49*, p. 2114..

[29] Willner, I., Zayats, M,, in *Angewandte Chemie International Edition* 2007, *Vol. 119*, p. 6528.

[30] Amatore, C., Arbault, S., Guille, M., Lemaitre, F., in *Chemical Reviews* 2008, *Vol. 108*, p. 2585.

[31] Stromberg, M., de La Torre, T. Z. G., Goransson, J., Gunnarsson, K., Nilsson, M., Svedlindh, P., Stromme, M., in *Analytical Chemistry* 2009, *Vol. 81*, p. 3398.

[32] Thévenot, D. R., Toth, K., Durst, R. A., Wilson, G. S., in *Biosensors and Bioelectronics* 2001, *Vol. 16*, p. 121.

[33] Patolsky, F., Zheng, G. F., Lieber, C. M., in *Analytical Chemistry* 2006, *Vol. 78*, p. 4260.

[34] Sajid, M. I., Jamshaid, U., Zafar, T. J. N., Fessi, H., Elaissari, A., in *International Journal of Pharmaceutics* 2016, *Vol. 501*, p. 278.

[35] Castro, V. G., Costa, I. B., Lopes, M. C., Lavall, R. L., Figueiredo, K. C. S., Silva, G. G. in *Journal of the Brazilian Chemical Society* 2017, *Vol. 28*, p. 1158.

[36] Zhang, Q., Huang, J., Qian, W., Zhang, Y., Wei, F., in *Small* 2013, *Vol. 9*, p. 1237.

[37] Chen, J., Chen, Q., Ma, Q., in *Journal of Colloid and Interface Science* 2012, *Vol. 370*, p. 32.

[38] Zeng, L., Alemany, L., Edwards, C., Barron, A., in *Nano Research* 2008, *Vol. 1*, p. 72.

[39] Medjo, R. E. *Characterization of Carbon Nanotubes, Physical and Chemical Properties of Carbon Nanotubes,* Dr. Satoru Suzuki (Ed.), ISBN: 978- 953-51-1002-6, 2013.

[40] Karousis, N., Tagmatarchis, N., Tasis, D., in *Chemical Reviews* 2010, *Vol. 110*, p. 5366.

[41] Park, H., Zhao, J., Lu, J. P., in *Nano Letters* 2006, *Vol. 6*, p. 916.

[42] Liu, Z., Tabakman, S., Welsher, K., Dai, H., in *Nano Research* 2009, *Vol. 2*, p. 85.

[43] Wildgoose, G. G., Hyde, M. E., Lawrence, N. S., Leventis, H. C., Jiang, L., Timothy G. J., Jones, T. G. J., Compton, R. G., in *Langmuir* 2005, *Vol. 21*, p. 4584.

[44] Urzua, J.; Carbajo, J.; Yanez, C.; Marco, J. F.; Squella, J. A., in *Journal of Solid State Electrochemistry* 2016, *Vol. 20*, p. 1131.

[45] Casero, E., Darder, M., Takada, K., Abruña, H. D., Pariente, F., Lorenzo, E., in *Langmuir* 1999, *Vol. 15*, p. 127.

[46] Lund. H., Hammerich, O., in *Cathodic Reduction of nitro and related compounds. Organic Electrochemistry, Fourth Edition.* ISBN: 9780824704308. 2001, 379-409.
[47] Rubinstein. I, in *Journal of Electroanalytical Chemistry* 1985, *Vol. 183*, p. 379.
[48] Squella, J. A., Bollo, S., L. Núñez-Vergara, L. J., in *Current Organic Chemistry* 2005, *Vol. 9*, p. 565.
[49] Hammerich, O., Reduction of nitrocompounds and related substrates. *In* Hammerich, O. and Speiser, B. *Organic Electrochemistry* ISBN: 978-1-4200-8402-3. 2015, p. 1151-1170.
[50] Squella, J. A.; Nunez-Vergara, L. J.; Campero, A.; Maraver, J.; Jara-Ulloa, P; Carbajo, J. in *Journal of The Electrochemical Society* 2007, *Vol. 154*, p. F77.
[51] Hernández-Munoz, L. S., González, F. J., González, I., Goulart, M. O. F., Abreu, F. C., Ribeiro, A. S., Ribeiro, R. T., Longo, R. L., Navarro, M., Frontana, C., in *Electrochimica Acta* 2010, *Vol. 55*, p. 8325.
[52] Brillas, E.; Farnia, G.; Severin, M. G.; Vianello, E., in *Electrochimica Acta* 1986, *Vol. 31*, p. 759.
[53] Taira, H., Nakano, K., Maeda, M., Takagi, M., in *Analytical Sciences* 1993, *Vol. 9*, p. 199.
[54] Tsutsumi, H., Furumoto, S., Morita, Y, Matsuda, M., in *Journal of Colloid and Interface Science* 1995, *Vol. 171*, p. 505.
[55] Mano, N., Kuhn. A., in *Journal of Electroanalytical Chemistry* 1999, *Vol. 477*, p. 79.
[56] Mano, N., Kuhn, A., in *Electrochemistry Communications* 1999, *Vol. 1*, p. 497.
[57] Santhiago, M.; Lima, P. R.; Santos, W. J. R.; Oliveira, A. B.; Kubota, L. T. in *Electrochimica Acta* 2009, *Vol. 54*, p. 6609.
[58] Moscoso, R., Inostroza, E., Bollo, Squella, J. A., in *Electrocatalysis* 2016, *Vol. 7*, p. 357.
[59] Paez, J. I.; Strumia, M. C.; Passeggi Jr., M. C. G.; Ferrón, J.; Baruzzi, A. M.; Brunetti, V., in *Electrochimica Acta*, 2009, *Vol. 54*, p. 4192.
[60] Munteanu, F. D.; Mano, N.; Kuhn, A.; Gorton, L., in *Bioelectrochemistry* 2002, *Vol. 56*, p. 67.
[61] Munteanu, F. D., Mano, N., Kuhn, A., Gorton, L., in *Journal of Electroanalytical Chemistry* 2004, *Vol. 564*, p. 167.
[62] Mano, N., Thienpont, A., Kuhn, A., in *Electrochemistry Communications* 2001, *Vol. 3*, p. 585.
[63] Mano, N.; Peyrou, P.; Kuhn, A., in *Electroanalysis* 2001, *Vol. 13*, p. 770.
[64] Mano, N., Kuhn, A., in *Journal of Electroanalytical Chemistry* 2001, *Vol. 498*, p. 58.
[65] Santhiago, M.; Lima, P. R.; Santos, W. J. R.; Kubota, L. T., in *Sensors and Actuators B* 2010, *Vol. 146*, p. 213.

[66] Giroud, F.; Sawada, K.; Taya, M.; Cosnier, S. in *Biosensors and Bioelectronics* 2017, *Vol. 87*, p. 957.

[67] Lima, P. R.; Santos, W. J. R.; Oliveira, A. B.; Goulart, M. O. F.; Kubota, L. T., in *Biosensors and Bioelectronics* 2008, *Vol. 24*, p. 448.

[68] Lima, P. R.; Santos, W. J. R.; Luz, R. C. S.; Damos, F. S.; Oliveira, A. B.; Goulart, M. O. F.; Kubota, L. T. *Journal of Electroanalytical Chemistry* 2008, *Vol. 612*, p. 87.

[69] Lima, P. R.; Santos, W. J. R.; Oliveira, A. B.; Goulart, M. O. F.; Kubota, L. T., in *Journal of Pharmaceutical and Biomedical Analysis* 2008, *Vol. 47*, p. 758.

[70] Lima, P. R., Miranda, P. R. B., Oliveira, A. B., Goulart, M. O. F., Kubota, L. T., in *Electroanalysis* 2009, *Vol. 21*, p. 2311.

[71] Lopes, C. B.; Silva, F. A. S.; Lima, P. R.; Freitas, J. D.; Sousa, J. S.; Kubota, L. T., Goulart, M. O. F., in *Journal of Solid State Electrochemistry* 2015, *Vol. 19*, p. 2819.

[72] Kovacic, P., Somanathan, R., in *Journal of Applied Toxicology* 2014, *Vol. 34*, p. 810.

[73] Manu B, Chaudhari S., in *Bioresource Technology* 2002, *Vol. 82*, p. 225.

[74] Bard, A. J., Faulkner, L. R. *Electrochemical Methods: Fundamentals and Applications,* 2nd Edition. ISBN: 978-0-471-04372-0 2001, p. 864.

[75] Radoi A., Compagnone D., in *Bioelectrochemistry* 2009, *Vol. 76*, p. 126.

[76] Reid, R. C., Minteer, S. D., Gale, B. K., in *Biosensors and Bioelectronics* 2015, *Vol. 68*, p. 142.

[77] Zhu, Z., Tam, T. K., Sun, F., You, C., Zhang, Y. H. P., in *Nature Communications* 2014, *Vol. 5*, p. 3026.

[78] Reuillard, B., Le Goff, A., Cosnier, S., in *Chemical Communications* 2014, *Vol. 50*, p. 11731.

[79] Fujita, S., Yamanoi, S., Murata, K., Mita, H., Samukawa, T., Nakagawa, T., Sakai, H., Tokita, Y., in *Scientific Reports* 2014, *Vol. 4*, p. 4937.

[80] Chenault, H., Whitesides, G., in *Applied Biochemistry and Biotechnology* 1987, *Vol. 14*, p. 147.

[81] Murray, R. W. Introduction to the Chemistry of Molecularly Designed Electrode Surfaces, *Techniques of Chemistry Series,* R. W. Murray (Ed.), ISBN: 978-0-471-55773-9, 1992.

[82] Santos, A. D., Gorton, L., Kubota, L. T., in *Electroanalysis* 2002, *Vol. 14*, p. 805.

[83] Santos, A. D., Gorton, L., Kubota, L. T., in *Electrochimica Acta* 2002, *Vol. 47*, p. 3351.

[84] Pereira, A. U., Santos, A. D., Kubota, L. T., in *Electrochimica Acta* 2003, *Vol. 48*, p.3541.

[85] Galus, Z., in *Fundamentals of Electrochemical Analysis*, Ellis Horwood, New York 1976, p. 313.

[86] Komiyama, T., Fujimori, K., in *Bioorganic Medicinal Chemistry Letters* 1997, *Vol. 7*, p. 175.

[87] Abreu F. C., Goulart, M. O. F., Brett, A. M., in *Biosensors and Bioelectronics* 2002, *Vol. 17*, p. 913.

[88] AL-Hadiya B. M., in *Profiles Drug Substances, Excipients and Related Methodology* 2005, *Vol. 32*, p. 67.

[89] WHO (World Health Organization) The Control of schistosomiasis: second report of the WHO Expert Committee, *WHO Technical Reports Series* 2003, Geneva, No. 830, p. 53.

[90] Pinczewska, A., Sosna, M., Bloodworth, S., Jeremy D. Kilburn, J. D., Bartlett, P. N. *Journal of the American Chemical Society* 2012, *Vol. 134*, p. 18022.

[91] Gutz, C., Klocher, B., Waldvogel, S. R. *Org. Process Res. Dev.* 2016, *Vol. 20*, p. 26-32

[92] Borgmann, S., Schuhmann, W. In *Integrated Analytical Systems;* Potyrailo, R. A., Mirsky, V. M., Eds.; Springer: New York, 2009.

[93] Gorton, L. in *Journal of the Chemical Society, Faraday Transactions* 1986, *Vol. 82*, p. 1245.

[94] Konig, B., Ed. *Chemical Photocatalysis*; De Gruyter: Berlin, 2013.

[95] Xu, K., in *Chemical Reviews* 2004, *Vol. 104*, p. 4303.

[96] Rasmussen, M., Abdellaoui, S., Minteer, S. D., in *Biosensors and Bioelectronics* 2016, *Vol. 76,* p. 91.

In: Advanced Materials and Systems …
Editor: Mauro Coelho dos Santos
ISBN: 978-1-53613-852-8

Chapter 5

ELECTROCHEMICAL NANOPLASMONIC SENSORS

Bruno P. Crulhas, Agnieszka E. Karpik, Gustavo R. Castro and Valber A. Pedrosa*

Department of Chemistry and Biochemistry, Institute of Bioscience, UNESP-Botucatu, SP, Brazil

ABSTRACT

We discuss nanoplasmonic sensors by analyzing different parameters in order to quantitatively assess their overall sensing performance. They have attracted much attention for their electrochemical potential to offer enhanced electrical and mechanical properties. These nanocomposites properties promote electron transfer between electroactive species and electrode surface with good chemical stability, high surface area, and excellent electronic properties. Finally, we discuss a series of considerations that should be addressed in order to bring the realization of nanoplasmonic sensor with competitive levels of sensitivity, robustness, and integration much closer to reality.

Keywords: plasmon, nanostructure, sensor, optical, electrochemistry

INTRODUCTION

The analytical chemistry has become a popular tool for the acquisition of valuable information about target systems over the years; moreover, it has been successfully used in a range of research fields such as medicine, pharmacy, food production and

* Corresponding Author Email: vpedrosa@ibb.unesp.br.

environmental protection. The quantitative and qualitative analyses of chemical and biological compounds are mainly implemented for clinical diagnosis, including diseases caused by viruses and bacteria, as well as for environmental monitoring, pathogen detection, and antigen-antibody binding. The biosensors are able to fulfill the gaps caused by the growing need for lowering costs, simplifying and minimizing the aforementioned technologies.

According to the International Union of Pure and Applied Chemistry, the term 'biosensor' refers to devices that transform biological or chemical information about the concentration of each component and the composition of the analyzed mixture into analytically useful signals. Biosensors were first reported in 1962, and they allow converting a biological response into a quantifiable signal by using a specific biochemical reaction involving a biologically active compound, i.e., enzymes, antibodies, tissues, DNA, RNA and cells, in order to detect a target substance coupled to a transducer (e.g., optical, electrochemical, piezoelectric, thermometric and mechanical transducers). The electrochemical and optical biosensors are of particular importance among several biosensor types due to their high sensitivity, simple design, and low cost.

The electrochemical sensors just require a small number of reagents; besides, they can be used in complex solutions (e.g., whole blood, plasma, culture medium and raw samples). Moreover, they are an attractive way to study biological sample contents due to the direct conversion of a biological event into an electronic signal and because they multiplex into multianalytic platforms able to achieve lower detection limits (values ranging from micromoles and nanomoles). The typical biosensing electrochemical techniques involve four different pathways: the generation of measurable currents (amperometric); the potential or charge accumulation (potentiometric); impedance (impedimetric) or the variations in the conductive properties between electrodes (conductometric). Voltammetry is a widely used electrochemical technique that measures the oxidation and reduction currents in target analytes. Its specificity is achieved through a range of biological recognition elements such as enzymes, DNA or RNA, antibodies, proteins, and even whole cells[1-4]) coupled to a redox cycling or to a fast scanning using small reagent and sample volumes.

In addition to the electrochemical biosensors, optical biosensors became a trend in the biosensing research field because they are based on detecting changes in the absorption of electromagnetic visible/infrared radiation, changes between reactants, and reactions produced by the measurements applied to the light emitted in luminescent processes. The measured parameters are absorption, adsorption, desorption, refractive and reflectivity index (RI), fluorescence, phosphorescence, and wavelength. The refractometric platforms based on surface plasmon resonance (SPR) or on localized surface plasmon resonance (LSPR) have been developed and improved because of their extraordinary properties. Nanomaterials have become the object of interest of many biosensor research fields due to the recent advances in nanotechnology. The main advantage of these nanomaterials lies in their exquisite sensitivity to both the biological and the chemical sensing. Many types of nanoparticles, including metal, oxide, and semiconductor nanoparticles, as well as the nanodimensional conducting polymers, have been used in biosensors.

Despite several reviews concerning the electrochemical [4-6] and optical biosensor topics [7-9], there is still a gap to be filled, namely: the use of different nanoparticles and the combination of these techniques. Therefore, the aim of the present review is to discuss the combination of nanoparticles, and electrochemical and optical techniques, in order to provide new insights into the development of new biosensors.

NANOPLASMONIC TECHNIQUES

The surface plasmons can be defined as charge density oscillations confined to a metal, which is excited by light at incident wavelength able to generate intense surface plasmon absorption bands [10]. Basically, the surface plasmon resonance (SPR) is the sensitivity of surface plasmon polaritons such as the changes in the RI of a gold film, which is measured through a reflected laser light passing across a prism and reflecting back. In addition, one part of the SPR technique is of particular interest due to its enormous potential for application in the surface plasmon resonance (LSPR) technique [11,12]. The LSPRs are restricted to the measurement of resonance waves from metal nanoparticles such as gold (AuNPs); silver (AgNPs); platinum (PtNPs) and palladium (PdNPs) [12,13]. The nanoscale resolution enables the detection of molecular interactions right on the nanoparticle surface by shifting the LSPR peak[11]. This resonance angle dependence on the surface is mainly caused by specific evanescent electromagnetic field interactions (evanescent wave) with the target molecules located close to the nanoparticles. Such phenomenon is accurately described by the classical Physics and is shown in the following plasma energy E_p expression:

$$E_p = h\sqrt{\frac{ne^2}{m\varepsilon_0}} \tag{1}$$

wherein: *m* is the electron mass, *e* is the electron charge, ε_0 is the permittivity of free space, and *n* is the electron density.

Metal nanoparticle suspensions can interact with light to produce a range of colored solutions due to light absorption by the oscillations of electrons in a UV-Vis range, and by their photon interaction with the conduction band of the metal nanostructures [8]. The LSPR has two intrinsic characteristics, namely: the electric field is larger on the metal nanoparticle surface, however, it decreases proportionally to the distance influencing the LSPR sensitivity; the optical extinction of the particle reaches maximum values at the plasmon resonant frequency in the visible wavelength of specific nanoparticulates and depends on the RI of the surroundings. The LSPR peaks are usually generated by spectral scattering measurements applied to the nanoparticles, or by the spectral extinction in a dense film. The agglomeration of nanoparticles generates a smaller spacing than their diameters; therefore, the electromagnetic interaction is triggered by the coupling mechanism between dipoles. This mechanism will lead to color shifts towards the red spectrum (lower energy) in the LSPR; however, when this distance increases 2.5 times in comparison to the diameter, the nanoparticles get more dispersed and behave as single species [7,14-16].

The LSPR response is strongly sensitive to changes near the surface and such characteristic is a prerequisite to measure the kinetic interaction between biomolecules. The additional properties of nanoparticles such as shape, size, distribution, shell thickness, dielectric constant of the surroundings, and spatial organization, can change band intensity and position. The shape of nanoparticles, nanorods, and nanostars significantly interfere in the experimental RI sensitivity [17-21]. Particles presenting sharp ends, nanoshells, and nanotriangles have higher RI sensitivity because the shape of these nanoparticles and it can create highly enhanced electric field intensity. The Mie theory was used and represented as follows, in order to analyze the optical properties at long wavelengths, which are limited by the electrostatic dipole of spherical nanoparticles:

$$E(\lambda) = \frac{24\pi N_A a^3 \varepsilon_m^{\frac{3}{2}}}{\lambda \ln(10)} \mathrm{X} \frac{\varepsilon_i}{(\varepsilon_r + \chi\varepsilon_m)^2 + \varepsilon_i^2} \quad (2)$$

wherein: *E(λ)* is the extinction (sum of scattering and absorption), *a* is the radius of metal sphere, *λ* is the wavelength of absorbed radiation, *m* is the dielectric constant of the medium, *NA* is the particle area density, ε_i is the imaginary part of the dielectric function of the metal particle, ε_r is the real part, and χ is related to the shape of the nanoparticle.

The selective absorption of the wavelength in LSPR occurs at high molar extinction coefficient (approximately 3 x 10^{11} $M^{-1}cm^{-1}$) due to the increased electromagnetic field near the surface of the nanoparticle [22]. The free valence shell electrons from the surface of the nanoparticles in the LSPR get in resonance with the incident light; this resonance gets confined to the surface and starts to continuously emit light. This process strongly depends on the nanoparticles' RI, as well as on changes in the RI, on length decay (δd) and on nanopattern thickness. The Drude model can be used to access the functional dependence form between LSPR peak wavelength and the dielectric properties of the medium.

$$\varepsilon_1 = 1 - \frac{\omega_{p^2}}{\omega^2 + y^2} \quad (3)$$

Wherein: y is the damping parameter of the bulk metal, and ωp is the plasma frequency. Moreover, the LSPR parameters can be achieved through the derivation of this equation, such as LSPR frequency (ωmax) and LSPR peak wavelength (λmax).

METAL NANOPARTICLES AND NANOPLASMONIC SENSORS

The nanoplasmonic sensors using metal nanoparticles are widely studied in analytics, biocatalytic processes, biomolecule analyses, and for monitoring biomaterials' local properties, and bioanalytical applications [19]. Prior to the use of metal nanoparticles to develop new biosensors, it is worth assessing what metals would be suitable for optical and electrical detection. Overall, most biosensing platforms use Au as active sensor surface, because it is a chemically stable metal of enhanced conductibility and of easy etching [23-25]. The only metal presenting Drude-like ideal electron behavior similar to the gold is the silver, which shows no interband transitions in the visible region and higher conductivity. The Ag's disadvantage lies on its oxidation susceptibility; however, several studies produced nanoplasmonic sensors using silver nanoparticles [26-28]. In addition to gold and silver, nanoplasmonic sensors made out of other metal types were tested to produce a range of new platforms for biosensing applications. Copper, for instance, has good conductivity as bulk metal, but its oxidation is more prominent than the Au and Ag platforms, even in face of efficient coupling with SPR and optical damping in visible wavelengths [29-31]. On the other hand, platinum and palladium present improved oxidative stability, but weak plasmon resonance of decreased intensity under the interband transitions. Their transitions generate lower signals than silver, but platinum and palladium achieve absolute extinction values 50 to 60% lower [32,33].

The gold nanoparticles (AuNPs) are the most promising and widely nanomaterial described in the literature due to their unique characteristics and extraordinary properties. The AuNPs can measure 1 to 100nm diameter; show high surface-to-volume ratio and high surface energy; therefore, it can stabilize biomolecule immobilization, as well as preserve biomolecules' bioactivity. Moreover, AuNPs enable fast and direct electron transfer between a wide range of electroactive species and electrode materials. In addition, the light-scattering properties and extremely large enhancement of the local electromagnetic field enable AuNPs to be used as signal amplification labels in different biosensors, including the optical and electrochemical techniques.

The AuNPs optical biosensor is based on the behavior of collective conduction band electron (plasmons) oscillations in response to the implemented electromagnetic radiation. The surface plasmon resonance is one of the most popular sensing modalities. The SPR is based on surface plasmon polaritons sensitivity to changes in the RI. These changes are measured as the changes in the reflected laser light pass through a prism and reflect back from the film. Accordingly, the molecules binding to the metallic film surface are responsible for deviations in the dielectric constant and, consequently, for changing the laser light reflection. The sensitive to the dielectric constant AuNPs is implemented in one of the SPR technique variations - the LSPR (localized on the surface of the plasmon resonance) – to generate good analytical characteristics. The main roles played by AuNP in optical biosensors are to amplify the signal and to improve the sensitivity of the device.

In addition to the optical sensors, AuNPs can be applied to electrochemical biosensing devices, because of their extraordinary properties, such as biocompatibility, conductivity, and catalytic assets, which derive from the enhanced electron transfer between the redox-molecule and the electrode. The AuNPs, for instance, presents more enhanced electron transfer than ferrocine derivatives such as quinones, and poly-2-aminoaniline polymer. Moreover, the AuNP can be coupled to several polymer matrices, and it results in increased biosensor stability and reusability, just as in biosensors based on nanocomposites including chitosan, poly(p-aminobenzene sulfonic acid) or Poly(ethylene)glycol. In addition, the immobilization of recognition elements in the electrode can be improved by AuNPs implementation. It prevents the degradation or deactivation of biological molecules after they get in contact with the naked surface of the bulk materials. The AuNPs provide effective, stable and reliable results because their biocompatibility, surface free energy (SFE) and surface area are higher than that of flat surfaces. The biosensing elements can be linked to AuNPs by a range of techniques such as thiol linkers, covalent bonds or amine groups. Apart from the aforementioned properties, and different from the bulk gold, the AuNPs also show extraordinary catalytic activity, which may result from their quantum-scale size and high surface-to-volume ratio. Some studies corroborate that this activity derives from the altered electronic core structure. Therefore, their overpotentials may be decreased and, as a result of it, some important redox reactions may become reversible, as well as the production of enzyme-free biosensors, e.g., the glucose biosensor combined with the self-assembling and sol-gel techniques for AuNPs immobilization. The AuNPs also show selective detection in the electrochemical analysis.

However, in order to combine the optical and electrochemical sensing in an electrochemical-plasmonic sensor, it is important assessing the AuNPs stability and the optical signals upon oxidation and reduction reactions because the metal could suffer from oxidative reactions and be dissolved or complexed with halide ions [34]. Moreover, the oxidation potential is particle-size dependent due to the increased surface energy for smaller particles [35]. It was assessed as reversible for the colloidal gold attachment to the indium tin oxide (ITO) [36,37]. Several studies have also assessed the AuNP changes resulting from responses to the electrochemical potentials caused by the restructuration and faceting of the gold nanorod edges [38]. The chemical interactions with ions can affect the optical response over the electrochemical sensing by changing the electron density, which leads to resonance wavelength shifts ([39].

Only a few sensing applications were developed due to the scarcity of information about the electrical-optical behavior of AuNPs. Novo et al. [40]) measured catalytic reactions in single particles using optical detection; the signals were generated by changes in the electron density. Wang and Lin [41] developed and electro-optically modulated the plasmonic biosensing frequency using AuNPs to improve the stability of the measurements. Heip et al. [42] used an electrochemical impedance spectroscopy (EIS) plasmonic sensor to detect the peptides responsible for forming pores in membranes wherein the EIS was used to assess changes in membrane resistance; the plasmonic signal was adopted to quantify protein accumulation in the membrane.

NANOWIRES ARRAYS AND NANOPLASMONIC SENSORS

The development of metallic nanowires became an area of particular interest due to their unique optical and electrochemical characteristics ([43-47]. Nanowires can be used in two singular plasmon resonances: first, the surface plasmon can propagate along the wire like 1D SPR, but this resonance is more related to plasmonic waveguides than to sensing [48]; the second plasmon resonance is a semi-localized or 2D plasmon, it works like nanoparticle plasmons - the transverse dipoles are perpendicular to the wire axis and the coupling effects occur in the arrays of the wires ([49,50]. Moreover, the resonance can be influenced by the incidence angle and cross-section shape [43,51,52], the refractometric sensing is similar to the metal nanoparticles presenting resonance wavelength sensitivity 17nm [52] or 104 nm [43], per RI increment.

In addition to the optical properties, nanowires can provide a substantial advantage to electrochemical experiments, because they open a conducting path along the wire [53,54]. The scattering of electrons over the surface lead to higher metal resistance in comparison to the bulk resistivity of the metal itself. It becomes an important information if at least one dimension of the nanowires is equal to or smaller than the mean Drude electro free path [55] and if it has optical analogy additional to the plasmon damping in small-sized nanoparticles (less than 10 nm) [56].

The resistance should be dominated by surface scattering in the sensing applications using nanowires; in this case, porous wires can be advantageous [57] for electrochemical sensing, because it enables assessing conductivity changes caused by the formation of alkanethiol monolayers on gold wires. The signal is proportional to the strength and chemical nature of the molecular interaction with the metal [57,58]. However, the way this conductivity will be able to provide valuable information for biosensing in a proper recognition interface remains unclear [59], as well as the measurements applied to biomolecular interactions through conductivity in metallic nanowires.

MacKenzie et al. [60,61] developed gold nanowire arrays for sensing applications produced by extreme UV interference lithography with optical and electrochemical detection. The nanowire arrays were similar to the interdigitated electrodes used for EIS experiments, so the implementation of the optical detection combined with EIS should be immediate. In addition, the system geometry in EIS is intrinsic to the improved performance, and it triggers the development of several electrode types, from nanogaps to 3D configurations [62,63] . The literature provides different LOD (for EIS in biosensing) showing values close to 1 fM; however, it is hard to find researches which detected target analytes in real biological samples through EIS. Moreover, there is no consensus about the presence of charge transfer species and about faradaic response enhancement [62,64-67]. Furthermore, Evans et al. [68]used the three-electrode setup standard (working, reference and auxiliary) for EIS sensing, with impressive LOD, in biological samples without any special asset on the gold electrode. They inferred the absence of differences between round or interdigitated gold electrodes; therefore, they could combine EIS and plasmonic measurement sensing in any geometric-shaped electrode.

Future Directions

The present review of new techniques and advances concerning the combination of electrochemical and optical techniques applied to nanoplasmonic sensing show that these methodologies have the potential to become a powerful instrument to enhance the sensitivity of biological detections. The LSPR provides the mechanisms to tune and control the optical behavior. The electrochemical measurements can be helpful to the understanding of the electron dynamics on the electrode surface. Optical and electrical methods are complementary; the optical techniques can provide the refractometric detection of small molecules. Voltammetry is able to detect any oxidation or reduction reactions on the electrode surface. Moreover, the use of the EIS techniques, in combination with optical sensors, is a new research topic, and it is still expanding. This is a topic of interest because these combinations can improve the understanding about charge resistance and plasmonic resonance, as well as help elucidating the EIS dynamics on the electrode surface and improving the quantitative interpretation of the sensor's responses.

References

[1] B. P. Crulhas, L. C. Recco, F. K. Delella, V. A. Pedrosa, *Electroanalysis* 2017, 29, 1-7.

[2] B. P. Crulhas, N. P. Ramos, G. R. Castro, V. A. Pedrosa, *Journal of Solid State Electrochemistry*, 2016, 20, 2427-2433.

[3] Q. Zhou, A. Rahimian, K. Son, D. S. Shin, T. Patel, A. Revzin, *Methods*, 2016, 97, 88-93.

[4] Q. Zhou, K. J. Son, Y. Liu, A. Revzin, *Annual Review of Biomedical Engineering*, 2015, 17, 165-190.

[5] J. Kirsch, C. Siltanen, Q. Zhou, A Revzin, A. Simonian, *Chemical Society Reviews*, 2013, 42, 8733-8768.

[6] D. Grieshaber, R. MacKenzie, J. Vörös,E. Reimhult, *Sensors*, 2008, 8, 1400–1458.

[7] M. C. Estevez, M. A. Otte, B. Sepulveda, L. M. Lechuga, *Analytica Chimica Acta*, 2014, 806, 55–73.

[8] K. M. Mayer, J. H. Hafner, *Chemical Reviews*, 2011, 111, 3828–3857.

[9] J. Homola, *Chemical Reviews*, 2008, 108, 462–493.

[10] I. Tokarev, *I.* Tokareva, V. Gopishetty, E. Katz, S. Minko, *Advanced Materials*, 2010, 22, 1412–1416.

[11] Y. Roiter, I. Minko, D. Nykypanchuk, I. Tokarev, S. Minko, *Nanoscale*, 2012, 4, 284-292.

[12] R. Lupitskyy, M. Motornov, S. Minko, *Langmuir*, 2008, 24, 8976–8980.

[13] C. Wu, X. Zhou, J. Wei, *Nanoscale Research Letters*, 2015, 10, 1–6.

[14] C. J. Murphy, A. M. Gole, S. E. Hunyadi, J. W. Stone, P. N. Sisco, A. Alkilany, B. E. Kinard, P. Hankins, *Chemical Communications*, 2008, 5, 544-557.

[15] B. Zhang, R. B. Kumar, H. J. Dai, B. J. Feldman, *Nature Medicine*, 2014, 20, 948–953.

[16] M. Sriram, K. Zong, S. R. C. Vivekchand, J. J. Gooding, *Sensors*, 2015, 15, 25774–25792.

[17] I. Tokareva, I. Tokarev, S. Minko, E. Hutter, J. H. Fendler, *Chemical Communications*, 2006, 31, 3343–3345.

[18] B. Nikoobakht, M. A. El-Sayed. *Chemistry of Materials*, 2003, 15, 1957–1962.

[19] Y. Zheng, X. Zhong, X. Li, Y. Xia, *Particle & Particle Systems Characterization,* 2014, 31, 266–273.

[20] M. R. Langille, M. L. Personick, J. Zhang, C. A. Mirkin, *Journal of American Chemistry Society*, 2012, 134,14542–14554.

[21] H. Y. Wu, H. C. Chu, T. J. Kuo, C. L. Kuo, M. H. Huang, *Chemistry of Materials*, 2005, 17, 6447–6451.

[22] T. R. Jensen, *The Journal of Physical Chemistry B,* 200, 104, 10549–10556.

[23] Y. Wang, X. Shan, H. Wang, S. Wang, N. Tao, *Journal of the American Chemistry Society*, 2017, 139, 1376-1379.

[24] W. Putzbach, N. J. Ronkainen, *Sensors*, 2013, 13, 4811-4840.

[25] A. B. Dahlin, T. Sannomiya, R. Zahn, G. A. Sotiriou, J. Vörös, *Nano Letters*, 2011, 11, 1337–1343.

[26] E. Abbasi, M. Milani, S. F. Aval, M. Kouhi, A. Akbarzadeh, H. T. Nasrabadi, P. Nikasa, S. W. Joo, Y. Haniferhpour, K. N. Koshki, M. Samiei, *Critical Reviews in Microbiology*, 2016, 42, 173-180.

[27] J. N. Anker, W. P Hall, O. Lyandres, N. C. Shah, J. Zhao, R. P. Van Duyne, *Nature Materials,* 2008, 7, 442–453.

[28] A. J. Haes, R. P. Van Duyne, *Analytical and Bioanalytical Chemistry*, 2004, 379, 920–930.

[29] H. Gao, J. Henzie, M. H. Lee, T. W. Odom, *Proceedings of the National Academy of Sciences*, 2008, 105, 20146–20151.

[30] G. H. Chan, J. Zhao, E. M. Hicks, G. C. Schatz, R. P. Van Duyne, *Nano Letters*, 2007, 7, 1947–1952.

[31] Zhang, W; Brongersma, SH; Richard, O; Brijs, B; Palmans, R; Froyen, L; Maex, K. *Microelectronic Engineering*, 2004, 76, 146–152.

[32] C. Langhammer, Z. Yuan, I. Zoric, B. Kasemo, *Nano Letters*, 2006, 6, 833–838.

[33] A. K. Bordoloi, S. J. Auluck, *Jounral of Physics F: Metal Physics,* 1988, 18, 237-247.

[34] B. H. Loo, *TheJournals of Physical Chemistry*, 1982, 86, 433–437.

[35] O. S. Ivanova, F. P. Zamborini, *Journal of the American Chemistry Society*, 2009, 132, 70–72.

[36] T. Sannomiya, H. Dermutz, C. Hafner, J. Voros, A. B. Dahlin, *Langmuir*, 2010, 26, 7619–7626.

[37] T. Miyazaki, R. Hasegawa, H. Yamaguchi, H. Oh-Oka, H. Nagato, I. Amemiya, S. Uchikoga, *The Journal of Physical Chemistry C*, 2009, 113, 8484–8490.

[38] C. Novo, P. Mulvaney, *Nano letters*, 2007, 7, 520–524.

[39] P. Mulvaney, *Langmuir*, 1996, 12, 788–800.

[40] C. Novo, A. M. Funston, P. Mulvaney, *Nature Nanotechnology*, 2008, 3, 598–602.

[41] T. J. Wang, W. S. Lin, *Applied Physics Letters*, 2006, 89, 173903.

[42] H. M. Hiep, T. Endo, M. Saito, M. Chikae, D. K. Kim, S. Yamamura, Y. Takamura, E. Tamiya, *Analytical Chemistry*, 2008, 80, 1859–1864.

[43] H. A. Chen, H. Y. Lin, H. N. Lin, *The Journal of Physical Chemistry C*, 2010, 114, 10359-10364.

[44] P. Shi, J. Y. Zhang, H. Y. Lin, P. W. Bohn, *Small*, 2010, 6, 2598–2603.

[45] V. Auzelyte, H. H. Solak, Y. Ekinci, R. MacKenzie, J. Voros, S. Olliges, R. Spolenak, *Microelectronic Engineering*, 2008, 85, 1131–1134.

[46] E. J. Menke, M. A. Thompson, C. Xiang, L. C. Yang, R. M. Penner, *Nature Materials*, 2006, 5, 914–919.

[47] Q. B. Xu, J. M. Bao, F. Capasso, G. M. Whitesides, *Angewandte Chemie [Applied Chemistry] International Edition*, 2006, 45, 3631–3635.

[48] S. Lal, S. Link, N. J. Halas, *Nature Photonics*, 2007, 1, 641–648.

[49] G. Della Valle, T. Sondergaard, S. I. Bozhevolnyi, *Optics Express*, 2008, 16, 6867–6876.

[50] G. Schider, J. R. Krenn, W. Gotschy, B. Lamprecht, H. Ditlbacher, A. Leitner, F. R. Aussenegg, *Journal of Applied Physics*, 2001, 90, 3825–3830.

[51] M. V. Sosnova, N. L. Dmitruk, A. V. Korovin, S. V. Mamykin, V. I. Mynko, O. S. Lytvyn, *Applied Physics B: Lasers and Optics*, 2010, 99, 493–497.

[52] K. M. Byun, S. J. Kim, D. Kim, *Applied Optics,* 2006, 45, 3382–3389.

[53] G. F. Zheng, F. Patolsky, Y. Cui, W. U. Wang, C. M. Lieber, *Nature Biotechnology*, 2005, 23, 1294–1301.

[54] Y. Cui, Q. Q. Wei, H. K. Park, C. M. Lieber, *Science*, 2001, 293, 1289–1292.

[55] L. Zhang, X. Jiang, E. Wang, S. Dong, *Biosensors and Bioelectronics*, 2005, 21, 337.

[56] V. Myroshnychenko, J. Rodriguez-Fernandez, I. Pastoriza-Santos, A. M. Funston, C. Novo, P. Mulvaney, L. M. Liz-Marzan, F. J. G. de Abajo, *Chemical Society Reviews*, 2008, 37, 1792–1805.

[57] Z. Liu, P. C. Searson, *The Journal of Physical Chemistry*, 2006, 110, 4318–4322.

[58] Y. M. Zhang, R. H. Terrill, P. W. Bohn, *Analytical Chemistry*, 1999, 71, 119–125.

[59] T. A. Bendikov, A. Rabinkov, T. Karakouz, A. Vaskevich, I. Rubinstein, *Analytical Chemistry*, 2008, 80, 7487–7498.

[60] R. MacKenzie, C. Fraschina, T. Sannomiya, J. Voros, *Nanotechnology*, 2011, 22, 055203.

[61] R. MacKenzie, C Fraschina, T. Sannomiya, V. Auzelyte, J. Vörös, *Sensors*, 2010, 10, 9808–9830.

[62] K. V. Singh, A. M. Whited, Y. Ragineni, T. W. Barrett, J. King, R. Solanki, *Analytical and Bioanalytical Chemistry*, 2010, 397, 1493–1502.

[63] A. Bratov, J. Ramon-Azcon, N. Abramova, A. Merlos, J. Adrian, F. Sanchez-Baeza, M. P. Marco, C. Dominguez, *Biosensors and Bioelectronics*, 2008, 24, 729–735.

[64] F. Lisdat, D. Schafer, *Analytical and Bioanalytical Chemistry*, 2008, 391, 1555–1567.

[65] J. S. Daniels, N. Pourmand, *Electroanalysis*, 2007, 19, 1239–1257.

[66] Y. X. Hou, S. Helali, A. D. Zhang, N. Jaffrezic-Renault, C. Martelet, J. Minic, T. Gorojankina, M. A. Persuy, E. Pajot-Augy, R. Salesse, F. Bessueille, J. Samitier, A. Errachid, V. Akimov, L. Reggiani, C. Pennetta, E. Alfinito, *Biosensors and Bioelectronics,* 2006, 21, 1393–1402.

[67] C. Berggren, B. Bjarnason, G. Johansson, *Electroanalysis*, 2001, 13, 173–180.

[68] D. Evans, S. Johnson, S. Laurenson, A. G. Davies, P. K. Ferrigno, C. Walti, *Journal of Biology*, 2008, 7, 1-11.

[69] Corrêa, C. C.; Santhiago, M.; Formiga, A. L. B.; Kubota, L. T., in *Electrochimica Acta*, 2013, *Vol. 90*, p. 309.

In: Advanced Materials and Systems …
Editor: Mauro Coelho dos Santos
ISBN: 978-1-53613-852-8

Chapter 6

RECENT MATERIALS AND RECOGNITION ELEMENTS APPLIED IN ELECTROCHEMICAL BIOSENSORS

Dênio E. P. Souto, Ana C. M. de Moraes, José T. C. Barragan, Everson T. S. G. da Silva, Lory C. Carossi and Lauro T. Kubota*

Department of Analytical Chemistry, Institute of Chemistry,
University of Campinas (UNICAMP), Campinas, São Paulo, Brazil

ABSTRACT

The development of biosensors has attracted increasing interest over the past decades, with a large number of publications on different applications, including in the fields of food quality control, environmental monitoring, and clinical diagnostics, among others. Although many transducers have been used in these fields, the electrochemical biosensor is by far the most popular transduction method in the biosensing area because it provides high sensitivity, simplicity of instrumentation, low cost, and the possibility of miniaturization for point-of-care (POC) analysis. The desired performance of electrochemical biosensors is intrinsically related to the recognition platform used for their construction; thus, in recent years, special attention has focused on the development of new materials and hybrid compounds. In this context, we highlight in this chapter some of the most recent materials and recognition elements, such as graphene and its derivatives, nanoparticles (e.g., metallic, magnetic, and quantum dots), dendrimers, and aptamers, applied in electrochemical biosensors. For each of these classes, a detailed discussion is presented on its definition, synthetic routes, properties, and applications in the electrochemical biosensing field. Finally, an overview of the current state and future

* Corresponding Author Email: kubota@iqm.unicamp.br.

trends of these recent materials and recognition elements in electrochemical biosensors is provided.

Keywords: biosensor, electrochemical biosensor, graphene, nanoparticles, dendrimer, and aptamer

INTRODUCTION

Biosensors are a class of chemical sensors that use biological components to selectively recognize certain analytes, thus ensuring sensitivity and specificity of the sensor. Although this description is acceptable for biosensors, in this chapter, we are going to follow the IUPAC (International Union of Pure and Applied Chemistry) definition, which considers biosensors as "chemical sensors in which the recognition system utilizes a biochemical mechanism" [1]. Biosensors are generally composed of three main components, (i) a bioreceptor, (ii) a transducer, and (iii) a signal-processing unit, in which a quantifiable signal can be generated when the bioreceptor interacts with a specific target analyte, thereby sending the processed analytical information to the display (Figure 1a) [2]. Research involving these devices and their applications has grown tremendously over the past decades, with a large number of publications in different areas, spanning the fields of clinical diagnostics, [3] food quality control, [4] and environmental monitoring, [5] among others. [6] Figure 1b shows a histogram of publications on biosensors over the years.

Although transducers, such as optical, [7] piezoelectric, [8] and thermometric, [9] among others, [10] have been used in biosensing applications, the electrochemical transducer is by far the most popular transduction method in this area because it provides high sensitivity, simplicity of instrumentation, low cost, and the possibility of miniaturization for point-of-care (POC) analysis. [11] The first electrochemical biosensor was introduced in 1962 by Leyland C. Clark in his seminal description of an "enzime electrode" [12]. In this work, Clark developed a biosensor based on a thin layer of glucose oxidase (GOx) on an oxygen electrode, with the amount of oxygen consumed during the enzymatic reaction with the target analyte (glucose) representing the analytical response. Since then, this electrochemical biosensor, known as the glucometer (or glucose meter), has become one of the most dominant biosensing devices worldwide. [13]. The glucometer is a glucose strip test based on a screen-printed electrode modified with GOx as the biorecognition element that is coupled to a portable electrochemical transducer for monitoring diabetic patients [14]. The development of this type of biosensor for medical applications was considered the main driving force in this research area, which resulted in an increasing number of research groups exploring this field. However, despite the tremendous success of the glucometer as an electrochemical biosensing device, to date, only a few examples of products involving this concept have reached the market, which is probably due to the multi-expertise required in the development of this type of device [4a] (more details on this topic will be discussed below).

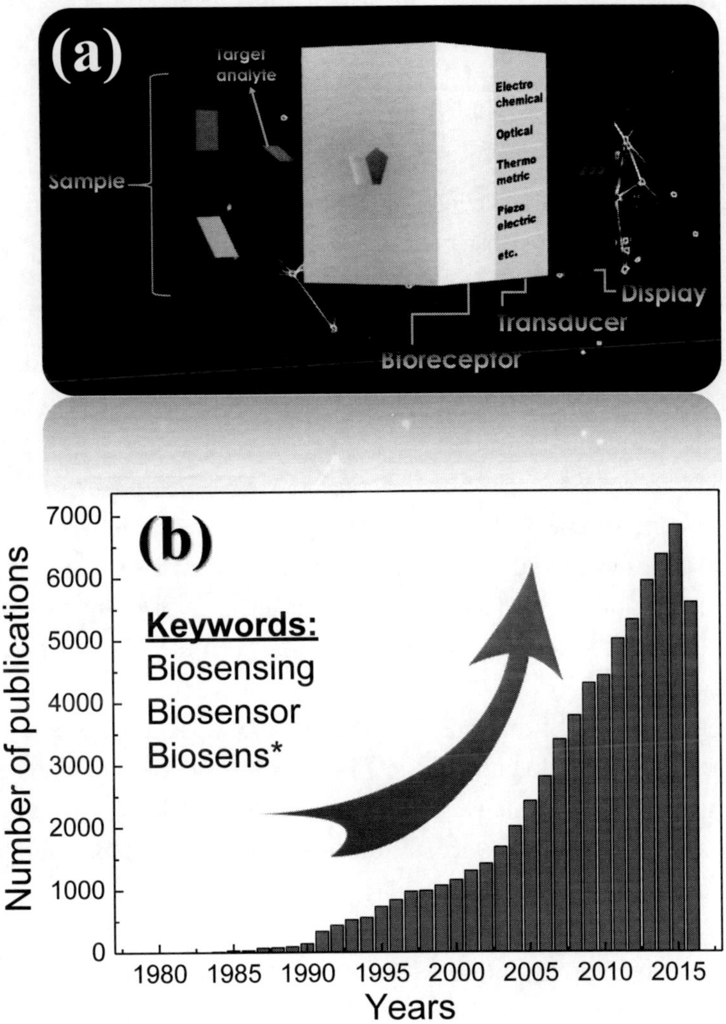

Figure 1. (a) Schematic illustration of the three main components of a biosensor, namely, the bioreceptor, transducer, and display. (b) General histogram of publications on biosensors over the years (data obtained from the Web Of Science™).

Electrochemical biosensors can be divided into two main categories according to the nature of the biorecognition process: catalytic biosensors and affinity biosensors [11, 15]. Catalytic biosensors are characterized by the incorporation of biological components, such as enzymes, whole cells or tissue slices, that can interact within the target analyte to generate electroactive species, which can be measured by the electrode [11]. Among all bioreceptors, enzymes are the most commonly used in biosensing applications due to their high specificity, which eventually eliminates the need for time-consuming and laborious pretreatment steps in complex samples such as blood and urine [16]. However, the lack of sufficient selective enzymes for every known biochemical analyte limits the use of this type of sensor; thus, in these cases, the affinity biosensor represents a better alternative. Affinity biosensors utilize the strong and selective binding abilities of some biomolecules with target species, which results in a measurable electrical signal [17]. Immunosensors are the main example of affinity biosensors, in which antibodies (Abs) bind specifically to an antigen (Ag), forming a complex (Ab-Ag) and generating an electrical signal that can be processed, recorded and displayed. In addition to the Ab used in immunosensors, membrane receptors, oligonucleotides and, more recently, nucleic acids (DNA, RNA and aptamers) have been used in affinity biosensors; however, these receptors have been used less frequently than antibodies [2].

Electrochemical transducers have been used in both catalytic and affinity biosensors, which use different detection strategies. Typically, biochemical reactions can generate different measurable signals, such as current, potential, conductivity, and resistance or reactance of the system. Thus, based on the measured electrical parameters, the electrochemical biosensors can be classified as amperometric, potentiometric, impedimetric or conductometric biosensors [18]. Amperometric biosensors are a class of electrochemical biosensors that evaluate current variations resulting from the electrochemical oxidation or reduction of a target analyte. In these cases, the resulting current is proportional to the concentration of the target molecule that interacts with the biorecognition element. This detection method is most commonly used in both catalytic and affinity electrochemical biosensors because of its simplicity and low limit of detection (LOD) [19]. Potentiometric biosensors represent another class of electrochemical biosensors, which use ion-selective electrodes (ISEs) to transduce the biochemical reaction between the target molecule and the bioreceptor into a measurable electrical signal (normally the electrode potential) [20]. Ion-selective field effect transistors (ISFETs) are an important variation of these potentiometric biosensors, and they can incorporate biorecognition elements to form enzymatic (ENFETs) [21] or immunological field-effect transistors (IMFETs) [22]. Conductometric [23] and impedimetric [24] biosensors are less common than the previously mentioned electrochemical biosensors. The latter, for example, is primarily used in affinity biosensors to monitor immunological binding events, which produce variations in the impedance of the system that are proportional to the concentration of the target analyte. By contrast, conductometric biosensors monitor variations in the electrical conductivity of sample solutions after the biorecognizing event; therefore, they are primarily used in catalytic biosensors that often utilize enzymes. More specific information about electrochemical biosensors and their definitions, architectures, syntheses, and applications can be found in a number of interesting reviews in the literature, such as that published by Ronkainen and co-workers (2010), [11] among others [6, 18, 25].

Although electrochemical biosensors exhibit great advantages, such as high specificity, sensitivity, and fast analysis with a relatively low cost, over traditional analytical methods, there are few drawbacks that can limit their use in some conditions [11, 18]. Fortunately, recent advances in nanotechnology and materials science as well as in micro- and nano-fabrication processes have pushed electrochemical biosensors to another level, overcoming some of the limitations of these devices, such as the reduced shelf life and poor stability of the biorecognition element and issues with non-specific binding. In this context, special attention is focused on new materials and hybrid compounds that can improve biosensing assays and eliminate their disadvantages [26]. In this chapter, we summarize some of the most recent materials and molecules (recognition elements) used in electrochemical biosensors and highlight their definitions, syntheses, properties, and applications in electrochemical biosensing. We pay special attention to graphene materials, and nanoparticles, in addition to some molecular recognition elements, such as aptamers and dendrimers. Thereafter, an overview of the advances and future prospects of these technologies will be presented.

GRAPHENE-BASED ELECTROCHEMICAL BIOSENSORS

Graphene, the thinnest known material, is composed of one-atom-thick two-dimensional (2D) sheets of carbon atoms with sp^2 hybridization, which are covalently bonded in honeycomb lattices [27]. Graphene is considered the basic building block of other carbon allotropes with different dimensionalities: 2D sheets can be wrapped into zero-dimensional fullerenes, [28] rolled up into one-dimensional carbon nanotubes [29] or stacked into three-dimensional graphite. Historically, as graphene is simply a single layer of graphite, the rapid rise of graphene in materials science cannot be explained without mentioning the abundant and naturally occurring of graphite. Graphite has been known for centuries, and its applications are varied, including in pencils, lubricants, electrodes, and reinforced composites. Graphite crystals are formed by layered graphene sheets separated by a distance of only 0.335 nm, in which the carbon atoms are strongly linked, without chemical bonding in the *c*-axis direction. In fact, the layers are weakly held together through van der Waals interactions, giving graphite anisotropic properties. For instance, graphite is highly conductive because electrons travel along the planes. However, the electrical conductivity is much lower through the planes [30]. The weak bonding between the layers also enables the easy detachment of graphite layers from each other. Essentially, a graphite crystalline flake is merely countless monolayers of graphene stacked together [31].

Graphene was firstly isolated in 2004 by Geim and co-workers at Manchester University [32]. Its experimental discovery resulted in an explosion of interest in materials science and condensed-matter physics, especially because 2D crystals were presumed to not exist due to their thermodynamic instability at finite temperatures [33]. Therefore, 2D graphene crystals were described as 'academic' materials until 2004 and were theoretically assumed to not exist in the free state because of the unstable formation of curved structures, such as fullerenes and nanotubes [32]. However, graphene crystals are not only stable but also structured to allow the charge carriers to move across relatively long interatomic distances without scattering, a phenomenon referred to as ballistic transport. The hexagonal lattices in graphene comprise two triangular interpenetrated sub-lattices of carbon atoms, which are linked through covalent sp^2 bonds with an interatomic distance of ~1.42 Å and 120° C–C–C bond angles (Figure 2a). Each carbon atom has s, p_x, and p_y atomic orbitals, which contribute to the mechanical stability of the sheet. The remaining p_z orbital overlaps with its neighboring carbon atoms to form a filled π (valence) band and an empty π^* (conduction) band, thus contributing to the delocalization of electrons in the network [34]. Single-layer graphene sheets present lateral dimensions in the micrometer range and are slightly flat with undulations that correspond to single-atom thick dimensions, as shown in Figure 2b.

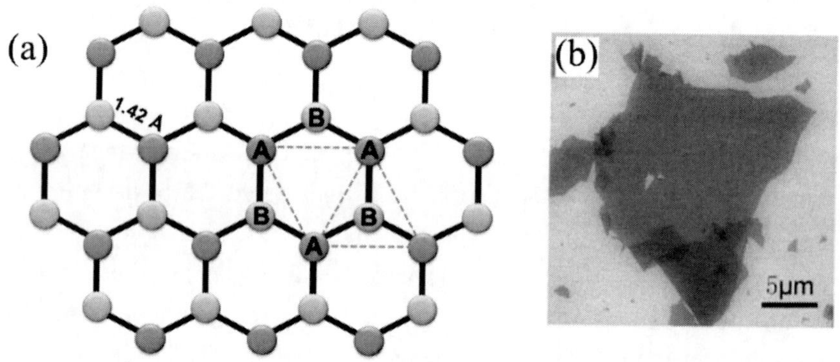

Figure 2. (a) Triangular sub-lattices of graphene showing that each carbon atom in one lattice (A) has three carbon atoms (B) as the nearest neighbors. (b) Scanning electron microscopy image of single-ayer graphene sheets deposited on Si/SiO_2 substrate (reproduced with permission from the Royal Society of Chemistry [35]).

Therefore, the long-range π-conjugation in the honeycomb network yields unusual and superlative electrical, mechanical, and thermal properties. Graphene presents high electron mobility, offering an exceptional electrical conductivity of ~10^4 S cm^{-1} and a high intrinsic mobility of 200,000 cm^2 V^{-1} s^{-1} at room temperature. It is a zero-band-gap semiconductor material because the conduction and valence bands meet at the Dirac points within the Brillouin zone, producing cone-shaped valleys. At this low energy limit, charge carriers behave as massless relativistic particles, which gives graphene an ambipolar electric field effect. Charge carriers can be modulated almost equally between electrons and holes at concentrations of 10^{13} cm^{-2}, which is suitable for developing field effect transistor (FET) devices. [32, 34b, 36] In particular, the electronic structure of graphene changes with the number of layers, and the 3D limit of graphite is reached at 10 layers. Single and bilayer graphene have simple electronic spectra and behave as zero-gap semiconductors. Thus, the higher number of layers, the more charge carriers appear, and the conduction and valence bands become overlapped. Consequently, structures with >10 layers should be considered as thin graphite films [32].

Due to the strong sp^2 covalent bonding between carbon atoms, graphene is the strongest known material – more than 200 times stronger than steel and aramid (Kevlar), which confers it a high elastic modulus of ~1.0 TPa and a fracture strength of 130 GPa [37]. These properties make graphene a desirable reinforcing material for polymer composites and coatings. Graphene is a good heat conductor, and its thermal conductivity of 5,000 W m^{-1} K^{-1} is superior to that of diamond and metals, such as copper [38]. It is not only strong but it is also a super-light material, being more than six times lighter than steel. Graphene possesses high transparency in the visible range, exhibiting an optical transmittance of ~97.7%, which is ideal for applications, such as transparent conductive films [39]. In addition to the aforementioned characteristics, graphene has a high specific surface area (2,630 m^2 g^{-1}) and is impermeable to standard gases, such as helium [40].

The combination of such properties in a single material enables a myriad of applications, including flexible electronics, efficient transistors, transparent protective coatings, optoelectronic devices such as touchscreens and organic light-emitting diodes, photovoltaic cells, energy storage, reinforced composites, and sensors. This list is ever growing, and the combined properties not yet explored might lead to new applications of graphene.

Graphene can be produced in bulk quantities and through a wide range of mechanical and chemical processes. As it is a single layer of graphite, it is reasonable to assume that it can be directly extracted from bulk graphite flakes. The top-down approach of mechanical exfoliation used by the Manchester group consists of isolating 2D single layers from 3D graphite flakes with an adhesive tape ('*Scotch tape*') in repeated and successive adhesion operations for bulk graphite and detaching the layers on a substrate of oxidized silicon wafer. This method enables the exfoliation into single sheets by overcoming weak van der Waals interactions between the layers without disturbing subsequent sheets. High-quality crystals can be obtained at a relatively low cost. However, this method is extremely laborious and is neither industrially scalable nor high yielding. Another limitation is the low probability of locating individual graphene sheets on the substrate [27, 31, 34b, 41].

Alternative routes to obtaining single-layer graphene include processes from non-graphitic sources. For instance, the epitaxial growth of graphene involves the thermal evaporation of silicon atoms from a monocrystalline silicon carbide (SiC) surface at temperatures of 1200 – 1500°C under high vacuum. The removal of silicon leaves a carbon-enriched surface, which undergoes reorganization and graphitization, giving rise to thin graphene sheets over the surface of the SiC wafers. High-quality graphene sheets are grown; however, they lack uniformity with respect to the size and number of layers, and the method requires high vacuum conditions [42]. Large-area graphene sheets can be generated through chemical vapor deposition (CVD). In this substrate-based technique, carbon atoms are deposited directly on transition metal surfaces (for example, nickel, copper, or ruthenium) from the decomposition of gaseous hydrocarbons (for example, methane or ethylene) at high temperatures. Typically, a gaseous hydrocarbon is exposed to a heated metal surface under vacuum, and the metal catalyzes the loss of hydrogen and dissolves carbon, forming a metal carbide layer. Upon cooling of the substrate, supersaturation is induced, which triggers the nucleation of graphene islands on the metal surface [43]. Then, a thin film of graphene with large crystalline domains is formed on the metal surface. The etching of the metal substrate allows the transferring of graphene to arbitrary substrates. Both CVD and epitaxial growth present prospects of producing graphene deposited in an entire wafer, which facilitate the integration of graphene as a new material for semiconductor devices.

Molecular bottom-up approaches for producing graphene have been explored by using small-sized and well-defined polycyclic aromatic hydrocarbons as building blocks. This method, pioneered by Mullen and Fasel, [44] involves the surface-assisted coupling of aromatic precursors and their subsequent cyclodehydrogenation to obtain graphene structures. More specifically, the authors exposed anthracene and triphenylene derivatives on a hot gold surface to promote intermolecular covalent interlinking through the radical addition of dehalogenated intermediates to generate fully aromatic graphene-like structures. This method is a clean chemical route; however, the generated graphene structures are limited in size, such as nanoribbons with a size of up to 12 nm. The carbon nanotubes unzipping method has also been demonstrated as a chemical approach for rendering graphene nanoribbons. Multiwalled carbon nanotubes subjected to strong solution-based oxidative treatment are capable of forming graphene nanoribbon structures from the lengthwise cutting of the side walls [45].

Individual graphene sheets can be exfoliated from bulk graphite by liquid-phase approaches. Solution-based processes involve the dispersion of graphite in a solvent, exfoliation, purification, and the separation of exfoliated sheets from unexfoliated graphite flakes [46]. The most commonly used methods for producing pristine graphene sheets are described as follows:

Liquid-Phase Exfoliation of Graphite

High quality and high-throughput graphene monolayers can be produced by ultrasound-assisted exfoliation. Hydrodynamic shear forces associated with cavitation cause the formation, growth, and collapse of bubbles or voids in the liquid, which are capable of "breaking" the layered graphite structure, thereby causing exfoliation into graphene sheets [47]. Graphene-solvent interactions are in equilibrium with the attractive forces between the graphene sheets and aggregation is minimized, which result in highly stable graphene dispersions. This method does not involve chemical reactions, and it is carried out in the presence of organic solvents that minimize the interfacial tension between graphene and the liquid. Low surface tension solvents, such as N-methylpyrrolidone (NMP), dimethyl formamide (DMF), and benzyl benzoate, enable the solvent-induced stabilization of graphene sheets [48]. Exfoliation may also be performed in the presence of surfactants or polymer solutions, resulting in electrostatically or sterically stabilized nanosheets [49].

Liquid-Phase Exfoliation of Graphene Oxide

This versatile method involves the production of chemically modified graphenes. The oxidative treatment of graphite in a mixture of strong acids and oxidizing agents partially disrupts the sp^2 network into sp^2–sp^3 lattices by introducing oxygenated functional groups, such as hydroxyl and epoxide, into the basal plane and carboxyl and carbonyl, which are attached to the edges [50]. These oxygen-bearing groups are essential for the subsequent liquid exfoliation, which results in the formation of oxidized graphene sheets that are highly stable in aqueous medium and organic polar solvents, the so-called graphene oxide (GO) sheets. The layered oxidized structure can be exfoliated to produce single GO sheets through sonication, mechanical stirring, and thermal expansion. However, large GO sheets are defective and insulating. The π-conjugation network can be partly restored by performing different reduction strategies, such as using chemical reducing agents, thermal annealing, UV irradiation, or electrochemical reduction. [51] The end material, reduced graphene oxide (rGO), is substially different from pristine graphene because the remaining defects cannot be totally removed from the subsequent reduction process, thus, rGO has inferior electronic properties compared to pristine graphene [52].

Liquid-Phase Exfoliation of Intercalated Graphite

graphite intercalation compounds (GICs) exhibit graphite crystal lattices that are intercalated with atoms or molecules. Donor or acceptor intercalants, such as alkali metals, alkali earth metals, lanthanides, halogens, halogen mixtures, metal chlorides, and acidic oxides, increase the graphite interlayer spacing, forming charge transfer complexes. Exfoliation into graphene sheets occurs spontaneously without ultra-sonication but mechanical stirring is still required [50, 53].

Table 1. Main production methods for synthesizing graphene from graphite and non-graphitic sources

Production method	Approach	Brief description
Graphene from graphite		
Mechanical exfoliation	Top-down	Straightforward, no need for specialized equipment, high-quality crystals, large-area sheets, time-consuming in locating sheets and identifying single and few layers, exfoliated sheets are often multilayer.
LPE of graphite		Stable graphene dispersions, inexpensive and easily scalable, thick graphite flakes must be removed, graphene sheets have a limited size.
LPE of GO		Inexpensive and scalable, versatile method for obtaining chemically modified graphenes, stable graphene dispersions, large-area sheets, rGO contains structural defects.
LPE of GIC		GICs tend to oxidize in air, requires a controlled ambient for processing, sheet size is limited by the size of the starting crystal.
Graphene from non-graphitic sources		
Epitaxial growth	Bottom-up	Relatively straightforward, high quality, high vacuum and temperature conditions, poor control of the size and number of layers, dependent on growth parameters and on the morphology of the SiC surface, requires highly specialized equipment, expensive.
CVD		High-quality and large-area sheets, compatible with CMOS technology, graphene growth is self-limiting, easy transfer to other substrates, transfer process can damage graphene sheets and/or introduce impurities, not suitable for applications that require significant quantities of graphene.
Molecular approach		Limited size range of graphene, clean chemical synthetic route.

Abbreviations: CVD, chemical vapor deposition; CMOS, complementary metal oxide semiconductor; GIC, graphite intercalation compound; GO, graphene oxide; LPE, liquid-phase exfoliation; rGO, reduced graphene oxide, SiC: silicon carbide.

A wide variety of synthetic methods for producing graphene sheets have been reported [54]. The advantages and disadvantages of the most exploited bottom-up (atom-by-atom growth) and top-down (from bulk graphite) routes are briefly illustrated in Table 1. In general, most synthetic strategies are highly scalable alternatives for preparing good-quality graphene, thereby paving the way toward industrial-scale production of graphene.

Beyond opto-electronic, energy, and material-based applications, graphene holds are great promise as platforms for a variety of bioapplications, one of which is biosensing. [55]. In particular, the electrochemical properties of graphene have attracted considerable interest with respect to its implementation into electrochemical biosensors. Graphene-based electrodes exhibit a low residual current, a renewable surface, a low charge resistance (in comparison to graphite and glassy carbon), a wide electrochemical potential window (e.g., 2.5 V in 0.1 M PBS and pH 7), and well-defined redox peaks [56]. Redox processes on their surfaces are primarily diffusion-controlled. The peak-to-peak potential separation values obtained from cyclic voltammetry indicate that the rapid single-electron transfer kinetics and electron transfer rates are much faster on graphene than on glassy carbon electrodes [57].

Graphene is an excellent electric charge conductor, and it exhibits high room temperature electron mobility. As the electron transfer processes occur at the edges of the graphene sheets or at the defects on the basal plane, its high surface area provides larger amounts of defects and consequently, multiple electroactive sites. Therefore, electrodes modified with graphene exhibit an improved and more uniform distribution of electrochemically active sites than those of graphite. Due to the 2D structure (flat surface) of graphene, all carbon atoms are exposed to the surrounding, which makes it a good platform for adsorbing and detecting molecules on the electrode surface, leading to increased sensitivity [55, 58]. In addition to the properties that make graphene an appealing transducer material in biosensors, it is produced from considerably clean methods (excluding epitaxial growth and carbon nanotubes unzipping methods), without the introduction of metallic impurities that remain electrochemically active, and hinder the biosensor's performance [59].

Graphene is also a biocompatible material. Biocompatibility assessments using mouse fibroblast cells showed good cell adhesion and proliferation on graphene papers, which indicates the excellent biocompatibility of graphene [60]. Graphene papers were found to be noncytotoxic toward three mammalian cell lines [61]. The Proliferation of osteoblasts on the rGO substrates was reported to be successful [62]. Therefore, graphene provides a highly biocompatible microenvironment for the immobilization of biocompounds, such as antibodies, enzymes, and DNA, while maintaining their bioactivity.

In terms of electrochemical applications, graphene produced from the reduction of GO (rGO) is the most extensively used form of graphene. Smaller-sized rGO sheets are easily obtainable in bulk quantities. rGO nanosheets still exhibit a degree of oxygenated functional groups and present abundant structural defects, which are potentially beneficial for fast heterogeneous electron transfer. In addition, the presence of the remaining oxygen-containing moieties can affect the local density of electronic states and may influence the adsorption of redox species, thereby catalyzing the redox reactions. Moreover, these oxygen-containing groups facilitate the immobilization/functionalization of the surface of graphene-modified electrodes with several biomolecules as recognition elements while improving hydrophilicity, which makes rGO highly advantageous for electrochemical biosensing applications [63].

Enzyme graphene-based biosensors take advantage of the good conductivity and large surface area of graphene, which favors enzyme immobilization and high electron transfer kinetics, which results in sensitive biosensors. For example, Liang et al. constructed a glucose biosensor based on glassy carbon electrodes modified with electrochemically reduced graphene oxide [64]. Glucose oxidase (GO_x) was immobilized over the graphene-modified electrode surface through self-assembly. The proposed biosensor exhibited a low detection limit along with efficient direct electron transfer between the immobilized GO_x and the electrode surface. Palanisamy et al. reported the modification of screen-printed carbon electrodes with electrochemically reduced graphene oxide and horseradish peroxidase (HRP) enzyme [65]. The remaining oxygen-containing groups on rGO played an important role in the immobilization of HRP on the graphene surface. The graphene-modified enzymatic amperometric biosensor showed electrocatalytic activity toward H_2O_2, and was capable to detect this analyte with appreciable accuracy.

Electrochemical graphene-based immunosensors are promising tools in biomedical analysis. These graphene immunosensors enhance electron transfer and also provide a larger surface area for the immobilization of antibodies. For instance, Kim et al. developed a FET-based immunosensor for the detection of prostate cancer biomarker (PSA) [66]. Reduced graphene oxide was used as the sensing channel to immobilize monoclonal PSA antibodies, which enabled the detection of protein-protein interactions down to the femtomolar level. Graphene also exhibits superior advantages as an electrode material in the catalytic oxidation of DNA bases. The high surface area of graphene provides a good platform for the immobilization of DNA base rings through π-π stacking interactions [67]. For example, Benvidi et al. reported a label-free DNA biosensor for the detection of the Amelogenin gene. DNA probes were immobilized on the surface of a glassy carbon electrode modified with rGO. The DNA biosensor presented an excellent performance, which can be attributed to the large surface area and high conductivity of rGO. These graphene-related properties enhance the DNA probe absorption and enable direct electron transfer between probe and the electrode [68].

The combination of graphene with other nanomaterials, such as nanoparticles, offers additional interesting properties to further expand and optimize the implementation of graphene in electrochemical biosensors. Graphene-hybrid structures usually display the individual properties of each nanomaterial and also exhibit synergistic properties that augment their use in electrochemistry [69]. For instance, graphene-metallic nanoparticle hybrids present excellent conductivity and catalytic properties, which makes them capable of decreasing the distance between the redox centers in proteins and electrode surfaces and acting as catalysts to increase the electrochemical reaction rates, thereby achieving direct electron transfer of enzymes on graphene-modified electrodes [70].

The direct wiring of enzymes to an electrode surface is essential to amplify the signal transduction of the biorecognition event [71]. In this case, due to the high conductivity of metal nanoparticles, they act as electron transfer mediators. Kavitha et al. pioneered a study on graphene nanosheets decorated with zinc oxide (ZnO) nanoparticles as an electrochemical platform for glucose biosensing. Graphene-ZnO hybrids presented a high surface area and excellent conductivity, which enable the immobilization of the GO_x enzyme in its native configuration on the electrode surface and the direct electron transfer between the redox center of GO_x and the graphene-hybrid electrode surface [72]. The direct electron transfer of GO_x was also efficiently achieved by a series of graphene-hybrids electrodes including TiO_2, [73] CdS, [74] Au, [75] Pt, [76] Pd, [77] and Ag [78] nanoparticles.

The design of graphene-hybrid immunosensors offers more opportunities for surface functionalization and an enhanced electron transfer, thereby improving the electrochemical performance of the immunosensors. For instance, Singal et al. used graphene nanosheets functionalized with PtNPs to construct an electrochemical impedance immunosensor to detect cardiac troponin I. The PtNPs were prepared using 3-Mercaptopropionic acid, which provided specific sites for the covalent immobilization of troponin antibodies. The graphene/PtNP hybrid provided a high surface area and a uniform distribution of PtNPs for loading more antibodies [79]. FET-based immunosensors using reduced graphene oxide decorated with AuNPs-antibody conjugates for protein detection were reported by Mao and co-workers [80]. The antibodies (anti-immunoglobulin G) were labeled on the graphene surface through the linking of AuNPs, preventing nonspecific-protein immobilization while providing stable binding for the probe proteins. Zhang and Jiang demonstrated a biosensor for the detection of sequence-specific DNA. Reduced graphene oxide sheets decorated with AuNPs on a glassy carbon electrode were used to improve the electrode surface. Short sequences of oligonucleotides as the DNA probes were covalently immobilized on the surface of the graphene-AuNPs hybrid (Figure 3). The biosensor was able to detect the complementary target DNA in the presence of large amounts of base mismatched target DNA [81]. Rasheed and Sandhyarani developed a graphene-DNA biosensor for the detection of the breast cancer-related BRCA1 gene by using a "sandwich" detection strategy in which the capture probe (DNA-c) and the reporter probe (DNA-r) conjugated with AuNPs DNAs were hybridized to the target probe DNA (DNA-t) on a graphene-modified electrode. The concentration detection of DNA-t was monitored from the electrochemical oxidation of AuNPs, which acted as electrochemical labels in this sensor [82].

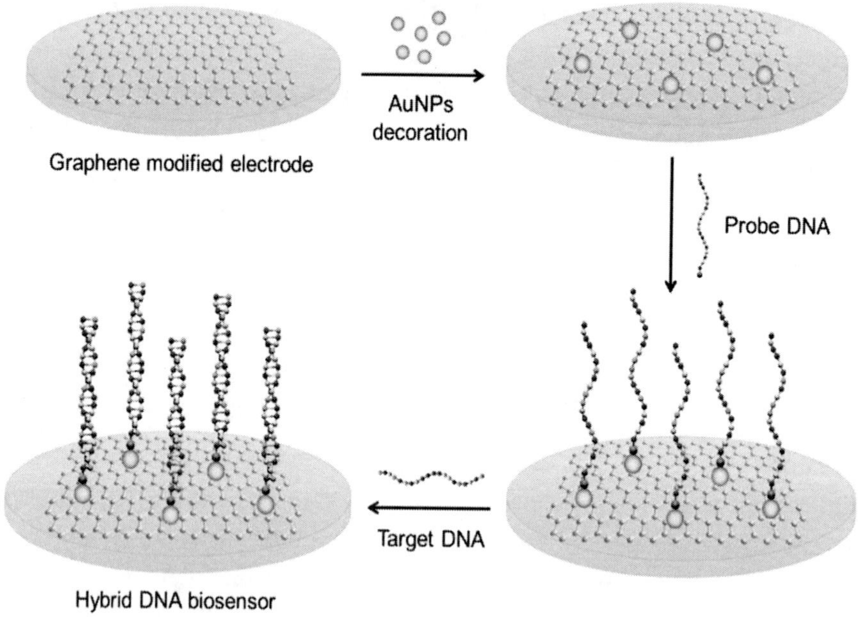

Figure 3. Scheme illustrating the fabrication of a graphene-AuNPs hybrid DNA biosensor. Based on Zhang and Jiang [81].

The combination of some polymers with graphene improves the solubility and processability of graphene nanosheets in water or organic solvents, and also prevents their restacking. With respect to enzymatic biosensors, Zhou et al. utilized a graphene-chitosan nanocomposite to form a film on the surface of glassy carbon electrodes to construct a H_2O_2 biosensor. The hybrid structure offered a more natural microenvironment for horseradish peroxidase (HRP) enzyme immobilization. The electrocatalytic response to H_2O_2 was also improved by introducing AuNPs over the electrode surface. The modification of the electrode by a graphene-hybrid greatly improved the biosensor sensitivity, allowing for an efficient direct electron transfer between HRP and the electrode [83]. Simirlarly, Xu et al. developed a H_2O_2 biosensor based on a graphene and chitosan nanocomposite film for immobilizing hemoglobin. The hybrid film offered a more biocompatible environment to ensure enzyme bioactivity while promoting direct electron transfer [84].

Conducting polymers, such as polypyrrole, polythiophene, polyaniline, polyphenol, present properties similar to those of metals and semiconductors, which make them excellent candidates for electrode materials. In particular, graphene-polymer electrodes improve the sensitivity of biosensors. For example, Wei et al. developed a label-free immunosensor for the detection of α-fetoprotein (AFP). Glassy carbon electrodes modified with graphene-thionine composite benefited from the large surface area of graphene, which improved the loading of specific antibodies. The good conductivity of graphene also enhanced the electroactivity of thionine. The proposed biosensor exhibited a low detection limit and long-term stability [85]. Graphene-based polyaniline (PANI) nanocomposites with different morphologies were used to construct DNA sensors. The nanocomposites, which presented higher surface area, provided a better surface coverage of single-stranded DNA and hybridization efficiency; this result implies that the morphology of the nanocomposites can also affect the sensitivity of the DNA biosensor [86].

Graphene-inorganic dye nanohybrids can also be used as electrode materials. Prussian blue (PB) behaves as an artificial peroxidase, and it is being a superior electrocatalyst for H_2O_2 reduction. Jiang et al. prepared a graphene-PB modified electrode that showed excellent electrocatalytic performance toward both H_2O_2 reduction and hydrazine oxidation. The increased amperometric response was attributed to the synergistic effect between graphene and PB, which resulted in a fast response, a wide linear range, and a long time of stability. [87] Organic dyes, including porphyrin, [88] methylene blue, [89] and methylene green, [90] exhibit an even larger potential in the modification of graphene-based electrodes. For example, Ferreira et al. reported the modification of glassy carbon electrodes with a graphene-DNA nanocomposite. This platform was efficient for entrapping methylene blue and it exhibited excellent electrocatalytic activity toward the oxidation of nicotinamide adenine dinucleotide (NADH) at lower potentials. The excellent performance of the graphene-MB hybrid DNA sensor can be related to the high conductivity and the large surface area of the graphene as well as the ability of the DNA to immobilize MB [89].

More recently, the modification of graphene electrodes with large biomolecules as recognition elements has been reported. In particular, aptamers, which are a new class of synthetic molecules that consist of artificial, short, single-stranded DNA/RNA oligonucleotides that can bind small molecules, proteins, or even cells with high specificity and affinity, have been explored [91]. Graphene-aptamer nanohybrid biosensors show additional advantages compared to antibody- or enzyme-based biosensors. For example, Feng et al. reported a label-free electrochemical biosensor based on the the first phase II clinical trial, which used aptamer AS1411 and a graphene-modified electrode for the detection of cancer cells [92]. The graphene-aptamer surface was used as an anchorage substrate to capture cells on the electrode surface through the specific binding between the cell surface and aptamer. Furthermore, this graphene aptasensor can be regenerated with a DNA hybridization technique, which ensures its reusability for cancer cell detection. Jiao et al. detected dopamine with an aptamer-graphene modified electrode [93]. The biosensor sensitivity was improved due to the excellent electronic properties of graphene, whereas its selectivity was guaranteed by the affinity of the RNA aptamer toward dopamine. In addition to the applications above, aptamers can also be combined with graphene-based hybrid materials [63a, 94].

Graphene has proven to be an excellent electrode material for fabricating various biosensing platforms for the determination of glucose, H_2O_2, DNA, biomarkers, cells, among others analytes. Due to the advantageous properties of graphene, such as the good conductivity, large surface area, and biocompatibility, this material has attracted considerable interest in the design of electrochemical biosensors with enhanced sensitivity. In terms of real applications, graphene produced from the solution-based approaches appers promising. rGO is a cost-effective option, and it can be considered as an appealing approach for large-scale graphene production. However, systematic studies on the synthetic routes are still needed to obtain graphene sheets with even better electrical properties. Graphene is also a highly versatile platform in the support of other nanomaterials, such as nanoparticles, polymers, and dyes. Hybrid systems, which combine graphene with other functional nanomaterials, yield synergistic properties that are beneficial for developing electrochemical biosensors with enhanced performance. Therefore, more effort should be invested on the synthesis and functionalization of graphene-hybrid materials coupled with a biological surface architecture for recognition elements. Graphene-based biosensors will connect to other technological advances, such as lateral-flow, point-of-care devices, and 3D printing techniques for the development of next-generation biosensors.

Nanoparticles-Based Electrochemical Biosensors

Nanoparticles (NPs) have been widely used because of their applications in several areas. The first report on NPs was in 1857, when Faraday verified a method for the synthesis of gold nanoparticles (AuNPs) based on the reduction of AuCl- in a biphasic medium of H_2O/CS_2 [95]. Since Faraday's initial study, several investigations have been carried out on nanoparticles to understand their properties, to obtain new nanoparticles, to improve/create new synthetic routes, and to analyze the practical applications of these materials. By definition, NPs are clusters of atoms with sizes ranging from 1 to 100 nm and have unique electronic, magnetic, physical, and chemical properties that are due to the small size and shape of these materials [96]. Currently, NPs can be subdivided into four classes: metallic, magnetic, quantum dots, and hybrid/semiconductor NPs; which will be discussed next.

Metal nanoparticles (MNPs) are the most widely used class of NPs due to specific characteristics, such as good biocompatibility, low-cost, high catalytic activity, and large surface area. In addition, MNPs may exhibit different colors according to their size and shape, and also due to the resonant interaction of free electrons on the MNPs surface with the incident light (see schematic illustration in Figure 4a).

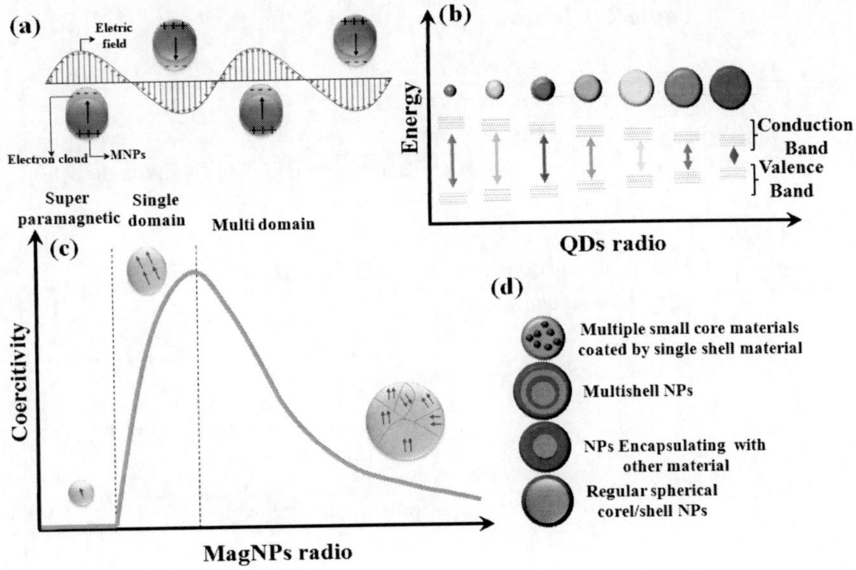

Figure 4. (a) Schematic illustration of the surface plasmon resonance phenomenon exhibited by the MNPs; (b) Graph illustrating the gap energy variation with the size of the QDs; (c) Graph showing the relationship between coercion and the MagNPs radius, and (d) Examples of different corel-shell structure types.

This phenomenon is the so-called surface plasmon resonance. Among those MNPs, AuNPs stand out due to specific characteristics, such as high chemical stability, low cytotoxicity, and good biocompatibility. In addition to simple and fast synthesis, AuNPs also present ease of chemical functionalization because of their high affinity for sulfur species. Depending on the method of synthesis, it is possible to obtain AuNPs with different shapes, size, and consequently with different colors. Reproducible and easily controlled synthetic methods for obtaining AuNPs are described in Table 2 [97].

AuNPs are mainly produced from the reduction of Au^{4+} ions using $HAuCl_4$ in solution in the presence of a reducing agent such as sodium citrate, [98] sodium borohydrate, ascorbic acid, among others [97]. It is possible to use stabilizing agents for the synthesis of AuNPs, which prevent the formation of agglomerates [108].

The most common method for obtaining AuNPs was proposed by Turkevich et al., [109] who described the reduction of $AuCl_4^-$ by using sodium citrate ($C_6H_5O_7^{-3}$) as reducing/stabilizing agent (Figure 5a). Equation 1 shows how this reduction reaction proceeds.

More specifically, the free conduction electrons present on the MNPs surface oscillate in response to external electromagnetic light field MNPs exposition, resulting in the collective resonant oscillation between the dipoles of the nanoparticle.

Table 2. Methods of obtaining different NPs

NPs		Main synthesis routes	Ref.
Metallic	*AuNPs*	Reduction of $HAuCl_4$ in sodium citrate.	[98]
		Reduction of $HAuCl_4$ in ascorbic acid and addition of cationic alkylammonium surfactant as stabilizing.	[99]
	AgNPs	*Chemical synthesis* Reduction of $AgNO_3$ in ethylene glycol.	[100]
		Physical synthesis Decomposition of complex Ag^{1+}-oleate.	[101]
		Biological synthesis Reduction of $AgNO_3$ by *Bacillus cereus* bacteria.	[102]
	PtNPs	Reduction of H_2PtCl_6 in $NaBH_4$.	[103]
Magnetic	*Fe_3O_4*	Coprecipitation of $FeCl_3$ and $FeSO_4$ in basic medium.	[104]
		Thermal decomposition of Fe(III) glucoronate.	[105]
QDs	*GaAs*	*Top-Down* Lithography	[106]
	CdS	*Bottow-Up* Wet chemical coprecipitation	[107]

This synthesis route results in negatively charged AuNPs surface, in the form of stable colloidal AuNPs [109]. The time of synthesis is proportional to the size and shape of the AuNPs, allowing for the obtainment of spherical NPs of up to 30 nm [108].

$$6Au^{3+} + C_6H_5O_7^{-3} + OH^- \rightarrow Au + 6CO_2 + 10H_2O \quad \text{(Eq. 1)}$$

Figure 5a also illustrates another method of obtaining AuNPs in which $AuCl_4^-$ ions are reduced by ascorbic acid, a weak reductant, in the presence of sufactants such as cationic alkylammonium salts.

In this method, the cationic surfactant can increase the concentration of the gold salt in solution, consequently forming cylindrical-shaped AuNPs with size ranging from 20 to 30 nm. This morphology can occur due to the balance among the surface of crystalline domains, growth kinetic parameters (e.g., time and reaction temperature), and interaction forces between the reducing agent, surfactant, and precursor metal salt [99].

Compared to AuNPs, silver nanoparticles (AgNPs) have higher extinction coefficient and lower cost. However, AgNPs have not been extensively used because they present certain limitations, such as easy oxidation and chemical degradation [110]. AgNPs can be synthesized via three primary methods: chemical, physical and biological (Table 2) [100]. The chemical method is the most widely used because it is considered the simplest way to obtain AgNPs. The formation of AgNPs occurs in two different stages: nucleation and

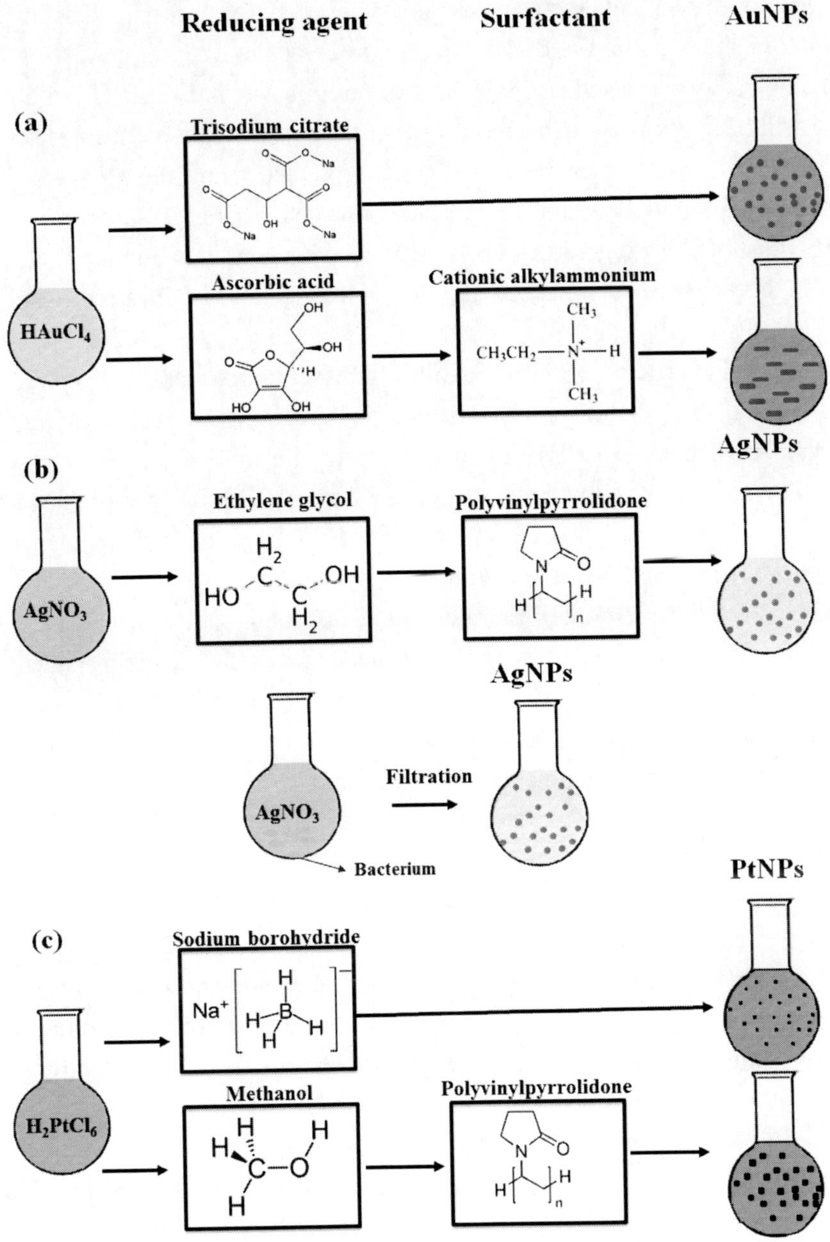

Figure 5. Synthesis methods of obtaining: (a) AuNPs, (b) AgNPs, and (c) PtNPs.

$6Au^{3+} + C_6H_5O_7^{-3} + OH^- \rightarrow Au + 6CO_2 + 10H_2O$ growth. Three main components are required: a metal precursor salt (typically $AgNO_3$), a reducing agent (for example, ethylene glycol) and a stabilizing agent (such as polyvinyl-pyrrolidone), as illustrated in Figure 5b [100]. For the physical method, the main technique for obtaining AgNPs is by thermal decomposition. From the reaction of $AgNO_3$ and sodium oleate in water, the complex Ag^{1+}-oleate is formed, which is then thermally decomposed to obtain AgNPs [101]. Finally, the biological method involves the use of living organisms to obtain

AgNPs, that is, biological agents such as bacteria, fungi, yeasts, algae or plants are responsible for the reduction and stabilization processes in obtaining AgNPs. Depending on the biological system used, AgNPs with different sizes and formats are obtained. For example, using *Bacillus cereus* bacteria, spherical nanoparticles with a size of 5 nm were obtained, [111] whereas when the AgNPs are produced using the *silver-tolerant strain MKY3*, the nanoparticles obtained were hexagonal shaped and 2-20 nm in size [112]. Figure 5b illustrates a biogenic route by using bacteria as a reducing agent. After the exposure of silver precursor ions to bacteria, the suspension is filtrated to isolate the as-produced AgNPs.

Platinum nanoparticles (PtNPs) exhibit unique antioxidant properties and high chemical and catalytic stability. These properties are even more interesting than those reported for AuNPs and AgNPs; however, due to the high cost, fewer investigations using PtNPs have been conducted [103]. In general, PtNPs are prepared from the reduction of a precursor metal ion (generally H_2PtCl_6) by a reducing agent such as $NaBH_4$. This method yields PtNPs with diameters smaller than 10 nm, [103, 113] as illustrated in Figure 5c. Another chemical route to obtain uniform and size-controlled PtNPs is by using methanol as reducing agent and a stabilizing agent such as polyvinylpyrrolidone (Figure 5c). In this case, methanol is oxidized to aldehyde, as shown in Equation 2. In general, this synthesis is mainly controlled by the reaction time, which allows a more homogeneous production of the PtNPs.

$$H_2PtCl_6 + 2\,CH_3OH \rightarrow Pt^o + 2HCHO + 6HCl \qquad \text{(Eq. 2)}$$

In terms of applications, quantum dots (QDs) have gained considerable attention as an alternative for replacing MNPs, mainly due to their exquisite optical properties. QDs are semiconductor nanoparticles with 2 - 10 nm in size that present quantum confinement effect. Bulk semiconductors present a minimum energy to excite electrons from the valence band to the conduction band - denominated the band gap energy. When an electron crosses this energy barrier, a hole arises in the valence band, creating a hole-electron pair (exciton) which has a finite size defined by the Bohr radius. Thus, if the particle size is equal or less than the dimensions of the Bohr radius for the hole-electron pair, the confinement occurs leading to a transition from continuous to discrete energy levels, and optoelectronic properties become strictly related to the particle size [114]. Figure 4b illustrates the relationship between the band gap and the diameter of QDs. It is possible to observe that the higher the band gap energy, the smaller the diameter of QDs, and consequently, different QD colors are obtained. Therefore, small differences in size result in distinct colors, e.g., CdS QDs have blue color when the diameter is 2.3 nm and red color when the diameter is 8 nm.

QDs are formed from the combination of 10 to 50 atoms of elements from groups II-VI, III-V or IV-VI of the periodic table, yielding, for example, CdS, CdSe, and CdTe

QDs. Historically, CdTe and CdSe were the first QDs synthesized, and they were obtained from mixing the precursors in a vitreous material [115]. New synthesis techniques have been investigated in recent years, and the syntheses are divided into two forms: top-down (based on lithographic processes) and bottom-up techniques (which are divided into wet-chemical and vapor-phase processes) (Table 2) [115]. For example, the fabrication process of QDs by lithography allows the obtainment of uniform QDs with low defect density; however, it is a cost-effective and time-consuming process [116]. Wet chemical methods represent bottom-up approaches which involve reduction and precipitation of QDs from precursor salts [117].

Magnetic nanoparticles (MagNPs) exhibit distinct characteristics of the MNPs and QDs due to their unique magnetic properties. These properties are also dependent on the size of the MagNPs, as illustrated in Figure 4c. In general, the coercivity of a given material decreases with increasing the size of the particles; however, in much smaller nanoparticles, these magnetic materials can exhibit different characteristics. MagNPs with diameters between 10 and 20 nm are in a superparamagnetic state, thus coercivity decreases with the size of the MagNPs nanoparticle. The superparamagnetism exhibited by MagNPs is an unstable magnetic state wherein the magnetization occurs only in the presence of an external magnetic field. Thus, MagNPs outside the magnetic field are no longer magnetized.

In addition to the distinct magnetic properties, the MagNPs have high surface area, good stability, low toxicity, and high separation efficiency. The most used magnetic cores are iron oxides, such as Fe_3O_4 and γ-Fe_2O_3. MagNPs are mainly obtained from the coprecipitation method (Table 2). It is possible to obtain MagNPs of Fe_3O_4 from the coprecipitation of iron salts (such as $FeCl_3$ and $FeSO_4$) in alkaline solution [104]. However, this synthesis method yields MagNPs with varied sizes. By contrast, the thermal decomposition method results in the formation of uniformly sized MagNPs due to the greater control of the synthesis phases of the nanoparticles. This methodology involves the thermal decomposition of coordination compounds in organic solvents containing surfactants (Table 2) [105].

It is noteworthy to mention that MagNPs have some disadvantages, such as a high tendency to aggregate, rapid biodegradation when exposed to certain biological systems, and nanoparticle instability, especially when exposed to external magnetic fields [118]. To improve these characteristics, it is possible to modify the structure of the MagNPs to obtain corel-shell type NPs (NPs-CS), providing nanoparticles with different characteristics, suach as an internal structure (core) and an outer layer (shell). Figure 4d shows examples of NPs-CS structures, which can be regular spherical NPs-CS, multishell NPs, NPs encapsulating, and multiple small core materials coated by single shell material. For example, corel-shell type structures for the MagNPs by adding a metallic outer layer avoid the nanoparticle degradation while improving stability [119].

Due to these unique properties, NPs have been used in a wide range of applications, including in microelectronics, [120] in drug delivery, [121] in catalysis, [122] as antibacterial agents, [123] in photovoltaic cells, [124] and in sensors and biosensors [96b]. The introduction of NPs in electrochemical biosensors has presented several improvements in these systems. For example, NPs have been conjugated to bioreceptors such as nucleic acids, proteins, enzymes, polymers, antibodies, etc., as illustrated in Figure 6a. In many applications, a more biocompatible microenvironment is required and the modification of NPs surface with silane, thiol, amine groups, and others, provides a more suitable surface to interact with the bioreceptor.

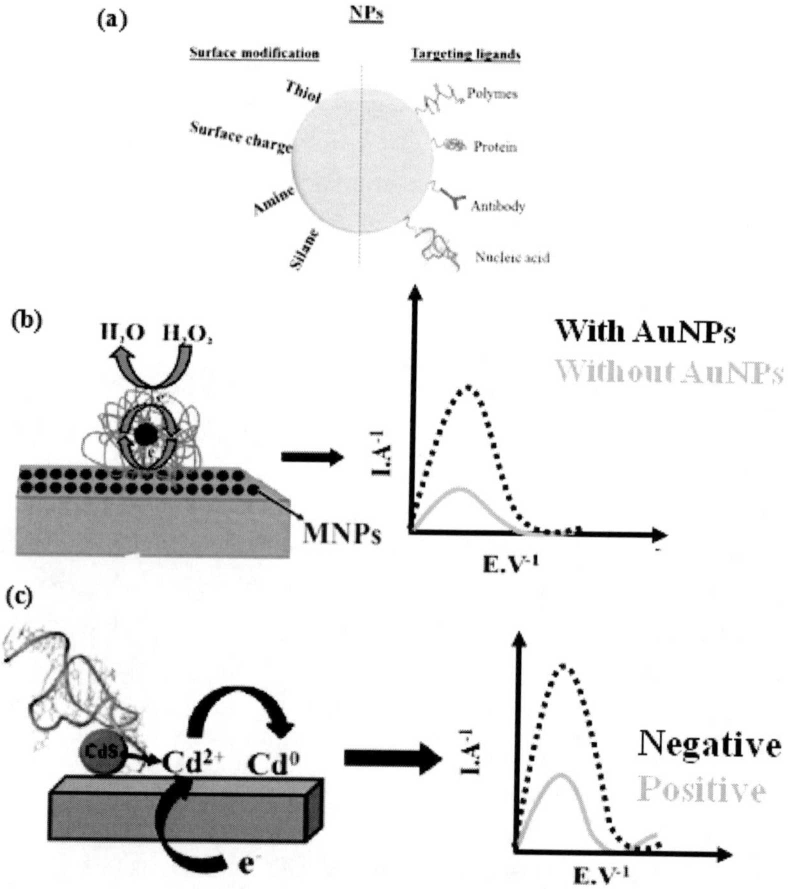

Figure 6. (a) Examples of NPs surface modification; (b) Schematic illustration of a biosensor using MNPs in which NPs act on the electron transfer between the electroactive center of the enzyme and the substrate; and (c) Illustration of a biosensor using QDs as electroactive marker.

MNPs have also been used in biosensors for biomolecules immobilization and as electron mediators. Depending on the biorreceptor type, it is possible that the electrical signal is non-existent. For example, in the case of some enzymes, the active centers are surrounded by proteins, which act as insulators, hindering electrical output signals. Using

MNPs, it is possible to improve the eletron transfer between the active center of the enzyme and the electrode, as shown in Figure 6b. In this example, MNPs act directly on the eletron transfer between the active center of the enzyme and the substrate, resulting in an increase in the electrochemical signal of the biosensor.

AuNPs are the most popular MNPs in electrochemical biosensors due to their attractive features such as the ability to catalyze the reduction and oxidation of H_2O_2, leading to increased electrochemical signals. For example, Barros and co-workers [125] developed a sensor for the detection of metallic ions and verified the influence of the AuNPs on the sensitivity of the sensors. In this study, nanocomposite Langmuir-Blodgett films with emeraldine salt polyaniline (PANI-ES), organophilic montmorillonite clay (OMt), and AuNPs (OMt/AuNPs/PANI-ES) were used. The authors determined the sensor response with and without the AuNPs. The utilization of the AuNPs enabled a better catalytic activity, which resulted in an increase in the conductivity of the films and, consequently, in the electrochemical signals. The LOD value for Pb^{2+} without AuNPs was 0.2324 μg L^{-1}. When the AuNPs were added to the Langmuir-Blodgett film, the LOD obtained was 0.0062 μg L^{-1}. This result demonstrates that the use of NPs can improve the electrochemical signals of biosensors. Although less commonly used, other MNPs applied in electrochemical biosensors include AgNPs and PtNPs, which also improve the conductivity, thereby facilitating electron-transfer pathways between the electrode surface and the bioreceptor.

QDs have shown to be promising NPs for applications in electrochemical biosensors, especially for multiplexed electrochemical sensors in which QDs act as efficient redox probes. Figure 6c illustrates the performance of CdS QDs in an electrochemical biosensor. The sensor detected the target analyte when CdS QDs were dissolved, thereby obtaining an electrochemical signal at a specific potential. The current density of the system increased as a function of the concentration of the analyte. In addition, QDs can enable the creation of devices with high sensitivity and specificity, high conductivity, and lower LOD with fast response. QDs formed by Cd^{+2} have been observed to be highly efficient in electrochemical detection [126]. For example, a multiplexed electro-chemical sensor for the simultaneous detection of ATP and cocaine was demonstrated by Zhang and co-workers [127] using aptamers conjugated with QDs. The aptamers for the identification of ATP and cocaine were labeled with PbS and QDs, respectively. CdS and PbS were dissolved in Pb^{2+} and Cd^{2+} ions when ATP and/or cocaine were added to the system. This process resulted in an increase in the current density at a specific potential and resulted in the creation of a QDs and aptamers sensor for simultaneous and multiple detection. Such systems are promising for identifying matrices of complex samples and may be useful in clinical diagnostics. The mechanism of this process is based on the display of well resolved voltammetric signals, which are obtained through the oxidation potentials of the metallic components of the QDs. This method resulted in a significant increase in the electrochemical signal, thus allowing ultrasensitive bioassays [128].

MagNPs can be used in electrochemical biosensors. One of the characteristics that distinguishes MagNPs from other NPs is their response to the application of an external magnetic field. For a biosensor, this feature may permit the rapid separation of the antigens immobilized on MagNPs by simply applying an external magnetic field. In general, MagNPs consist of a polymeric outer layer, which is effective for immobilizing biomolecules, with a paramagnetic or supermagnetic core. For their application in biosensing, it is necessary to coat the MagNPs with a protective layer, usually a metallic layer, to obtain a MagNP core@shell type, which preserves the properties and composition of the MagNPs. This effect is due to the addition of the outer layer, which makes it impossible for O_2 to diffuse to the surface of NPs. With this modification, it is possible to obtain NPs with characteristics of both MagNPs and MNPs, which results in a system with excellent biocompatibility, a large surface area, an ease of regeneration using the external magnetic field, and immobilization of the bioreceptor. In addition, core@shell type NPs enable improvements in the sensitivity of electrochemical biosensors by contacting the core@shell NPs with the surface of the electrode and/or by transporting redox species to the surface of the electrode [129].

In summary, NPs have unique optical, electrical, and magnetic properties, and have been used in several applications. NPs are used in the construction of electrochemical biosensors, mainly to amplify the electrochemical signals and to immobilize biomolecules (enzymes, for example), which improve the sensitivity and selectivity of these biosensors. Additionally, MNPs are the most commonly applied NPs for biosensors, especially AuNPs, because of their high conductivity and excellent biocompatibility. However, QDs, MagNPs, and hybrid NPs are promising for electrochemical sensor applications because they can enable, in addition to improvements in the electrochemical signals, multiple detection systems, and real sample analysis. Therefore, due to their very high potential, there has been an increasing interest over the years in using NPs mainly in the fabrication of miniaturized electrochemical biosensors, such as the point-of-care devices.

Dendrimers-Based Electrochemical Biosensors

Dendrimers represents an interesting class of nanoparticles; dendrimers are supramolecular three-dimensional architectures consisting of hyperbranched organic units with a well-defined size and number of functional groups [130]. The term "dendrimer" was created in a sequence of articles by Tomalia, and it has its origins in the word "dendron" (which is a Greek word meaning "tree") [131]. Typically, a dendrimer is structurally symmetric around a nucleus and exhibits a globular three-dimensional morphology (Figure 7). The high versatility of the chemistry of these molecules enables the synthesis of ordered structures of different generations, which not only provide varied sizes of nanoparticles but also defines the terminal groups, morphology, solubility, stability, flexibility, and viscosity of these systems [131d].

Independent of the synthetic route selected (a divergent or convergent approach), dendrimers are molecules produced by iterative sequences of reaction steps, in which each additional iteration leads to a higher generation, and the eventual formation of robust structures with a high density of terminal groups (e.g., $-NH_2$, -COOH, -OH, -CHO, and others) [131-132].

According to the literature, the synthesis of dendrimers began in the 1980s through the divergent approach [131a-c]. Subsenquently, in 1990, Hawker and Fréchet also demonstrated that it is possible to construct dendrimers through a convergent approach, [133] as observed in the schematic illustration in Figure 7.

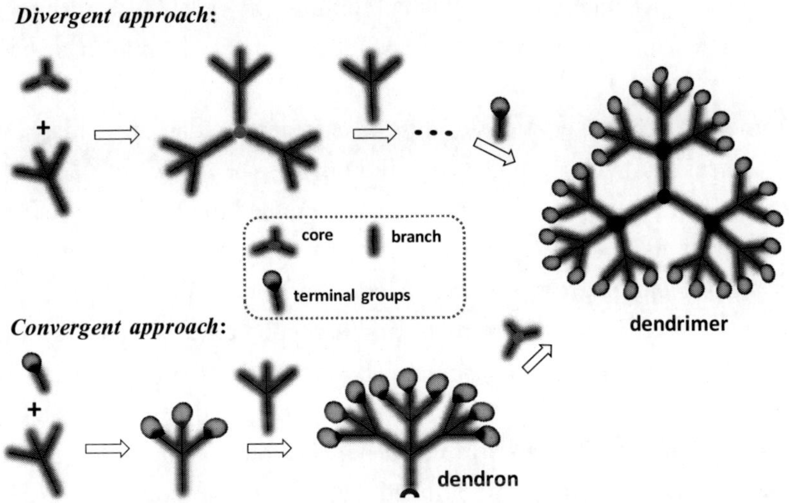

Figure 7. Schematic illustration of the steps involved in the synthesis of dendrimers by the convergent and divergent approaches.

Regarding the details of the synthesis using the divergent approach developed by Tomalia et al. (1985) [130a] and Newkome et al. (1985) [130b] through the repetition of a series of coupling and activation steps, the core grows from the inside out by the even increase of the number of branches at each branching point (Figure 7). In the coupling steps, the core reacts with the complementary reactive group (unprotected) of other monomers, and other reactive groups of the monomers are protected to produce a controlled growth. In the activation steps, the protected group can be chemically activated due its conversion to a reactive chemical group for coupling to a second set of molecules, which produces the next dendrimer generation [130-131]. The divergent approach has the advantage of enabling the large-scale synthesis of dendrimers, and it has been used to construct numerous classes of these macromolecular systems, such as poly(propylene imine) (PPI), [134] poly(amidoamine) (PAMAM), [135] polyphenylene, [136] organosilicon, [137] and phosphorous dendrimers [138]. However, several studies have demonstrated that in determined types of reactions the number of defects in dendrimers increases as the generation number increases. These events can occur due, for example, to incomplete reactions, fragmentation, intramolecular cyclization, and other effects [139].

By contrast, in the convergent approach, the synthesis starts from the periphery by the construction of dendrons, followed by its coupling to the core; a globular dendritic molecule (Figure 7) is constructed in the last step [133]. As each growth step requires a relatively small number of coupling reactions, this approach leads to fewer defects in the dendritic structure and offers greater structural control. Furthermore, the convergent approach can accurately place functional groups throughout the structure of the molecule. However, a significant disadvantage is the steric constrain on the focal point of large dendrons, which prevents the utilization of this strategy in the construction of dendrimers with high-generation numbers [140]. In this context, through the dendrimer's design of interest, it is important to consider the advantages and disadvantages of each synthetic approach.

The numerous characteristics of these supramolecular systems, such as the nanometric dimensions with monodispersed diameters (below 20 nm), biocompatibility, existence of dendritic crevices, and stability, have enabled the utilization of dendrimers in numerous applications [141]. Recently, the use of dendrimers has exhibited impressive results in systems ranging from sensing to drug delivery, e.g., sensors, light-harvesting devices, catalytic systems, protein mimics, and nanomedicine [132, 140b]. In addition to their applications in biosensors, dendrimers have been reported as ideal platforms for the immobilization of biomolecules [142]. The three-dimensional nanometric architectures of these supramolecular systems have provided better control of the sensor array thickness and the spacing between the immobilized recognition molecules. These properties preserve the functionality of the immobilized binding groups and reduce the non-specific interactions, which has led to reported improvements in the analytical performance of dendrimer-based biosensors by several researches [143]. Among the variety of dendrimers used in biosensors, an extensive and established family is the structures based on PAMAM (Figure 8) [144].

The PAMAM type dendrimer was first developed by Tomalia and co-authors, which was synthesized through an addition sequence of methyl acrylate (Michael addition) on a diamino nucleus, followed by amidation with ethylenediamine [131a]. A particularly simple approach to increase the reactivity of the terminal groups of PAMAM dendrimers is through the reaction of the free amines with coupling agents, such as glutaraldehyde (GLU) or a diester-type molecule, e.g., BS3 (bis (sulfosuccinimidyl) suberate) [145]. The schematic illustration in Figure 8 of the approach using glutaraldehyde shows the formation of an imine type molecule and the exposition of terminal aldehyde-groups for the subsequent covalent immobilization of biomolecules [146]. Currently, PAMAM

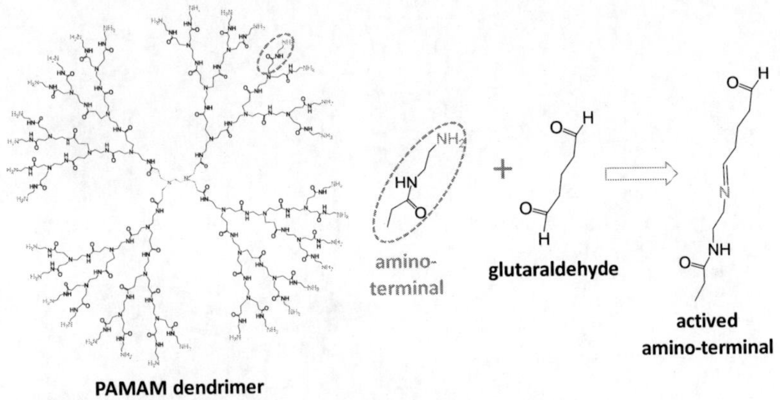

Figure 8. Activation of the terminal amino-groups of a PAMAM dendrimer of generation 3 (PAMAM-G3) via a reaction with glutaraldehyde. The formation of an imine type molecule, and terminal aldehyde-groups for subsequent reactions with the biomolecules can be observed.

dendrimers are commercially available, which is a very important factor in the dissemination of these molecules in studies involving biosensing and has consequently culminated in a better understanding of these nanostructures [144b].

Due to the unique synthetic chemical structure and the varied properties discussed above, dendrimers are advantageous when combined in electrochemical biosensors, in which dendrimers are used with the aim of amplifying the transducer interface in these systems. However, because dendrimers present poor electrical conductivity, applications of these macromolecules in electrochemistry are possible through their combined use with materials or other molecules that exhibit electroactivity [147]. Therefore, the combination of dendrimers with metal nanoparticles, redox probes, and carbon materials will be highlighted in this section. Some recent successful applications of dendrimers in electrochemical biosensors will be discussed to emphasize the strategies used to overcome the limitations of dendrimers in electrochemistry and their enormous potential in this field.

Dendrimers of higher generations can usually encapsulate metal complexes, metallic nanoparticles (MNPs), and other organic and inorganic molecules, which enable the extension of their properties for different applications (Figure 9a) [148].

The metal-dendrimer nanocomposite has attracted interest in catalysis because the conductivity of the dendrimer can be significantly improved by its combination with MNPs. Recently, various MNPs, e.g., gold, platinum, copper and, palladium, have been successfully incorporated on the dendrimer surface or encapsulated within its structure. [147a, 149]. Some studies have demonstrated that the incorporation of MNPs by the electrostatic adsorption of noble metal cations into a dendrimer layer, followed by chemical reduction, resulted in high electric and catalytic activities (Figure 9b) [150]. This very interesting approach, enables the dendrimer to act as a protective coating of

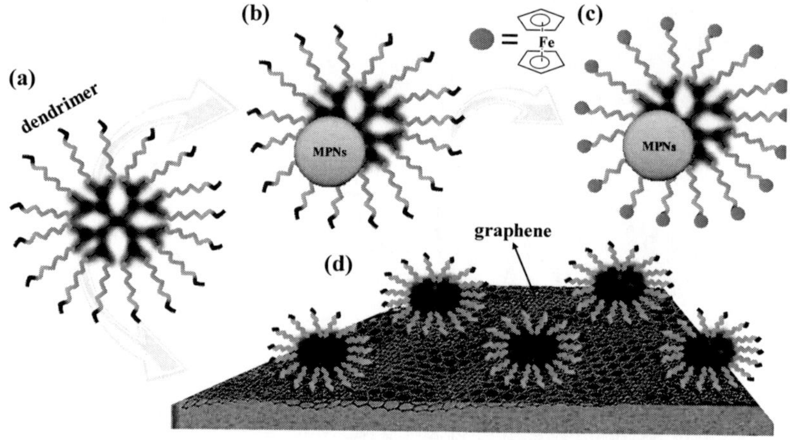

Figure 9. Schematic structural representation of dendrimers and different application approaches. (a) Generic structure of a dendrimer; (b) Incorporation of MNPs into a dendrimer layer; (c) Covalent bond of ferrocene on the terminal groups of the dendrimer; and (d) Formation of a hybrid by the dendrimer-graphene combination.

ultra-small particles of free metals, allowing one to obtain a controlled size of nanoparticles, which are homogeneously distributed within the organic matrix.

It is important to consider that although aggregates of metal nanoparticles with dendrimers find applications in electrocatalysis, in some cases the dendrimers present low activity as mediators in electron transfer. This feature may restrict the use of dendrimers in electrochemical biosensors, whereby the limiting step is often the electron transfer from the biological component. Therefore, researchers synthesized functionalized dendrimers, in which the terminal groups are covalently bound to electrochemically active fragments, which can perform reversible redox reactions on the electrode surface, and thus, improving the electrical communication of the recognition platform (Figure 9c). Among the variety of available redox probes, the most widely known are the ferrocene derivatives, which differ due to their high efficiency of electron transfer and chemical stability [151, 152]. For example, the ferrocenyl dendrimers make up a well-known class that has been used in combination with dendrimers to construct electrochemical biosensors [153]. Usually, the presence of guest molecules in close proximity to the ferrocenyl groups causes significant changes in their electric potentials during the redox reactions. In addition, due to the intrinsic properties of the receptor of the dendrimers, this effect is much more pronounced in this configuration than in monomeric ferrocenyl [154]. In a study reported by E. Çevik et al. (2016), [155] an electrochemical immunosensor was developed for prostate specific antigen (PSA) detection. In this work, the biosensor recognition platform was constructed using the layer-by-layer method. First, a self-assembled monolayer (SAM) was formed on a gold electrode by the adsorption of cysteamine (CYS), followed by the binding of the ferrocene-cored PAMAM dendrimer (Fc-PAMAM) and, finally, the binding of the PSA monoclonal antibody (anti-PSA). The differential pulse voltammetry technique was used to quantify the PSA levels and a detection limit of 1.00 pg mL^{-1} was obtained. Furthermore, the authors evaluated the analytical performance and specificity of the assays using human serum, and other proteins.

After researching recent publications in this area, it is possible to observe that a dual signal amplification strategy has been developed through the formation of hybrid materials for electrochemical biosensors, which illuminates the importance of the combination of dendrimers with multi-walled carbon nanotubes (MWCNTs) and graphene derivatives [156]. The unique physical and chemical properties of MWCNTs, previously discussed in the text, have enabled new and improved biosensing devices, mainly for electrochemical biosensors [157]. Additionally, the use of MWCNT-PAMAM hybrids has effectively improved the immobilization of biomolecules on the sensor substrates, resulting in better analytical performance of the electrochemical biosensors for the detection of numerous biomolecules [147a]. In turn, the interesting properties of graphene-based materials, discussed in detail in the text, such as the good thermal properties, excellent conductivity, and high mechanical strength, have also attracted considerable attention in the electrochemistry field [158]. The combination of PAMAM dendrimers with graphene sheets has enabled the expansion of the electroactive area of the sensor to grasp more capture probes and amplified the signal for electrochemical biosensing, which results in an improved sensitivity [159] (Figure 9d). A novel nanostructured architecture for the construction of enzymatic electrochemical biosensors was described by Borisova et al. (2015) [160]. In the methodology adopted by the authors, the PAMAM dendrimer was bound on a glassy carbon electrode coated with a hybrid nanomaterial formed by graphene oxide-carboxymethylcellulose. This functionalized surface was used for the covalent immobilization of the enzyme tyrosinase using glutaraldehyde as a coupling agent. The developed biosensor enabled the amperometric detection of catechol with a detection limit of 0.9 nmol L^{-1}. In addition, the biosensor retained over 93% of the initial enzymatic activity at 48°C after 40 days.

In general, according to the literature, dendrimer-based electrochemical biosensors have been most widely used in applications with oxidoreductases. This fact likely is due to the intrinsic characteristic of the dendrimers, which allows the more favorable orientation, excellent biocompatibility, and a large density of active sites for the immobilization of enzymes [158a]. Furthermore, based on the discussion above, it is possible to observe that the combined use of dendrimers with metallic nanoparticles, redox probes, and carbon materials, presents a variety of applications in electrochemical biosensors. However, it is important to mention that the number of published studies using these methodologies in miniaturized devices, such as point-of-care devices, is still very low. There presumably, there are still challenges in exploring new protocols and strategies for these applications, and more platforms and electrochemical techniques should be applied in these systems.

APTAMERS-BASED ELECTROCHEMICAL BIOSENSORS

The use of aptamers as biological recognition elements has gained a prominent role in recent years due to their stability, specificity and the affinities achieved for different targets. Aptamers are single-stranded DNA or RNA oligonucleotides that bind specifically to a particular target, which may be from a biological tissue to a metal ion. The term aptamer originates from the Latin word *aptus*, which means adapting, fitting, or adjusting. This ability to adapt and adjust is one of the fundamental characteristics of aptamers. Because of the degree of freedom of the intramolecular nucleotide bonds, aptamers can exhibit different conformations and be enveloped in a three-dimensional fashion, and this ability allows certain sequences to fit into the target and interact with it at various sites. This feature allows aptamers to be used in a similar way as antibodies to recognize the three-dimensional structure of the targets. The selection of the single strand of DNA or RNA can be performed *in vitro*, amplified by polymerase chain reaction (PCR), and fragments with a high affinity can be purified by electrophoretic and chromatographic methods. If desired, the selected aptamers can also be sequenced, and their synthesis can be reproduced with high purity [161].

Although the possibility of using fragments of DNA and RNA as binders had already been previously reported, [162] the introduction of a synthetic technology called Systematic Evolution of Ligands by Exponential Enrichment (SELEX) played a major role in the development and dissemination of aptamers. The SELEX principle is based on successive replications of fragments with higher affinity for the target [161]. Figure 10 shows an adapted scheme of the hybrid SELEX principle [163]. Initially, a synthetic library containing approximately 10^{15} different sequences is maintained in contact with the target in which specific aptamers are desired. The unbound portion of the sequences is eluted and discarded, whereas the portion bound to the target is subjected to a first PCR amplification. Compared to traditional SELEX, SELEX hybrid includes an additional step to improve the specificity of the selected aptamers through the use of one or more control targets. Thus, the enriched sequences are subjected to the presence of the control and contrary to the previous step, the portion that binds to the control is discarded and the unbound fragments are subjected to a second PCR amplification. After this step, fragments can be selected or monitored by separation methods. This entire process is generally repeated 6 to 20 times, resulting in an exponential enrichment of high affinity aptamers by the selected target. High-affinity aptamers can be obtained for a wide variety of targets, such as proteins, carbohydrates, lipids, nucleotides, viruses, cells, tissues, drugs, organic dyes, antibiotics or even ions, such as mercury, [164] and arsenic [165].

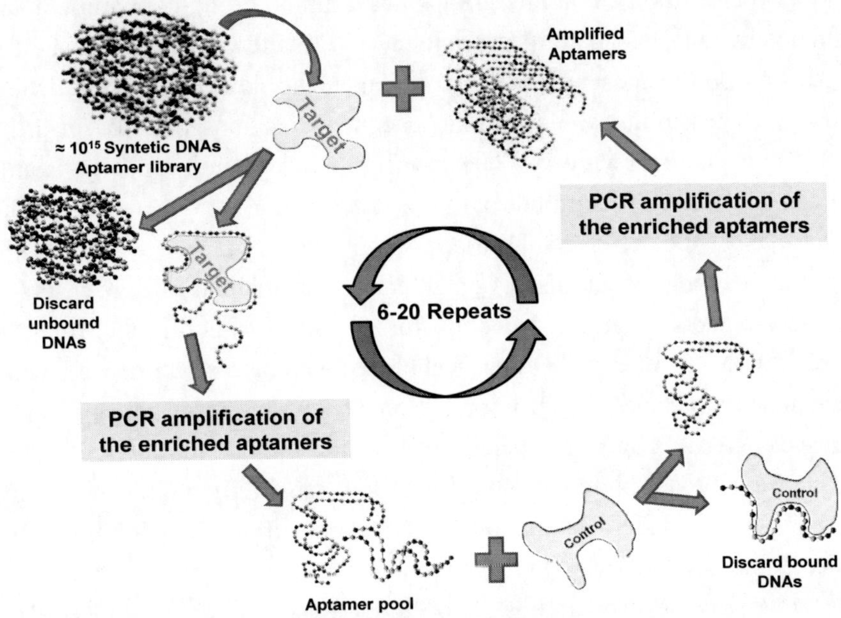

Figure 10. Schematic diagram of the hybrid SELEX methodology to obtain specific aptamers for a selected target. Based on [163].

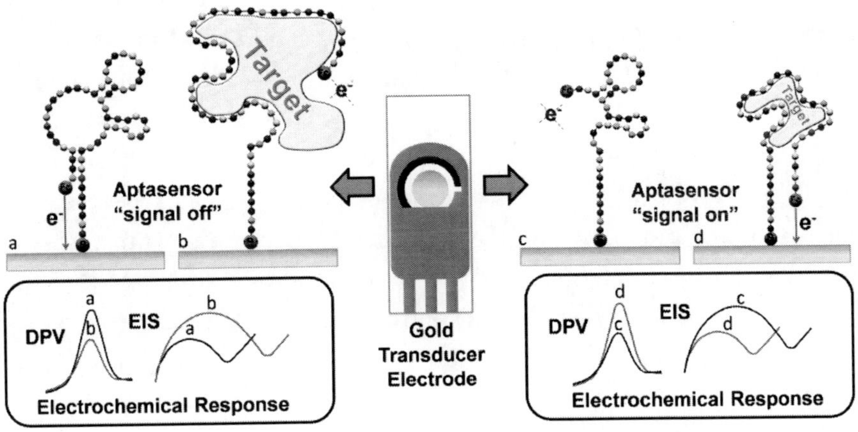

Figure 11. Schematic illustration of the electrochemical biosensor created using an aptamer as a recognition element. The "signal off" type detection (a and b) is shown on the left side, and the "signal on" type detection (c and d) is shown on the right side.

Despite the requirement of high affinity and purity, aptamers currently do not present major difficulties and can be obtained from commercial suppliers (SomaLogic, Noxxon Pharma AG, Gilead Sciences, AptSci, AptaMatrix, Base Pair Biotechnologies, NeoVentures Biotechnology and AMBiotech). The effective combination of the aptamers with transducers represents an initial challenge in the development of biosensors. Some surface immobilization strategies involve the use of physical adsorption (electrostatic or hydrophobic interactions), streptavidin-biotin interactions using biotinylated aptamers, reactions between carboxylic amine groups using amine group-containing aptamers, and gold-thiol interactions using thiolated aptamers. Despite the higher cost of gold, this metal seems to be the most widely used metal in the development of electrochemical biosensors. The possibility of performing chemical reactions directly on the molecular structure of the aptamers, without a decrease in the recognition capacity, is an important advantage in comparison to antibodies. In this context, thiolation reactions at one of the 3' or 5' ends can easily be performed [166].

A second challenge in the use of aptamers in electrochemical biosensors lies in the fact that these binders are not good conductors of electrons and do not offer redox sites suitable for the use of electrochemical techniques such as voltammetric techniques and electrochemical impedance spectroscopy. Some alternating current and potentiometry techniques can be used in non-labeled aptasensors without the aid of a redox probe; however, these approaches generally result in low sensitivity and accuracy because the responses are more prone to fluctuations in the composition of the solutions and samples used in detection. Thus, to overcome this inconvenience, some other extracts are typically used. One involves the use of redox probes such as $Fe(CN)_6^{4-/3-}$ or $Ru(NH3)_6^{2+/3+}$, in which electrochemical sensing results from changes in the permeability of the aptamer film and subsequent electronic transfer. Although this approach is used to a relative extent, the need to submit the biosensor to other solutions in addition to the sample presents an inconvenience, especially when the biosensor is used in diagnostics both by doctors and lay people. Considering this limitation, the use of labeled aptamers may be a better option. One possibility is the use of enzyme-labeled aptamers, in which the aptamer becomes responsible for generating an electro-active product, such as hydrogen peroxide [167]. However, in this case special care must be taken because the enzyme activity loss can limit the lifetime of the biosensor. Another strategy that is gaining more attention is the labeling of the aptamer with a redox group at one of its 3' or 5' ends. Redox groups such as methylene blue or ferrocene, are in these cases the most widely used. Figure 11 shows a schematic illustration of this more elegant type of biosensor. In one of these examples, the aptamer is previously modified with a thiol group (S) and labeled with a ferrocene group (Fc) at opposite ends. On the left side of the scheme, the immobilization of this element on the surface of the gold electrode allows the labeling to approach the surface. The presence of the target changes the conformation of the aptamer and hinders the electrochemical reaction of the ferrocene group. Thus, the current decrease observed in the voltammograms or the charge transfer resistance increase in the impedance spectra can be correlated with the concentration of the target. This type of approach is called a "signal off" approach. Some electrochemical aptasensors may also exhibit a reciprocal behavior, as shown on the right side of the scheme. In this case, in the absence of the target, the redox label remains far from the surface. The presence of the target brings the ferrocene group closer to the surface and facilitates the electrochemical reaction. The increase in the current or in the charge transfer resistance is correlated with the concentration of the target. This type of approach is called the "signal on" [166].

Aptamers are becoming promising recognition elements for the development of electrochemical biosensors. This promise becomes evident when we relate their characteristics to those of antibodies. Unlike antibodies, the synthesis of aptamers does not require animals or cell lines, and aptamers can be obtained using instruments. This process facilitates automation, decreases costs and significantly reduces the time of production, which in the case of antibodies can take months. Furthermore, aptamers can be immobilized directly on different surfaces without losing their ability to recognize the target. By contrast, direct immobilization on surfaces is avoided for antibodies because they generally cause denaturation of the protein. Another important characteristic is that aptamers are more resistant to chemical reactions because they exhibit greater stability when subjected to aggressive ionic strength, temperature and pH conditions. These perturbations, in many cases, only result in conformational changes that can easily be reversed without losing their recognition capability. This higher stability also makes them more tolerant to the electric field gradient that arises when electrochemical techniques are used in detection. Due to its higher stability, solutions with temperatures as high as 95°C, conditions that easily denature an antibody, may be used to regenerate the aptasensor, which allows successive and reproducible detection to be performed on the same device [168]. It is also important to highlight that given the nature of the oligonucleotides, the modifying and labeling reactions may be better directed at one of their 3 'or 5' ends, which results in better control of the orientation of those elements on the surface. The proper orientation of the antibodies is more difficult to control. The aptamer molecules, in general, also have a lower mass and smaller volume than antibodies. These characteristics, in addition to the possibility of the direct immobilization on the surface of the electrodes, facilitate the use of electrochemical techniques, as they exhibit a lower dielectric thickness and a higher coverage density. Consequently, all these properties result in a larger number of detection sites, an increased sensitivity and a miniaturization capability that is compatible with modern microfabrication technologies [169].

The possibility of synthesis through chemical routes and direct immobilization on different electrode surfaces, as well as the simplicity, miniaturization, low cost, the possibility of analyzing turbid samples, and the speed of data acquisition for electrochemical devices, make this junction relatively promising for the development of electrochemical biosensors using aptamers. Table 3 lists selected examples of aptamer-based electrochemical biosensors, with the aptamer as the recognition element. The typical targets of aptasensors include protein markers of diseases (Mucin 1 protein - MUC1), toxins (Aflatoxin B1 - AFB1), enzymes (Lysozymes -LYZ), hormones (17β-estradiol), toxic ions (Hg^{2+}) and controlled substances (e.g., cocaine). Many other examples of electrochemical aptamers applied to biologically relevant targets can be found in the following reviews [166, 170]. Studies involving RNA aptamers are also found, [171] however, these studies are not as numerous and less widespread, which is probably attributed to the greater difficulty in obtaining and reusing these aptamers and their being prone to attacks by nucleases [172].

The main electrode materials are gold and carbon. The most commonly used approach for gold electrodes involves the use of thiolated aptamers at the 5' end. Another relatively new approach uses the immobilization of the thiolated complementary tape, and the aptamers are immobilized by interactions with portions of the complementary tape, resulting in an H-shaped structure [179]. In carbon electrodes, the immobilization strategies are more varied. The deposition of streptavidin on the surface of the electrode and the immobilization by the use of biotinylated aptamers are the most commonly found strategies. Some other approaches use amide formation between amine group-modified aptamers at one of the 5 'ends and carbon electrodes modified with carboxylic groups [174]. Another less frequently used technique employs carbon surfaces containing electropolymerized films, which are functionalized with carboxyl groups for attachment to aptamers modified with amine groups; [178] magnetic glassy carbon electrodes and interaction with magnetic nanoparticles previously immobilized with aptamers are also other new approaches [164].

The construction of aptasensors using hybrid materials with gold nanoparticles, oxidized graphene, quantum dots, carbon nanotubes and silica are approaches that have been recently explored, [180] and they may contribute to improvements in the recognition kinetics, electronic conduction and amplification of the analytical signal. Regarding the use of redox compounds, the approaches include the use of a soluble redox probe, such as $Fe(CN)_6^{4-/3-}$ or $[Ru(NH3)_6]^{3+}$, and using aptamers labeled with methylene blue and alkaline phosphatase.

A different approach is used for the mercury target, [164] in which the thiamine-Hg^{2+} complex is electroactive and its redox reaction is used in the quantification of this metal ion.

Electrochemical techniques such as differential pulse voltammetry (DPV), square wave voltammetry (SWV), alternating current voltammetry (ACV), and electrochemical impedance spectroscopy (EIS) are the most used techniques for this type of biosensor. Voltammetric techniques are particularly interesting because they enable a better selectivity by the selection of the potential range investigated, and they may allow multiple signaling, as was recently demonstrated [181]. The use of EIS is also very frequent in these applications, and in these cases, the charge transference resistance (Rct) is usually the most used parameter in obtaining the analytical curves. The detection limits found depend on the architecture used in the construction of this type of biosensor and the target used as the analyte. However, the detection limits are usually less than 10^{-6} mol L^{-1}. Detection limits as low as 10^{-15} mol L^{-1} for the 17β-estradiol target are also found, [178] in addition, their relative simplicity attests to the promising nature of the use of these recognition elements in the development of electrochemical biosensors

Table 3. Examples of electrochemical biosensors based on aptamers as recognition elements

Target/ Aptamer	Cell	Iimmobilization	Redox compounds	LOD/ Techinique	Reference
MUC1 (DNA)	WE: Au RE:Ag\AgCl AE:Pt wire	5' End thiolaed; chemisorption	3' End methylene blue label aptamer	50 nM (SWV)	[173]
AFB1 (DNA)	WE: Carbon RE:Ag/AgCl AE:Carbon	5' End amino modified Carboxyl-amine bound	$Fe(CN)_6^{4-/3-}$ soluble redox probe	380 nM (EIS)	[174]
CRP (DNA)	WE: Carbon RE:Ag AE:Carbon	5' End biotinylated streptovidin-biotin interaction	Aptamer conjugated alkaline phosphatase	2 nM (DPV)	[175]
VEGF (DNA)	WE: Au RE:Ag\AgCl AE:Pt wire	5' End thiolaed; chemisorption	3' End methylene blue label aptamer	5 pM (ACV)	[176]
LYZ (DNA)	WE: Au RE: $Hg\backslash Hg_2Cl_2$ AE:Pt wire	5' End thiolaed; chemisorption	3' End ferrocene label aptamer and $[Ru(NH3)_6]^{3+}$ soluble redox probe	0.8 pM (SWV)	[177]
17β-estradiol (DNA)	WE: GC RE:Ag\AgCl AE:Pt wire	5' End amino modified poly(Py-co-PAA) Carboxyl-amine bound	$Fe(CN)_6^{4-/3-}$ soluble redox probe	1 fM (EIS)	[178]
Hg^{2+} (DNA)	WE: MGC RE:$Hg\backslash Hg_2Cl_2$ AE:Pt wire	5' End biotinylated streptovidin-biotin interaction	Hg^{2+}-thymine probe	0.33 nM (DPV)	[164]
Cocaine (DNA)	WE: Au RE:Ag AE:Au	H-shape complementary aptamer	$Fe(CN)_6^{4-/3-}$ soluble redox probe	0.27 nM (DPV)	[179]

MUC1, Mucin 1 protein; AFB1, Aflatoxin B1; LYZ, Lysozymes; MGC, magnetic glassy carbon; WE, working electrode; RE, reference electrode; AE, auxiliary electrode; LOD, limit of detection DPV, differential pulse voltammetry; SWV, square wave voltammetry; ACV, alternating current voltammetry; EIS, electrochemical impedance spectroscopy.

There is no doubt that aptamers have sufficient merits as promising biological recognition elements in electrochemical biosensors. The trends point to a replacement of antibodies by aptamers in biosensor science. However, compared to more mature technologies, such as immunoassays, it is clear that there is still much to be explored. Recent studies have described different ways of building electrochemical aptasensors, and, in many cases, only well-defined synthetic solutions are used as proofs of concept. Due to the significant diversity of biomolecules in living organisms and the need to analyze real samples, the opportunity for ongoing research in this area is enormous. Fundamental studies, the compilation and comparison of affinity constants, and the search for other thermodynamic/kinetic parameters should be further explored in the future. The possibility of miniaturization coupled with microfluidic technology can lead to the construction of devices that are capable of analyzing thousands of targets. These perspectives are striking because they can provide a rapid diagnosis of numerous diseases using only a small sample volume. This would be an unprecedented development in the history of electrochemical biosensors.

Conclusion and Future Trends

The proper functioning and desirable performance of electrochemical biosensors are directly associated with the materials and/or recognition elements selected for their construction. In this sense, there has been a recent focus on the technology of new materials and composites in the design of biosensors, especially carbon-based materials (e.g., graphene and its derivatives) and metallic nanoparticles. There is also an increasing trend in the synthesis of new molecules as modifiers and recognition elements to improve the sensitivity, biocompatibility, specificity, and stability of electrochemistry biosensors, and this is exemplified by the use of dendrimers and aptamers coupled to another materials and biomolecules.

Concerning the main perspectives in the development of electrochemical biosensors, we strongly believe that the motivation is focused on nanotechnology, miniaturization, and the design of multisensory arrays for POC analysis in biomedical applications. Although some studies have addressed the use of nanoparticles and carbon-based materials in miniaturized devices, the use of dendrimers and aptamers is still rare in these systems. Therefore, advances in the use of these materials and recognition elements, as discussed in this chapter, in miniaturized electrochemical biosensors will require a multidisciplinary approach that involves different fields, ranging from analytical chemistry, and electronics to molecular biology.

In terms of the materials for future prospects, hydrogel nanoparticles are considered promising materials for application in electrochemical biosensors. Hydrogels, by definition, are cross-linked polymer networks swelled with water in the presence of nanostructures or nanoparticles [182]. The unique physicochemical properties of these systems enable their application in several fields and reinforce their use in the development of biosensors with improved performance, biocompatibility, and minimized costs [183]. Therefore, we anticipate that hydrogels will also play a significant role in the development of portable electrochemical biosensors in biomedical applications, and more.

REFERENCES

[1] B. Nagel, H. Dellweg, L. M. Gierasch, in *Pure and Applied Chemistry, Vol. 64*, 1992, p. 143.

[2] P. Mehrotra, *Journal of Oral Biology and Craniofacial Research* 2016, *6*, 153-159.

[3] a) E. B. Bahadır, M. K. Sezgintürk, *Analytical Biochemistry* 2015, *478*, 107-120; b) S. Murugaiyan, R. Ramasamy, N. Gopal, V. Kuzhandaivelu, *Advanced Biomedical Research* 2014, *3*, 67-67.

[4] a) E. K. Witkowska Nery, E. T. S. G. da Silva, L. T. Kubota, in *Food Biosensors*, The Royal Society of Chemistry, 2017, pp. 22-43; b) L. D. Mello, L. T. Kubota, *Food Chemistry* 2002, *77*, 237-256.

[5] K. R. Rogers, *Analytica Chimica Acta* 2006, *568*, 222-231.

[6] S. Vigneshvar, C. C. Sudhakumari, B. Senthilkumaran, H. Prakash, *Frontiers in Bioengineering and Biotechnology* 2016, *4*.

[7] S. M. Borisov, O. S. Wolfbeis, *Chemical Reviews* 2008, *108*, 423-461.

[8] P. Skládal, *TrAC Trends in Analytical Chemistry* 2016, *79*, 127-133.

[9] K. Ramanathan, B. Danielsson, *Biosensors and Bioelectronics* 2001, *16*, 417-423.

[10] R. Fogel, J. Limson, Ashwin A. Seshia, *Essays in Biochemistry* 2016, *60*, 101-110.

[11] N. J. Ronkainen, H. B. Halsall, W. R. Heineman, *Chemical Society Reviews* 2010, *39*, 1747-1763.

[12] L. C. Clark, C. Lyons, *Annals of the New York Academy of Sciences* 1962, *102*, 29-45.

[13] E. Witkowska Nery, M. Kundys, P. S. Jeleń, M. Jönsson-Niedziółka, *Analytical Chemistry* 2016.

[14] J. Wang, *Biosensors and Bioelectronics* 2006, *21*, 1887-1892.

[15] F. Xu, G. Zhen, F. Yu, E. Kuennemann, M. Textor, W. Knoll, *Journal of the American Chemical Society* 2005, *127*, 13084-13085.

[16] Z. Zhao, H. Jiang, *Enzyme-based Electrochemical Biosensors*, 2010.

[17] K. R. Rogers, *Molecular Biotechnology* 2000, *14*, 109-129.

[18] Jules L. Hammond, N. Formisano, P. Estrela, S. Carrara, J. Tkac, *Essays in Biochemistry* 2016, *60*, 69-80.

[19] S. J. Sadeghi, in *Encyclopedia of Biophysics* (Ed.: G. C. K. Roberts), Springer Berlin Heidelberg, Berlin, Heidelberg, 2013, pp. 61-67.

[20] N. Verma, A. K. Singh, P. Kaur, *Journal of Analytical Chemistry* 2015, *70*, 1111-1115.

[21] K. Yao, Y. Zhu, P. Wang, X. Yang, P. Cheng, H. Lu, *Materials Science and Engineering: C* 2007, *27*, 736-740.

[22] W. Muangsuwan, C. Promptmas, W. Jeamsaksiri, W. Bunjongpru, A. Srisuwan, C. Hruanun, A. Poyai, P. Wongchitrat, M. Yasawong, *Heliyon* 2016, *2*, e00188.

[23] O. O. Soldatkin, I. S. Kucherenko, V. M. Pyeshkova, A. L. Kukla, N. Jaffrezic-Renault, A. V. El'skaya, S. V. Dzyadevych, A. P. Soldatkin, *Bioelectrochemistry* 2012, *83*, 25-30.

[24] J.-G. Guan, Y.-Q. Miao, Q.-J. Zhang, *Journal of Bioscience and Bioengineering* 2004, *97*, 219-226.

[25] D. Grieshaber, R. MacKenzie, J. Vörös, E. Reimhult, *Sensors-Basel* 2008, *8*, 1400.

[26] a) Z. Wang, J. Yu, R. Gui, H. Jin, Y. Xia, *Biosens Bioelectron* 2016, *79*, 136-149; b) S. Kumar, W. Ahlawat, R. Kumar, N. Dilbaghi, *Biosensors & Bioelectronics* 2015, *70*, 498-503; cJ. Wang, *Analyst* 2005, *130*, 421-426.

[27] K. S. Novoselov, A. K. Geim, S. V. Morozov, D. Jiang, Y. Zhang, S. V. Dubonos, I. V. Grigorieva, A. A. Firsov, *Science* 2004, *306*, 666-669.

[28] H. W. Kroto, J. R. Heath, S. C. Obrien, R. F. Curl, R. E. Smalley, *Nature* 1985, *318*, 162-163.

[29] S. Iijima, *Nature* 1991, *354*, 56-58.

[30] a) P. Delhaes, *Graphite and precursors, Vol. 1*, CRC Press, 2000; b) T. Enoki, M. Suzuki, M. Endo, *Graphite intercalation compounds and applications*, Oxford University Press, 2003.

[31] M. J. Allen, V. C. Tung, R. B. Kaner, *Chemical Reviews* 2010, *110*, 132-145.

[32] A. K. Geim, K. S. Novoselov, *Nat Mater* 2007, *6*, 183-191.

[33] a) L. D. Landau, *Phys. Z. Sowjetunion* 1937, 26–35; b) R. E. Peierls, *Ann. I. H. Poincare* 1935, 177–222.

[34] a) C. N. R. Rao, A. K. Sood, K. S. Subrahmanyam, A. Govindaraj, *Angewandte Chemie-International Edition* 2009, *48*, 7752-7777; b) C. Soldano, A. Mahmood, E. Dujardin, *Carbon* 2010, *48*, 2127-2150.

[35] J. Zheng, C.-a. Di, Y. Liu, H. Liu, Y. Guo, C. Du, T. Wu, G. Yu, D. Zhu, *Chemical Communications* 2010, *46*, 5728-5730.

[36] Y. W. Zhu, S. Murali, W. W. Cai, X. S. Li, J. W. Suk, J. R. Potts, R. S. Ruoff, *Advanced Materials* 2010, *22*, 3906-3924.

[37] C. Lee, X. Wei, J. W. Kysar, J. Hone, *science* 2008, *321*, 385-388.

[38] A. A. Balandin, S. Ghosh, W. Bao, I. Calizo, D. Teweldebrhan, F. Miao, C. N. Lau, *Nano letters* 2008, *8*, 902-907.

[39] W. Cai, Y. Zhu, X. Li, R. D. Piner, R. S. Ruoff, *Applied Physics Letters* 2009, *95*, 123115.

[40] J. S. Bunch, S. S. Verbridge, J. S. Alden, A. M. Van Der Zande, J. M. Parpia, H. G. Craighead, P. L. McEuen, *Nano letters* 2008, *8*, 2458-2462.

[41] S. Unarunotai, Y. Murata, C. E. Chialvo, N. Mason, I. Petrov, R. G. Nuzzo, J. S. Moore, J. A. Rogers, *Advanced Materials* 2010, *22*, 1072-1077.

[42] a) C. Berger, Z. M. Song, T. B. Li, X. B. Li, A. Y. Ogbazghi, R. Feng, Z. T. Dai, A. N. Marchenkov, E. H. Conrad, P. N. First, W. A. de Heer, *J Phys Chem B* 2004, *108*, 19912-19916; b) C. Berger, Z. M. Song, X. B. Li, X. S. Wu, N. Brown, D.

Maud, C. Naud, W. A. de Heer, *Physica Status Solidi a-Applications and Materials Science* 2007, *204*, 1746-1750; cF. Kim, J. Y. Luo, R. Cruz-Silva, L. J. Cote, K. Sohn, J. X. Huang, *Advanced Functional Materials* 2010, *20*, 2867-2873.

[43] a) C. Mattevi, H. Kim, M. Chhowalla, *Journal of Materials Chemistry* 2011, *21*, 3324-3334; b) H. Marsh, A. Warburton, *Journal of Applied Chemistry* 1970, *20*, 133-142.

[44] J. Cai, P. Ruffieux, R. Jaafar, M. Bieri, T. Braun, S. Blankenburg, M. Muoth, A. P. Seitsonen, M. Saleh, X. Feng, *Nature* 2010, *466*, 470-473.

[45] D. V. Kosynkin, A. L. Higginbotham, A. Sinitskii, J. R. Lomeda, A. Dimiev, B. K. Price, J. M. Tour, *Nature* 2009, *458*, 872-876.

[46] J. N. Coleman, *Accounts of chemical research* 2012, *46*, 14-22.

[47] V. Nicolosi, M. Chhowalla, M. G. Kanatzidis, M. S. Strano, J. N. Coleman, *Science* 2013, *340*, 1226419.

[48] J. N. Coleman, M. Lotya, A. O'Neill, S. D. Bergin, P. J. King, U. Khan, K. Young, A. Gaucher, S. De, R. J. Smith, *science* 2011, *331*, 568-571.

[49] M. Lotya, Y. Hernandez, P. J. King, R. J. Smith, V. Nicolosi, L. S. Karlsson, F. M. Blighe, S. De, Z. Wang, I. McGovern, *Journal of the American Chemical Society* 2009, *131*, 3611-3620.

[50] S. Park, R. S. Ruoff, *Nat Nanotechnol* 2009, *4*, 217-224.

[51] G. Eda, M. Chhowalla, *Advanced materials* 2010, *22*, 2392-2415.

[52] F. Bonaccorso, A. Lombardo, T. Hasan, Z. Sun, L. Colombo, A. C. Ferrari, *Materials Today* 2012, *15*, 564-589.

[53] K. E. Whitener, P. E. Sheehan, *Diamond and related materials* 2014, *46*, 25-34.

[54] A. C. Ferrari, F. Bonaccorso, V. Fal'Ko, K. S. Novoselov, S. Roche, P. Bøggild, S. Borini, F. H. Koppens, V. Palermo, N. Pugno, *Nanoscale* 2015, *7*, 4598-4810.

[55] M. Pumera, *Materials Today* 2011, *14*, 308-315.

[56] a) M. Zhou, Y. Zhai, S. Dong, *Analytical Chemistry* 2009, *81*, 5603-5613; b) Y. Shao, J. Wang, H. Wu, J. Liu, I. A. Aksay, Y. Lin, *Electroanalysis* 2010, *22*, 1027-1036.

[57] M. S. Artiles, C. S. Rout, T. S. Fisher, *Advanced drug delivery reviews* 2011, *63*, 1352-1360.

[58] A. T. Lawal, *Talanta* 2015, *131*, 424-443.

[59] M. Pumera, *The Chemical Record* 2009, *9*, 211-223.

[60] H. Chen, M. B. Müller, K. J. Gilmore, G. G. Wallace, D. Li, *Advanced materials* 2008, *20*, 3557-3561.

[61] S. Park, N. Mohanty, J. W. Suk, A. Nagaraja, J. An, R. D. Piner, W. Cai, D. R. Dreyer, V. Berry, R. S. Ruoff, *Advanced materials* 2010, *22*, 1736-1740.

[62] S. Agarwal, X. Zhou, F. Ye, Q. He, G. C. Chen, J. Soo, F. Boey, H. Zhang, P. Chen, *Langmuir* 2010, *26*, 2244-2247.

[63] a) H. Ma, D. Wu, Z. Cui, Y. Li, Y. Zhang, B. Du, Q. Wei, *Analytical Letters* 2013, *46*, 1-17; b) N. Celik, W. Balachandran, N. Manivannan, *IET Circuits, Devices & Systems* 2015, *9*, 434-445.

[64] B. Liang, L. Fang, G. Yang, Y. Hu, X. Guo, X. Ye, *Biosensors and Bioelectronics* 2013, *43*, 131-136.

[65] S. Palanisamy, B. Unnikrishnan, S.-M. Chen, *Int. J. Electrochem. Sci* 2012, *7*, 7935-7947.

[66] D.-J. Kim, I. Y. Sohn, J.-H. Jung, O. J. Yoon, N.-E. Lee, J.-S. Park, *Biosensors and Bioelectronics* 2013, *41*, 621-626.

[67] Y. Song, Y. Luo, C. Zhu, H. Li, D. Du, Y. Lin, *Biosensors and Bioelectronics* 2016, *76*, 195-212.

[68] A. Benvidi, N. Rajabzadeh, M. Mazloum-Ardakani, M. M. Heidari, A. Mulchandani, *Biosensors and Bioelectronics* 2014, *58*, 145-152.

[69] P. T. Yin, S. Shah, M. Chhowalla, K.-B. Lee, *Chemical reviews* 2015, *115*, 2483-2531.

[70] R. A. Marcus, *Angewandte Chemie [Applied Chemistry] International Edition in English* 1993, *32*, 1111-1121.

[71] a) J. F. Rusling, *Accounts of chemical research* 1998, *31*, 363-369; b) Y. Xiao, F. Patolsky, E. Katz, J. F. Hainfeld, I. Willner, *Science* 2003, *299*, 1877-1881.

[72] T. Kavitha, A. I. Gopalan, K.-P. Lee, S.-Y. Park, *Carbon* 2012, *50*, 2994-3000.

[73] H. D. Jang, S. K. Kim, H. Chang, K.-M. Roh, J.-W. Choi, J. Huang, *Biosensors and Bioelectronics* 2012, *38*, 184-188.

[74] K. Wang, Q. Liu, Q.-M. Guan, J. Wu, H.-N. Li, J.-J. Yan, *Biosensors and Bioelectronics* 2011, *26*, 2252-2257.

[75] C. Shan, H. Yang, D. Han, Q. Zhang, A. Ivaska, L. Niu, *Biosensors and Bioelectronics* 2010, *25*, 1070-1074.

[76] H. Wu, J. Wang, X. Kang, C. Wang, D. Wang, J. Liu, I. A. Aksay, Y. Lin, *Talanta* 2009, *80*, 403-406.

[77] L.-M. Lu, H.-B. Li, F. Qu, X.-B. Zhang, G.-L. Shen, R.-Q. Yu, *Biosensors and Bioelectronics* 2011, *26*, 3500-3504.

[78] W. Lu, Y. Luo, G. Chang, X. Sun, *Biosensors and Bioelectronics* 2011, *26*, 4791-4797.

[79] S. Singal, A. K. Srivastava, A. M. Biradar, A. Mulchandani, *Sensors and Actuators B: Chemical* 2014, *205*, 363-370.

[80] S. Mao, G. Lu, K. Yu, Z. Bo, J. Chen, *Adv Mater* 2010, *22*, 3521-3526.

[81] Y. Zhang, W. Jiang, *Electrochimica Acta* 2012, *71*, 239-245.

[82] P. A. Rasheed, N. Sandhyarani, *Sensors and Actuators B: Chemical* 2014, *204*, 777-782.

[83] K. Zhou, Y. Zhu, X. Yang, J. Luo, C. Li, S. Luan, *Electrochimica Acta* 2010, *55*, 3055-3060.

[84] H. Xu, H. Dai, G. Chen, *Talanta* 2010, *81*, 334-338.

[85] Q. Wei, K. Mao, D. Wu, Y. Dai, J. Yang, B. Du, M. Yang, H. Li, *Sensors and Actuators B: Chemical* 2010, *149*, 314-318.

[86] T. Yang, L. Meng, J. Zhao, X. Wang, K. Jiao, *ACS applied materials & interfaces* 2014, *6*, 19050-19056.

[87] Y. Jiang, X. Zhang, C. Shan, S. Hua, Q. Zhang, X. Bai, L. Dan, L. Niu, *Talanta* 2011, *85*, 76-81.

[88] L. Wu, L. Feng, J. Ren, X. Qu, *Biosensors and Bioelectronics* 2012, *34*, 57-62.

[89] G. M. M. Ferreira, F. M. de Oliveira, F. R. F. Leite, C. M. Maroneze, L. T. Kubota, F. S. Damos, R. d. C. S. Luz, *Electrochimica Acta* 2013, *111*, 543-551.

[90] H. Liu, J. Gao, M. Xue, N. Zhu, M. Zhang, T. Cao, *Langmuir* 2009, *25*, 12006-12010.

[91] C. Tuerk, L. Gold, *science* 1990, *249*, 505-510.

[92] L. Feng, Y. Chen, J. Ren, X. Qu, *Biomaterials* 2011, *32*, 2930-2937.

[93] L. Jiao, G. Lu, C. Zhang, Y. Yuan, J. Wang, Q. Zhang, P. Li, *IET Micro & Nano Letters* 2013, *8*, 903-905.

[94] a) H. Wang, Y. Wang, S. Liu, J. Yu, W. Xu, Y. Guo, J. Huang, *Rsc Advances* 2014, *4*, 60987-60994; b) B. Jiang, M. Wang, Y. Chen, J. Xie, Y. Xiang, *Biosensors and Bioelectronics* 2012, *32*, 305-308.

[95] M. Faraday, *Philosophical Transactions of the Royal Society of London* 1857, *147*, 145-181.

[96] a) X. Luo, A. Morrin, A. J. Killard, M. R. Smyth, *Electroanalysis* 2006, *18*, 319-326; b) H. Kang, L. Wang, M. O'Donoghue, Y. C. Cao, W. Tan, in *Optical Biosensors (Second Edition)*, Elsevier, Amsterdam, 2008, pp. 583-621.

[97] R. Britto Hurtado, M. Cortez-Valadez, L. P. Ramírez-Rodríguez, E. Larios-Rodriguez, R. A. B. Alvarez, O. Rocha-Rocha, Y. Delgado-Beleño, C. E. Martinez-Nuñez, H. Arizpe-Chávez, A. R. Hernández-Martínez, M. Flores-Acosta, *Physics Letters A* 2016, *380*, 2658-2663.

[98] C. Li, D. Li, G. Wan, J. Xu, W. Hou, *Nanoscale Research Letters* 2011, *6*, 440.

[99] M. S. Bakshi, S. Sachar, G. Kaur, P. Bhandari, G. Kaur, M. C. Biesinger, F. Possmayer, N. O. Petersen, *Crystal Growth & Design* 2008, *8*, 1713-1719.

[100] L. Quang Huy Tran and Van Quy Nguyen and Anh-Tuan, *Advances in Natural Sciences: Nanoscience and Nanotechnology* 2013, *4*, 033001.

[101] D. K. Lee, Y. S. Kang, *ETRI Journal* 2004, *26*, 252-256.

[102] S. Sunkar, C. V. Nachiyar, *Asian Pacific Journal of Tropical Biomedicine* 2012, *2*, 953-959.

[103] Y. Wu, W. Xu, L. Bai, Y. Yuan, H. Yi, Y. Chai, R. Yuan, *Biosensors and Bioelectronics* 2013, *50*, 50-56.

[104] U. Tamer, Y. Gündoğdu, İ. H. Boyacı, K. Pekmez, *Journal of Nanoparticle Research* 2010, *12*, 1187-1196.

[105] V. Patsula, L. Kosinová, M. Lovrić, L. Ferhatovic Hamzić, M. Rabyk, R. Konefal, A. Paruzel, M. Šlouf, V. Herynek, S. Gajović, D. Horák, *ACS Applied Materials & Interfaces* 2016, *8*, 7238-7247.
[106] S. Lüscher, A. Fuhrer, R. Held, T. Heinzel, K. Ensslin, M. Bichler, W. Wegscheider, *Microelectronics Journal* 2002, *33*, 319-321.
[107] J. Zhang, Z. Sun, J. Fang, *MRS Proceedings* 2006, *942*.
[108] C. Siti Rabizah Makhsin and Khairunisak Abdul Razak and Rahmah Noordin and Nor Dyana Zakaria and Tan Soo, *Nanotechnology* 2012, *23*, 495719.
[109] J. Turkevich, P. C. Stevenson, J. Hillier, *Discussions of the Faraday Society* 1951, *11*, 55-75.
[110] G. Bülbül, A. Hayat, S. Andreescu, *Sensors* 2015, *15*.
[111] M. M. Ganesh Babu, P. Gunasekaran, *Colloids and Surfaces B: Biointerfaces* 2009, *74*, 191-195.
[112] a) W. V. a. J. U. a. S. K. K. a. K. M. P. Meenal Kowshik and Shriwas Ashtaputre and Sharmin Kharrazi and, *Nanotechnology* 2003, *14*, 95; b) L. Sintubin, W. Verstraete, N. Boon, *Biotechnology and Bioengineering* 2012, *109*, 2422-2436.
[113] P. L. Suryawanshi, S. P. Gumfekar, P. R. Kumar, B. B. Kale, S. H. Sonawane, *Colloid and Interface Science Communications* 2016, *13*, 6-9.
[114] A. M. Smith, S. Nie, *Accounts of chemical research* 2009, *43*, 190-200.
[115] L. Wu, E. Xiong, X. Zhang, X. Zhang, J. Chen, *Nano Today* 2014, *9*, 197-211.
[116] V. B. Verma, J. J. Coleman, *Applied Physics Letters* 2008, *93*, 111117.
[117] L. C. d. S. Viol, F. O. Silva, D. L. Ferreira, J. L. A. Alves, M. A. Schiavon, *Química Nova* 2011, *34*, 595-600.
[118] R. Ghosh Chaudhuri, S. Paria, *Chemical Reviews* 2012, *112*, 2373-2433.
[119] A. S. de Dios, R. Badía-Laíño, M. E. Díaz-García, *Sensors and Actuators B: Chemical* 2013, *184*, 48-53.
[120] M. Homberger, U. Simon, *Philosophical Transactions of the Royal Society A: Mathematical, Physical and Engineering Sciences* 2010, *368*, 1405.
[121] a) C. Saraiva, C. Praça, R. Ferreira, T. Santos, L. Ferreira, L. Bernardino, *Journal of Controlled Release* 2016, *235*, 34-47; b) R. Singh, J. W. Lillard, *Experimental and molecular pathology* 2009, *86*, 215-223.
[122] E. Gross, H.-C. LiuJack, F. D. Toste, G. A. Somorjai, *Nat Chem* 2012, *4*, 947-952.
[123] a) H. H. Lara, N. V. Ayala-Núñez, L. d. C. Ixtepan Turrent, C. Rodríguez Padilla, *World Journal of Microbiology and Biotechnology* 2010, *26*, 615-621; b) G. A. Sotiriou, S. E. Pratsinis, *Current opinion in chemical engineering* 2011, *1*, 3-10.
[124] B. R. Saunders, M. L. Turner, *Advances in Colloid and Interface Science* 2008, *138*, 1-23.
[125] A. De Barros, C. J. L. Constantino, J. R. R. Bortoleto, N. C. Da Cruz, M. Ferreira, *Sensors and Actuators B: Chemical* 2016, *236*, 408-417.

[126] a) I. L. Medintz, M. H. Stewart, S. A. Trammell, K. Susumu, J. B. Delehanty, B. C. Mei, J. S. Melinger, J. B. Blanco-Canosa, P. E. Dawson, H. Mattoussi, *Nat Mater* 2010, *9*, 676-684; b) F. Tan, L. Cong, X. Li, Q. Zhao, H. Zhao, X. Quan, J. Chen, *Sensors and Actuators B: Chemical* 2016, *233*, 599-606; c) G. Bhanjana, N. Dilbaghi, R. Kumar, A. Umar, S. Kumar, *Electrochimica Acta* 2015, *169*, 97-102.

[127] H. Zhang, B. Jiang, Y. Xiang, Y. Zhang, Y. Chai, R. Yuan, *Analytica Chimica Acta* 2011, *688*, 99-103.

[128] J.-J. Zhang, T.-T. Zheng, F.-F. Cheng, J.-R. Zhang, J.-J. Zhu, *Analytical Chemistry* 2011, *83*, 7902-7909.

[129] E. Hamidi-Asl, F. Dardenne, S. Pilehvar, R. Blust, K. De Wael, *Chemosensors* 2016, *4*, 16.

[130] a) D. A. Tomalia, H. Baker, J. Dewald, M. Hall, G. Kallos, S. Martin, J. Roeck, J. Ryder, P. Smith, *Polymer Journal* 1985, *17*, 117-132; b) G. R. Newkome, Z. Q. Yao, G. R. Baker, V. K. Gupta, *Journal of Organic Chemistry* 1985, *50*, 2003-2004.

[131] a) D. A. Tomalia, H. Baker, J. Dewald, M. Hall, G. Kallos, S. Martin, J. Roeck, J. Ryder, P. Smith, *Macromolecules* 1986, *19*, 2466-2468; b) D. A. Tomalia, V. Berry, M. Hall, D. M. Hedstrand, *Macromolecules* 1987, *20*, 1164-1167; c) D. A. Tomalia, M. Hall, D. M. Hedstrand, *Journal of the American Chemical Society* 1987, *109*, 1601-1603; d) D. A. Tomalia, A. M. Naylor, W. A. Goddard, *Angewandte Chemie-International Edition in English* 1990, *29*, 138-175.

[132] C. Ornelas, *Macromolecular Chemistry and Physics* 2016, *217*, 149-174.

[133] a) C. J. Hawker, J. M. J. Frechet, *Journal of the American Chemical Society* 1990, *112*, 7638-7647; b) C. J. Hawker, J. M. J. Frechet, *Journal of the Chemical Society-Chemical Communications* 1990, 1010-1013.

[134] C. Worner, R. Mulhaupt, *Angewandte Chemie-International Edition in English* 1993, *32*, 1306-1308.

[135] P. R. D. D. A. Tomalia, Dendritic Polymers, Divergent Synthesis (Starburst Polyamidoamine Dendrimers), in Polymeric Materials Encyclopedia (Ed: J. C. Salamone), CRC Press, New York 1996.

[136] F. Morgenroth, E. Reuther, K. Mullen, *Angewandte Chemie-International Edition in English* 1997, *36*, 631-634.

[137] J. Ruiz, G. Lafuente, S. Marcen, C. Ornelas, S. Lazare, E. Cloutet, J. C. Blais, D. Astruc, *Journal of the American Chemical Society* 2003, *125*, 7250-7257.

[138] a) C. Larre, D. Bressolles, C. Turrin, B. Donnadieu, A. M. Caminade, J. P. Majoral, *Journal of the American Chemical Society* 1998, *120*, 13070-13082; b) C. Larre, B. Donnadieu, A. M. Caminade, J. P. Majoral, *Journal of the American Chemical Society* 1998, *120*, 4029-4030.

[139] M. A. van Dongen, A. Desai, B. G. Orr, J. R. Baker, M. M. B. Holl, *Polymer* 2013, *54*, 4126-4133.

[140] a) S. M. Grayson, J. M. J. Frechet, *Chemical Reviews* 2001, *101*, 3819-3867; b) G. Y. Chen, I. Roy, C. H. Yang, P. N. Prasad, *Chemical Reviews* 2016, *116*, 2826-2885.

[141] a) K. Oberg, J. Ropponen, J. Kelly, P. Lowenhielm, M. Berglin, M. Malkoch, *Langmuir* 2013, *29*, 456-465; b) H. Tokuhisa, J. a. Liu, K. Omori, M. Kanesato, K. Hiratani, L. A. Baker, *Langmuir* 2009, *25*, 1633-1637.

[142] G. Jiang, S. Deng, A. Baba, C. Huang, R. C. Advincula, *Macromolecular Chemistry and Physics* 2010, *211*, 2562-2572.

[143] M. J. L. Felipe, R. R. Ponnapati, R. B. Pernites, P. Dutta, R. C. Advincula, *Acs Applied Materials & Interfaces* 2010, *2*, 3401-3405.

[144] a) X. Li, K. Takeda, E. Yuba, A. Harada, K. Kono, *Journal of Materials Chemistry B* 2014, *2*, 4167-4176; b) E. B. Bahadir, M. K. Sezginturk, *Talanta* 2016, *148*, 427-438.

[145] B. S. Day, L. R. Fiegland, E. S. Vint, W. Shen, J. R. Morris, M. L. Norton, *Langmuir* 2011, *27*, 12434-12442.

[146] D. E. P. Souto, A. M. Fonseca, J. T. C. Barragan, R. d. C. S. Luz, H. M. Andrade, F. S. Damos, L. T. Kubota, *Biosensors & Bioelectronics* 2015, *70*, 275-281.

[147] a) M. Hasanzadeh, N. Shadjou, M. Eskandani, J. Soleymani, F. Jafari, M. de la Guardia, *Trac-Trends in Analytical Chemistry* 2014, *53*, 137-149; b) P. P. Pednekar, K. R. Jadhav, V. J. Kadam, *Expert Opinion on Drug Delivery* 2012, *9*, 1273-1288; c) P. Iqbal, F. J. Rawson, W. K. W. Ho, S.-F. Lee, K. C.-F. Leung, X. Wang, A. Beri, J. A. Preece, J. Ma, P. M. Mendes, *ACS Applied Materials & Interfaces* 2014, *6*, 6264-6274.

[148] T. J. Cho, R. I. MacCuspie, J. Gigault, J. M. Gorham, J. T. Elliott, V. A. Hackley, *Langmuir* 2014, *30*, 3883-3893.

[149] X. W. Fu, C. L. Zhao, X. H. Liu, J. H. Zhao, X. F. Niu, L. Zheng, Y. H. Yang, *International Journal of Environmental Analytical Chemistry* 2016, *96*, 284-297.

[150] a) X. N. Zhang, J. Z. Shen, H. L. Ma, Y. X. Jiang, C. Y. Huang, E. Han, B. S. Yao, Y. Y. He, *Biosensors & Bioelectronics* 2016, *80*, 666-673; b) A. K. Deb, S. C. Das, A. Saha, M. B. Wayu, M. H. Marksberry, R. J. Baltz, C. C. Chusuei, *J Appl Electrochem* 2016, *46*, 289-298.

[151] M. Frasconi, D. Deriu, A. D'Annibale, F. Mazzei, *Nanotechnology* 2009, *20*.

[152] D. Astruc, J. Ruiz, *Journal of Inorganic and Organometallic Polymers and Materials* 2015, *25*, 2-11.

[153] a) M. Karadag, C. Geyik, D. O. Demirkol, F. N. Ertas, S. Timur, *Materials Science & Engineering C-Materials for Biological Applications* 2013, *33*, 634-640; b) D. Astruc, Y. L. Wang, A. Rapakousiou, A. Diallo, R. Djeda, J. Ruiz, C. Ornelas, *Polyhedron* 2015, *86*, 24-30.

[154] A. F. Wang, C. Ornelas, D. Astruc, P. Hapiot, *Journal of the American Chemical Society* 2009, *131*, 6652-+.

[155] E. Cevik, O. Bahar, M. Senel, M. F. Abasiyanik, *Biosensors & Bioelectronics* 2016, *86*, 1074-1079.

[156] J. N. Tiwari, V. Vij, K. C. Kemp, K. S. Kim, *Acs Nano* 2016, *10*, 46-80.

[157] a) F. Y. Li, J. Peng, Q. Zheng, X. Guo, H. Tang, S. Z. Yao, *Analytical Chemistry* 2015, *87*, 4806-4813; b) J. V. da Silva, D. M. Pimentel, D. E. P. Souto, R. D. S. Luz, F. S. Damos, *Journal of Solid State Electrochemistry* 2013, *17*, 2795-2804; c) N. D. Pereira, F. M. de Oliveira, N. R. Pereira, R. M. Verly, D. E. P. Souto, L. T. Kubota, A. A. Tanaka, F. S. Damos, R. D. S. Luz, *Analytical Sciences* 2015, *31*, 29-35.

[158] a) B. Borisova, A. Sanchez, S. Jimenez-Falcao, M. Martin, P. Salazar, C. Parrado, J. M. Pingarron, R. Villalonga, *Sensors and Actuators B-Chemical* 2016, *232*, 84-90; b) G. Y. Shen, X. Hu, S. B. Zhang, *J Electroanal Chem* 2014, *717*, 172-176.

[159] L. Z. Liu, S. T. Jiang, L. Wang, Z. Zhang, G. M. Xie, *Microchim Acta* 2015, *182*, 77-84.

[160] B. Borisova, J. Ramos, P. Diez, A. Sanchez, C. Parrado, E. Araque, R. Villalonga, J. M. Pingarron, *Electroanalysis* 2015, *27*, 2131-2138.

[161] I. Bazin, S. A. Tria, A. Hayat, J. L. Marty, *Biosens Bioelectron* 2016, *87*, 285-298.

[162] A. D. Ellington, J. W. Szostak, *Nature* 1990, *346*, 818-822.

[163] H. Sun, X. Zhu, P. Y. Lu, R. R. Rosato, W. Tan, Y. Zu, *Molecular therapy. Nucleic acids* 2014, *3*, e182.

[164] D. Wu, Y. Wang, Y. Zhang, H. Ma, X. Pang, L. Hu, B. Du, Q. Wei, *Biosens Bioelectron* 2016, *82*, 9-13.

[165] L. Cui, J. Wu, H. Ju, *Biosensors & bioelectronics* 2016, *79*, 861-865.

[166] S. G. Meirinho, L. G. Dias, A. M. Peres, L. R. Rodrigues, *Biotechnol Adv* 2016, *34*, 941-953.

[167] B. Prieto-Simón, M. Campàs, J.-L. Marty, *Bioanalytical Reviews* 2010, *1*, 141-157.

[168] Y. Lu, X. Li, L. Zhang, P. Yu, L. Su, L. Mao, *Anal Chem* 2008, *80*, 1883-1890.

[169] a) M. Jarczewska, L. Gorski, E. Malinowska, *Analytical Methods* 2016, *8*, 3861-3877; b) S. G. Kim, J. S. Lee, J. Jun, D. H. Shin, J. Jang, *ACS Appl Mater Interfaces* 2016, *8*, 6602-6610.

[170] a) C. H. van den Kieboom, S. L. van der Beek, T. Mészáros, R. E. Gyurcsányi, G. Ferwerda, M. I. de Jonge, *TrAC Trends in Analytical Chemistry* 2015, *74*, 58-67; b) M. Labib, E. H. Sargent, S. O. Kelley, *Chem Rev* 2016, *116*, 9001-9090.

[171] S. G. Meirinho, L. G. Dias, A. M. Peres, L. R. Rodrigues, *Biosensors & bioelectronics* 2015, *71*, 332-341.

[172] A. Sassolas, L. J. Blum, B. D. Leca-Bouvier, *Electroanal* 2009, *21*, 1237-1250.

[173] F. Ma, C. Ho, A. K. H. Cheng, H.-Z. Yu, *Electrochimica Acta* 2013, *110*, 139-145.

[174] K. Yugender Goud, G. Catanante, A. Hayat, S. M, K. Vengatajalabathy Gobi, J. L. Marty, *Sensors and Actuators B: Chemical* 2016, *235*, 466-473.

[175] S. Centi, L. Bonel Sanmartin, S. Tombelli, I. Palchetti, M. Mascini, *Electroanalysis* 2009, *21*, 1309-1315.

[176] S. Zhao, W. Yang, R. Y. Lai, *Biosensors & bioelectronics* 2011, *26*, 2442-2447.

[177] X. Cao, J. Xia, H. Liu, F. Zhang, Z. Wang, L. Lu, *Sensors and Actuators B: Chemical* 2017, *239*, 166-171.

[178] B. Zhu, O. A. Alsager, S. Kumar, J. M. Hodgkiss, J. Travas-Sejdic, *Biosensors & bioelectronics* 2015, *70*, 398-403.

[179] K. Abnous, N. M. Danesh, M. Ramezani, S. M. Taghdisi, A. S. Emrani, *Sensors and Actuators B: Chemical* 2016, *224*, 351-355.

[180] F. Charbgoo, F. Soltani, S. M. Taghdisi, K. Abnous, M. Ramezani, *TrAC Trends in Analytical Chemistry* 2016, *85*, 85-97.

[181] P. Yu, Y. Liu, X. Zhang, J. Zhou, E. Xiong, X. Li, J. Chen, *Biosens Bioelectron* 2016, *79*, 22-28.

[182] a) M. Hasanzadeh, N. Shadjou, M. de la Guardia, *Trac-Trends in Analytical Chemistry* 2014, *62*, 11-19; b) S. S. Babu, V. K. Praveen, A. Ajayaghosh, *Chemical Reviews* 2014, *114*, 1973-2129; c) M. A. Krepker, E. Segal, *Analytical Chemistry* 2013, *85*, 7353-7360.

[183] S. W. Han, W. G. Koh, *Analytical Chemistry* 2016, *88*, 6247-6253.

ELECTROCHEMICAL ADVANCED OXIDATION PROCESSES

In: Advanced Materials and Systems …
Editor: Mauro Coelho dos Santos
ISBN: 978-1-53613-852-8

Chapter 7

SYNTHESIS OF SELF-ORGANIZED NANOSTRUCTURED OXIDE FOR PHOTOCATALYSIS AND PHOTOELECTROCATALYSIS APPLICATION

***Carlos Ponce de León Albarrán*[1]**
***and Christiane de Arruda Rodrigues*[2,*]**

[1]Electrochemical Engineering Laboratory, Energy Technology Research Group, Department of Mechanical Engineering, University of Southampton, Highfield, Southampton, UK

[2]Departamento de Engenharia Química, Instituto de Ciências Ambientais, Químicas e Farmacêuticas, Universidade Federal de São Paulo (UNIFESP), Diadema, SP, Brazil

ABSTRACT

Photocatalytic processes using self-organized TiO_2 nanotubes semiconductor are considered together with the factors that affect the growth mechanism and formation of the nanotubes during the anodization of titanium. The chapter analyses the morphology of the nanotubes when they are modified with metals, oxides and wall decoration. Illustrated examples of the applications of TiO_2 nanotubes in photocatalysis, photoelectrocatalysis, solar cells, semiconductors and biomedical areas are presented and the main characteristics required for these applications are highlighted.

---

* Corresponding Author Email: chris.arruda72@gmail.com.

Keywords: degradation of organic compounds, photocatalyst, photoelectrocatalysis, photocurrent, solar cells, titanium dioxide nanotubes

List of Symbols

A	Kubelka-Munk constant
A	anatase phase
AES	Auger Electron Spectroscopy
ACL	anion contaminated layer
BOL	barrier oxide layer
CB	conduction band
CP-Ti	commercially pure titanium
DSSC	Dye-Sensitized Solar Cell
EFB	flat band potential
E_{bg}	band gap energy
E_F	Fermi level
E_F*	quasi Fermi level
E_{ph}	photopotential
E_F(redox)	electrochemical potential of the electrolyte
E_F(sc)	Fermi energy level in a semiconductor
h	Planck constant
i_{ph}	photocurrent
NHE	normal hydrogen electrode
N-doping	Nitrogen doping
n	type of electron transition
nE_F*	quasi-Fermi level of electrons for the conduction band
pE_F*	quasi-Fermi level of holes for the valence band
R	absolute reflectance for a given radiation energy
R	rutile phase
SC	semiconductor
SEM	Scanning Electron Microscopy
VB	valence band
XPS	X-ray Photoelectron Spectroscopy
XRD	X-Ray Diffraction
α,	absorption coefficient
v	incident light frequency
ΔU	applied potential

INTRODUCTION

The fundamental requirement for efficient photocatalysis and electrochemical treatment is strongly based on the choice of anode and cathode materials. At the anode, the main requirements are: high corrosion resistance, physical and chemical stability at high positive potentials, large surface area electrodes, and that the material can be sourced at low cost and is abundant. Some examples include: semiconductor with low band gap energy (E_{bg}) or reduced recombination rate of excited electron-hole pairs, such as the anatase form of TiO_2 particles [1-6], nanotubular oxides from Ti, Zr, Nb, Ta and W together with the alloys of these metals, and dimensionally stable anodes (DSA) [7, 8]. Similar characteristics are required for the cathode electrode, and carbonaceous materials such as reticulated vitreous carbon (RVC) [9-11], carbon felt, carbon fibre and carbon nanotubes have been successfully employed to generate oxidizing species such as hydrogen peroxide and the Fenton reactant [12, 13].

This chapter considers the importance of the photocatalytic processes and their large applications in different areas (biomedical and technological) especially for processes using self-organized TiO_2 nanotubes as semiconductor. The chapter will present a review of the fundamental aspects of semiconductor as photocatalyst, the growth mechanism of self-organized TiO_2 nanotubes obtained by the anodization method and the factors that affect its growth and morphology. Additional aspects discussed here include: modification of the TiO_2 nanotubes layer with other metals, oxides and introduction of other elements (doping) or by tube wall decoration, the influence of heat treatment in the crystallization of anodized nanotubular oxide layers and some applications of self-organized nanotubular oxides in photocatalysis, solar cell and biomedical areas.

SEMICONDUCTOR AS PHOTOCATALYSTS: FUNDAMENTAL ASPECTS

The electrical properties of any material are based on its electronic band structure and the way these bands are filled with electrons. The position of the Fermi level (E_F) relative to the conduction band is an important parameter to determine the electrical properties of a particular material. Figure 1 presents the electronic bands of metal (a), insulator (b) and semiconductor (c) at 0 K temperature. In the case of metals, all energy levels in the conduction band (CB) below the Fermi energy level are filled with electrons, while all levels above E_F are empty, and the electrons are free to move into "empty" states of conduction band with only a small electric field, leading to high electrical conductivity. Insulators and semiconductors (SC) show lower valence band (VB) filled with electrons while the upper conduction band is empty, and these bands are separated by a band gap energy (E_{bg}). For insulators, the Fermi energy level is at the midpoint of a large energy gap (> 4 eV) between the conduction and the valence bands. The electrons cannot “be excited” from the valence to the conduction band, which leads to zero conductivity. In semiconductors, the Fermi energy is at the midpoint of a small energy gap (≤ 4 eV) between the conduction and the valence bands and the electrons can “be excited” from the valence to the conduction band under irradiation of energy higher than E_{bg}, leading to a measurable conductivity [14].

The electrical behavior of a semiconductor is determined by its impurities, which, even at very low concentrations will promote an excess of electrons or holes. Holes are positive charges that arise when an electron is excited from the valence to the conduction band. Under the excited condition, electrons and holes move in opposite directions and they spread throughout the materials following the imperfections of the semiconductor lattice.

When a semiconductor electrode is irradiated by a photon beam (light) with an energy equal or greater than the energy band gap of the semiconductor, the electrode absorbs the energy of the photon in order to excite the electrons from the valence to the conduction band, producing photoexcited pairs of electrons in the conduction band and holes in the valence band as shown in Figure 2. In general, the photoexcitation increases the concentration of minority charge carriers (holes in the *n*-type and electrons in the *p*-type semiconductors) in semiconductor electrodes, but has little influence in the concentration of the majority charge carriers (electrons in the *n*-type and holes in the *p*-type semiconductors). In the case of *n*-type semiconductors, the photoexcitation has little influence in the concentration of electrons in the conduction band (majority charge carriers), but the concentration of holes in valence band increase significantly by the photoexcited holes. The holes have a very positive potential (+1.0 to +3.5 V *vs.* NHE depending on the semiconductor and pH) that enables the generation of hydroxyl radicals from the oxidation of water which are able to oxidize organic compounds [15]. The efficiency of the photocatalytic process will depend on the competition between the process in which the electrons are removed from the semiconductor surface and the recombination process of the electron/hole pair, which also results in heat release.

Fermi Energy Level

In the dark, thermal equilibrium is established between the electrons in the conduction band and the holes in the valence band so that the electrochemical potential is called Fermi energy level of the semiconductor (E_F). Under a photoexcitation, the Fermi energy level of semiconductors is divided in two levels: the quasi-Fermi level of electrons for the conduction band (nE_F*) and the quasi-Fermi level of holes for the

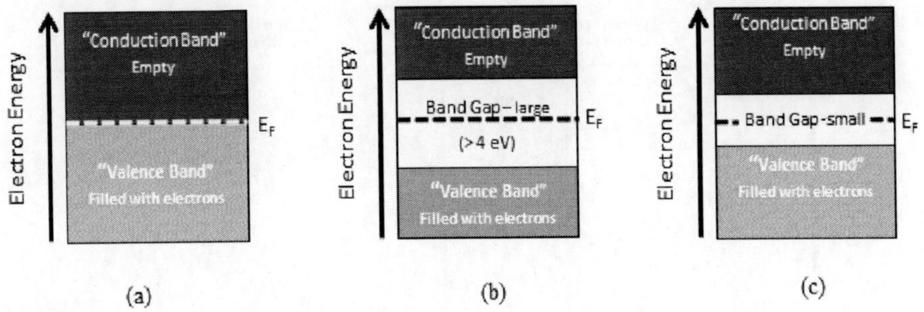

Figure 1. Schematic representation of the electronic bands for: (a) metal, (b) insulator and (c) semiconductor, at 0 K temperature.

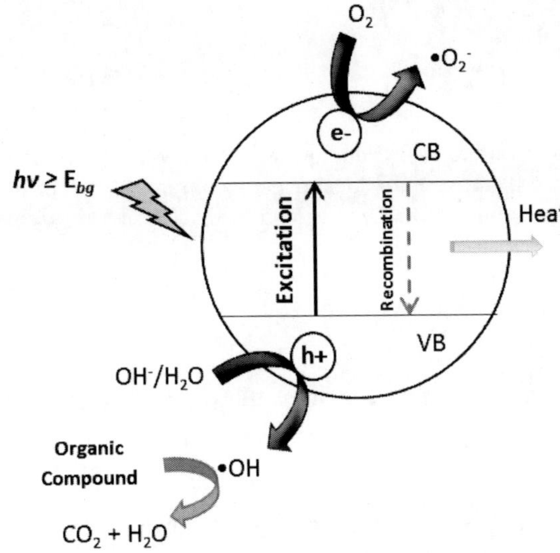

Figure 2. Schematic representation of a semiconductor in a photocatalys process. The figures shows the photoexcitation of electrons from the valence to the conduction band, producing photoexcited pairs of electrons in the conduction band and holes in the valence band. Hydroxyl radicals will be generated in the valence band and the process of recombination of the electron/hole pair occurs.

valence band (pE_F*) as shown in Figure 3. Under illumination, the quasi Fermi level (E_F*) depends on the local concentration of the minority carriers. Usually, the quasi-Fermi level of majority charge carriers remains close to the original Fermi level; but the quasi-Fermi level of minority charge carriers shifts away from the original Fermi level [16].

In the *n*-type extrinsic semiconductors, the electrons are the major charge carriers, the original Fermi level is shifted to near the conduction band and the quasi-Fermi level of the majority of the charge carries, is close to the original Fermi energy level.

On the other hand for *p*-type semiconductors, the holes are the major charge carriers, the original Fermi level is shifted to near the valence band and the quasi-Fermi level of the majority of the charge carries, is close to the original Fermi level.

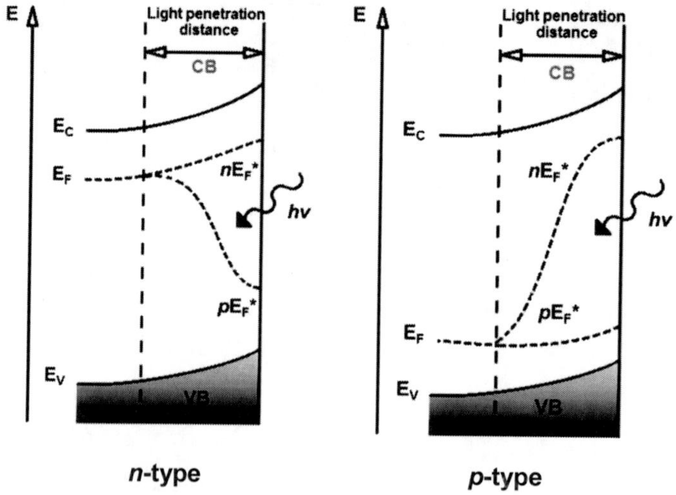

Figure 3. Fermi level of semiconductor (E_F) and quasi-Fermi levels of electrons (nE_F^*) and holes (pE_F^*) for *n*-type and *p*-type semiconductors under local photoexcitation. A shift of quasi-Fermi levels from original Fermi Level is greater for minority charge carries than for majority charge carries. Adapted from [16].

Interface Semiconductor/Electrolyte

When the semiconductor (SC)/electrolyte interface is formed, the Fermi level of the SC electrode E_F(sc) and the electrochemical potential of the electrolyte E_F(redox) are not in an state of equilibrium. In order to establish a balance at the interface, there should be a flow of charges from one phase to another one, followed by the "depletion of bands" of the semiconductor because the Fermi level will move down until it has reached the same energy level as the other side of the interface. This depleted region is called 'space charge region' (SCR) characterized by the accumulation of charges (electrons or holes) on the surface of the semiconductor (for *n*-type semiconductor, holes will be accumulated on the surface). At the equilibrium state, the Fermi energy level in a semiconductor, E_F(sc), and the electrochemical level in the electrolyte, E_F(redox), are equal.

In the dark, the Fermi energy level E_F(sc) of the *n*-type semiconductor electrode is located between $E_F(H_2O/H_2)$ and $E_F(O_2/H_2O)$ so that both electron transfer of the cathodic hydrogen reaction and hole transfer of the anodic oxygen reaction are thermodynamically impossible. Therefore, if light is absorbed, the Fermi level of the semiconductor electrode shifts by an energy equivalent to the photopotential (ΔE_{ph}), and it also splits into the two quasi-Fermi levels, $nE_F^*$and pE_F^*, for interfacial electrons and holes. The transfer reactions of both, electrons for the hydrogen reaction and holes for the oxygen reaction, becomes thermodynamically possible, provided that the nE_F^* is above the H_2O/H_2 electrochemical level of water reduction ($nE_F^* > E_F(H_2O/H_2)$) and the pE_F^* is below the

O_2/H_2O electrochemical level of water oxidation ($pE_F^* < E_F(O_2/H_2O)$) [16], as shown in Figure 4.

In the presence of a space charge layer, however, the photoexcited electrons and holes are separated; each move in opposite direction under the influence of an electric field in the space charge layer as shown in Figure 5. The separation of the charges will depend on the control of the Fermi levels and, therefore, the depletion of bands. The greater the bending of the bands the faster the separation of the electron/hole pair is and thus the recombination of charges is minimized.

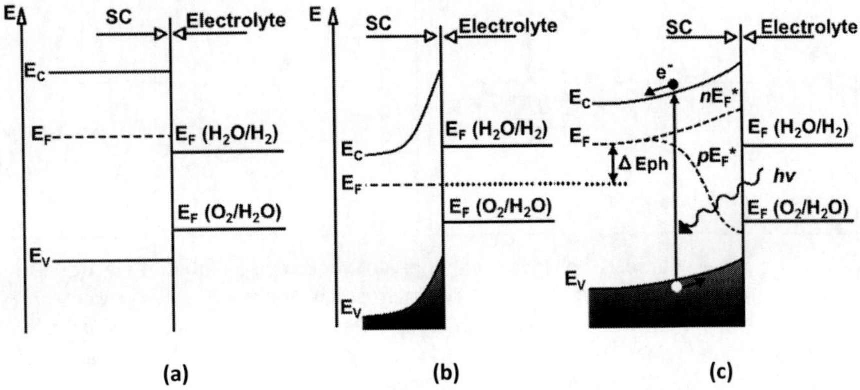

Figure 4. Energy Diagrams for an *n*-type semiconductor electrode: (a) in dark and before the equilibrium of energy level in semiconductor/electrolyte interface, (b) in the dark and after the equilibrium of energy level in semiconductor/electrolyte interface, (c) in a photoexcited state. Adapted from Ref. [16].

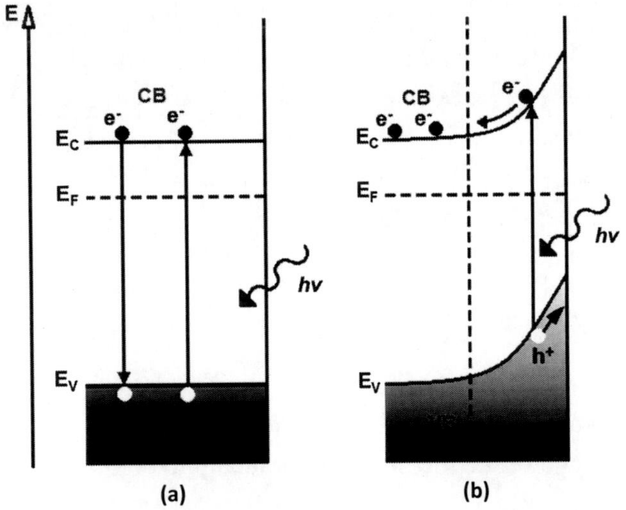

Figure 5. Energy diagrams for an *n*-type semiconductor electrode: (a) generation and recombination of photoexcited electron/hole pairs in a flat band, (b) generation and separation of photoexcited electron/hole pairs in space charge layer. Adapted from Ref. [16].

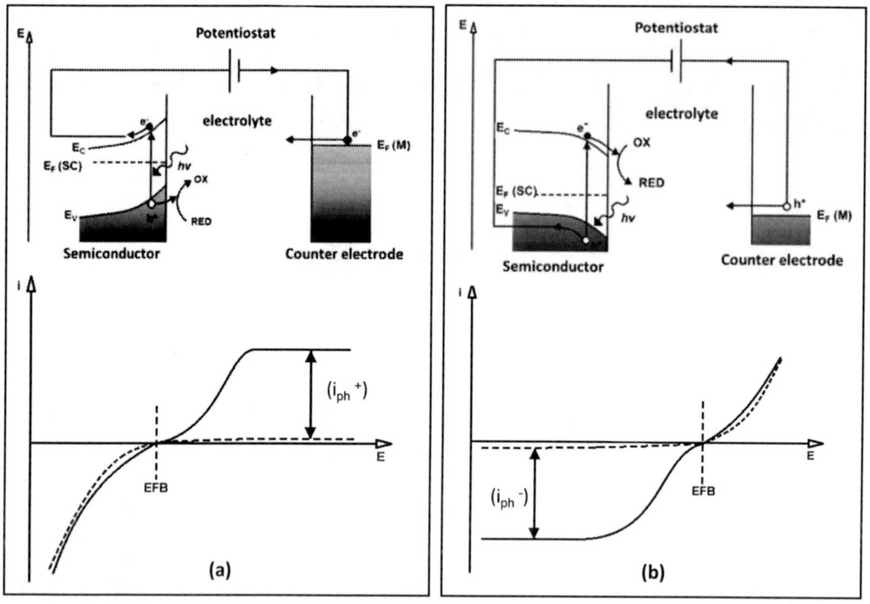

Figure 6. Schematic representation of photocurrent at photoexcited semiconductor electrodes: (a) photoexcited reaction of anodic hole transfer at *n*-type semiconductor and its polarization curves in the dark (dashed curve) and under illumination (solid curve), (b) photoexcited reaction of cathodic electron transfer at *p*-type semiconductor and its polarization curves in the dark (dashed curve) and under illumination (solid curve). Adapted from Ref. [16].

One way to control the depletion band is by applying a potential. However for a given semiconductor and electrolyte, there is a unique condition where the potential drop between the surface and the bulk of solution is zero. In this situation there is not space charge layer and the photoexcited electrons and holes are rapidly recombined; this is the flat band potential, EFB. From the point of view of the photoelectrochemical process it is desired to have large and flat band potential. The flat band potential of the electrodes can be estimated by measuring the potential at which the sign changes, since at this point the photopotential corresponds to the flat band potential. Applying a potential greater than the flat band potential in an *n*-type SC, the electrons will be drained and the holes will enrich the surface of the semiconductor (Figure 6).

Photocurrent

In order to separate the photoexcited electrons from the holes, a combination of the electrochemical and photocatalytic processes, called photoelectrocatalysis, which creates an electrical potential gradient can be implemented. The system is composed by one semiconductor (working electrode) connected to a counter electrode, as illustrate in Figure 6. Under illumination and at an appreciable electric field the photoexcited electrons or holes migrate within space charge layers towards the electrode surface and participate in the charge transfer reactions (cathodic or anodic). These reactions produce a photocurrent (i_{ph}).

The photocurrent of electrons and holes is significant for minority charge carriers whose concentration increases greatly by photoexcitation, compared with the concentration of the minority charge carriers in the dark. Whereas, for the majority of charge carriers, the same photocurrent is insignificant. When no electric field is present to separate the photogenerated charges, no photocurrent occurs and the flat band potential can be obtained. Then, the flat band potential may be regarded as the potential for the onset for the photocurrent.

In a photoelectrocatalytic system, the greater the lamp intensity, the greater the potential applied for an effective separation of electron-hole pairs. If the potential is high enough for an effective charge separation, the interface reaction is the rate-limiting step and, in this case, the rate of degradation is proportional to the light intensity [17].

Band Gap Energy

When a semiconductor is illuminated, the electrons absorb the photons and undergo transition to larger level of energy. If the incident light energy (hv) is greater than the band gap energy (E_{bg}) of the semiconductor ($hv > E_{bg}$), then the electrons of the valence band are excited towards the conduction band. The optical absorption from this transition is known as *fundamental absorption*, and the minimum energy in which this absorption occurs is called the *fundamental absorption limit*. The fundamental absorption limit energy is determined by the band gap energy of the semiconductor [18].

Some methods can be used to determine the band gap energy (E_{bg}), such as differential capacitance [19] and optical [20]. Among the optical methods, the UV-Vis diffuse reflectance spectroscopy is the most commonly used, since it does not depend on temperature variations or uncertainties due to the surface state, such as electrical conductivity, Hall constant and photoconductivity techniques. The Hall constant is defined as the ratio of the induced electric field to the product of the current density and the applied magnetic field for a known semiconductor thickness. It is a characteristic property of each material and depends on the type, number, and properties of the charge carriers (electrons or holes) that create the current. Several other methods to calculate E_{bg} have been developed and applied over the years, for example the atomic absorption spectra and diffuse reflectance, E_{bg} can be estimated from the absorbance *vs.* energy plot, from the Kubelka-Munk function (F(R)) based on the following equation:

$$\boldsymbol{F}\,(\boldsymbol{R}) = \frac{(\boldsymbol{1}-\boldsymbol{R})^2}{\boldsymbol{2R}} \qquad \text{(Eq.1)}$$

where ($F(R)$) is proportional to α, absorption coefficient, and R is the absolute reflectance for a given radiation energy. Taking into account the predominant type of electron transition in the absorption region close to band gap (n) energy, equation (1) can be rewritten as:

$$(\propto h\nu)^n = A(h\nu - E_{bg}) \qquad \text{(Eq.2)}$$

where h is the Planck constant, ν is the incident light frequency, A is the Kubelka-Munk constant and n reflects the type of transition. For direct transitions, n may assume values of 2 for allowed transitions or 2/3 for forbidden transitions whereas for indirect transitions, n may be considered as ½ for allowed transitions or 1/3 for forbidden transitions [20-22]. Because it is difficult to establish the type of transition experimentally, the literature reports controversial values for the band gap energy of TiO_2, which may differ by up to 0.5 V for P25 powder for example, depending on the extrapolation method used [20].

SELF-ORGANIZED TiO_2 NANOTUBULAR AS PHOTOCATALYST

The efficiency of the photocatalysis process depends on the photoactivity of the semiconductor that is a function of its surface area, crystal morphology, adsorption capacity and crystalline microstructure. The latter should be able to optimize charge separation and minimize the recombination of the electron/hole pair [23]. Other characteristics of the semiconductor are also required, such as the possibility to combine photocatalytic and electrochemical processes, reducing the recombination of photogenerated charges, chemical and physical stability, and excitation on visible light.

Among several semiconductors, TiO_2 is one of the most investigated and promising materials, because it is abundant, non-toxic, has low cost and physical and chemical stability in a wide range of pHs [24].

TiO_2 nanotubes of different geometrical shapes and microstructures have been manufactured by various research groups using techniques such as templates [25-27], sol-gel synthesis [28], seeded growth [29] and hydrothermal [30, 31]. While many of these processes are complicated due to the use of templates or the chemical technique involved, many studies [32-35] have demonstrated that self-organized titanium oxide nanotube arrays can be fabricated using a simple anodization technique. The anodization process is simple and fast. Besides, it allows growing self-organized nanoestructures from their metallic substrate that can be used directly as back-contacted oxide electrode.

The nanotubes provide a more optimized geometry for photocatalytic applications with significantly shorter carrier-diffusion paths in the walls (10-15 nm), including lower trapping and recombination kinetics of light-generated electron-hole pairs in comparison to nanoparticulate layers, where these pairs have to travel between single nanoparticles [36]. Also, the tubular geometry provides a short (straight) diffusion path for organic molecules in the solution to reach the active surface area whereas the porous structure resulting from nanopowder fixed in a film contains a broad spectrum of diffusion paths (including very long diffusion lengths). Cardoso et al. [37] investigated the structural effects of nanotubes, nanowires and nanoporous Ti/TiO_2 electrodes on photoelectrocatalytic oxidation of 4,4-oxydianiline and shown that the complete mineralization of 5×10^{-6} mol L^{-1} solution was reached in 120 minutes using Ti/TiO_2 nanotubes, while for nanowires and nanoporous the mineralization was around 70% and 50%, respectively, for same time of treatment.

A Brief History

The first reports on self-ordered TiO_2 nanotube formation date back to 1984 by Assefpour-Dezfuly [38] and later by Zwilling et al. in 1999 [39] who discussed the importance of the presence of dissolved fluoride in solution for the growth of a nanoporous oxide layer in Ti substrate by anodizing process. Nevertheless a significant amount of studies have been reported on the anodic formation of TiO_2 nanotubular layer, aiming to investigate the influence of anodization conditions (pH, aqueous or organic electrolytes, applied potential, anodization time, fluoride concentration and the water content in the organic electrolyte) and the mechanism of the growth of nanotubes oxides. These reports can be organized in four generations of formation of TiO_2 nanotubes described below [35]:

1. In the first generation, TiO_2 nanotubes were formed in fluoride containing aqueous electrolyte. The use of HF containing solution led to tubes length limited to 600 nm due to high chemical dissolution of TiO_2 in the acid media, with rough and ribbed walls [40-45], as shown in Figure 7;
2. By reducing the acidity of the medium and using fluoride salts instead of HF in the second generation, the rate dissolution of TiO_2 was reduced and self-organized nanotube TiO_2 layers with thickness around 5 μm with ripples on the side of the walls could be grown [46-51]. Different salts (NH_4F, NaF, CsF and KF) were investigated [48] and the importance of selecting an appropriated cation was detected. While electrolytes containing NH_4F and NaF produced a nanotubular oxide layer, the presence of CsF and KF caused more precipitation and non-uniform TiO_2 dissolution phenomena;
3. The third generation of nanotubes were grown in organic fluoride electrolytes (mostly glycerol, ethylene glycol, DMSO – dimethyl sulfoxide, ionic liquids) with small amount of water (0.1 to 5 wt.%) [52-64], as shown in Figure 7. In the

studies using ethylene glycol, NH_4F and small amount of water, self-aligned hexagonally closed-packed titania nanotube arrays of over 1000 μm in length with smooth walls and faster oxide growth rates were obtained [58, 59]. While the hydroxyl incorporation negatively affects the ion transportation through the oxide, the incorporation of small amount of hydroxyl ions and oxygen leads to sub-stoichiometry (oxygen vacancies) of the oxide and enhances the ionic transport through the barriers layer. Besides, the size and viscosity of the cation in the fluorine salt affects the length and growth rate of the nanotubes. Due to the possibly of an inhibitory steric effect, the presence of quaternary ammonium cations maintained the thickness of the oxide at the tube base (barrier oxide) below a critical value, facilitating ionic migration [35];

4. In the fourth generation, hexagonal and high degree of self-ordered nanostructured were achieved using organic electrolytes containing fluoride salts with small amount of water applying two steps of anodization [65] or fine tuning the anodization potential [66]. In such conditions, it was allowed to establish smooth walled tubes layers of several 100 μm thickness and diameter range from 10 to 800 nm [34, 65]. Besides, a porous layer is observed on the top of the nanotubes, due to limited dissolution of the oxide in the organic solution. This layer is a remnant of the oxide formed in previous stage of the anodization and the nanotubes being developed underneath [35].

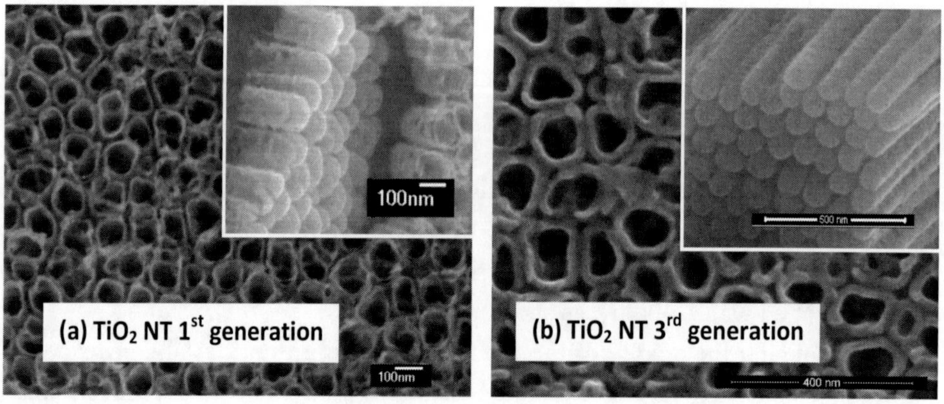

Figure 7. Images of nanotubes obtained in (a) 1st generation grown in HF containing aqueous electrolytes and (b) 3rd generation grown in fluoride containing organic electrolytes.

The Growth Mechanism of Self-Organized TiO_2 Nanotube

The models of TiO_2 nanotubes growth (Figure 8) are based on the initial formation of a barrier layer (metal/oxide interface) at the surface of the metal due to interaction of Ti^{4+}, resulting for metal oxidation (Eq. 4), with OH^- and O^{2-} species provided by water decomposition at the oxide/electrolyte interface (Eq. 3) [67]. This results in the formation of hydrated oxide (Eq. 5) and in formation of the oxide (Eq. 6) ($\Delta G°_{298}$ = - 821.3 KJ mol^{-1}) [68]. More oxide is produced when the hydrated anodic layer releases water by condensation reaction (Eq. 7):

$$H_2O(aq) \leftrightarrow OH^-(aq) + H^+(aq) \leftrightarrow O_2\text{-}(ox) + 2H^+(aq) \qquad \text{(Eq. 3)}$$

$$Ti \rightarrow Ti^{4+} + 4e- \qquad E°red = -0.88V \text{ vs. NHE } (1 \text{ mol dm}^{-3} \text{ HCl}) \qquad \text{(Eq. 4)}$$

$$Ti^{4+} + 4OH^- \rightarrow Ti(OH)_4 \qquad \text{(Eq. 5)}$$

$$Ti^{4+} + 2O^{2-} \rightarrow TiO_2 \qquad \text{(Eq. 6)}$$

$$Ti(OH)_4 \rightarrow TiO_2 + 2H_2O \qquad \text{(Eq. 7)}$$

These equations represents the possible oxidation reactions under field-assisted, since the potential controls the migration of Ti^{4+} species (in direction to electrolyte) and O^{2-} and OH^- species (in direction to metal) within the metal/oxide interface. At the cathode, the hydrogen evolution takes place according to Eq. 8.

$$4H^+ + 4e^- \rightarrow 2H_2 \quad E°_{red} = 0 \text{ V } \textit{vs}. \text{ NHE} \qquad \text{(Eq. 8)}$$

The overall reaction for oxide formation can be represented by the following reaction:

$$Ti + 2H_2O \rightarrow TiO_2 + 2H_2 \qquad \text{(Eq. 9)}$$

At the same time, the fluoride ions present in the electrolyte promote the chemical dissolution of the hydrated layer (Eq. 10), oxide (Eq. 11) or react with Ti^{4+} (Eq. 12). Due to applied electric field and action of fluoride ions the Ti-O bond is weaken promoting dissolution of the cation Ti^{4+}. Fluoro-complexes are formed, the most stable being TiF_6^{2-} ($\Delta G°_{298}$ = - 2118.4 KJ mol $^{-1}$) [68]. It is noteworthy that the fluoride ions also migrate towards the anode by the applied potential.

$$TiO_2 + 6F^- + 4H^+ \rightarrow TiF_6^{2-} + H_2O \qquad \text{(Eq. 10)}$$

$$Ti(OH)_4 + 6F^- \rightarrow TiF_6^{2-} + 4OH^- \qquad \text{(Eq. 11)}$$

$$Ti^{4+} + 6F^{-} \rightarrow TiF_6^{2-} \quad \text{(Eq. 12)}$$

A determining factor for anodic titanium oxide nanostructure production is to stablish an oxide formation/dissolution equilibrium. In order for this to occur, the steady-state oxide layer must be thin enough. Thus, a moderately degree of solubility of the oxide in the electrolyte is needed to provide growth of porous/nanotubular oxide films [34, 69]. When the anodic film is insoluble the oxide layer is thicker and the process is self-limiting (the field drops and becomes too weak to assist the ion migration). However, if the dissolution rate of the oxide layer is too high, no steady-state oxide layer is formed, and a complete oxide dissolution occurs (electropolishing of Ti) [70].

After some irregular nanoscale pores that penetrate the initial compact oxide are initially formed (the current increases as the reactive area increases), the pores spread over the surface and a competition between the rate of oxide formation at the metal/oxide interface and the rate of dissolution at the oxide/electrolyte interface can be observed. When these rates are equal the steady state is reached [32]. The thickness of the tubular structure ceases to increase when the chemical dissolution rate of the oxide at the mouth of the tube (top surface) becomes equal the rate of inside movement of the oxide layer at the base of the tube [32]. Under steady-current conditions, the magnitude of the current is significantly higher than when a compact oxide (at the same voltage) is formed [34]. However, this explanation is compatible for aqueous media, where the nanotube thickness is constant over time. Considering that in organic media the maximum thickness is reached after a long time of anodization, others factors need to be taken into account to explain the growth of the nanotubes.

The "field assisted flow" mechanism has been suggested for some studies [71-75] in order to explain the initial formation of porosity and growth of the nanotubes in organic electrolytes. In these cases, the effect of dissolution is considered secondary. The nanotubes are formed by a mechanical stress induced by volume expansion and electrostriction. The development and growth of pores in the oxide layer is driven by the oxide displacement from the barrier layer towards the pore wall in order to relieve the stress created by volume expansion when converting metal to oxide. Additionally it may be enhanced by the voltage induced electrostrictive forces [39, 76]. Similar mechanism was proposed for anodized aluminum oxide (AAO) by Skeldon et al. [71] and Garcia-Vergara et al. [72] who carried out experiments with traces of W inserted in an Al substrate.

Berger et al. [74, 75] verified that the film initially evolves as a barrier layer with a volume expansion of 2.4, known as the Pilling- Pilling-Bedworth ratio (RBR). At the point of pore initiation the volume expansion decreases to 1.7-1.9 and during the growth of the nanotube the volume expansion increases again to 2.7-3.1. The expansion factor of TiO_2 nanotubes is affected by the water contained in the electrolyte and the applied potential [77]. At higher water content there is a decrease in the expansion factor due to the higher dissolution of Ti and TiO_2. In this case, lower growth efficiency is observed. At lower water content there is an increase of expansion factor due to incorporation of impurities and the dissolution is reduced, leading to higher growth efficiency.

The transition from pores to tubes is also favored by an increase in the concentration of fluoride ions. Although fluorine is present in the oxide film, it is not homogeneously distributed [78]. A fluoride rich layer is formed at the metal/oxide interface as consequence of the fast migration rate of fluoride ions in relation to oxygen ions, about twice higher [79]. The existence of this fluoride-rich layer was shown by Albu et al. using XPS and SEM analysis.[80] Berger et al. [81] also showed evidence of the presence of a fluoride layer between the nanotubes walls using high resolution Scanning Auger Electron Spectroscopy (AES). Fluoride accumulated at the metal/oxide interface moving upwards with tube growth as an outer cladding (decorating the hexagonal cell walls) [34, 81] Taking in account that Ti-fluoride compounds are easily dissolved in aqueous media, the fluoride rich regions between the TiO_2 tubes (or cells) are dissolved. As the attack at the top of the structure is more resistant to chemical attack relative to the lower part of the tube, a tubular morphology is more apparent in the top, whereas in the bottom generally a porous and hexagonal arrangement is observed [68]. According Wei et al. [82], if the amount of water in solution is drastically reduced, the etching of cell boundaries is difficulty and the porous layer can be observed. Then the fluoride ions have two roles: maintain a thinner bottom oxide layer by chemical etching of the oxide layer and immediate complexation (solvatization) of Ti^{4+} species arriving at the oxide–electrolyte interface [70].

The efficiency of nanotube growth can also be affected by the oxygen evolution reaction at the anode (Eq.13). In aqueous electrolyte, there is intense oxygen evolution which affects the formation of nanotubes, producing oxide rings and ribs on the wall of the nanotubes [35].

$$2H_2O \rightarrow O_2 + 4e^- + 4H^+ \qquad \text{(Eq. 13)}$$

In a photoelectrocatalytic system, the greater the lamp intensity, the greater the potential applied for an effective separation of electron-hole pairs. If the potential is high enough for an effective charge separation, the interface reaction is the rate-limiting step and, in this case, the rate of degradation is proportional to the light intensity [17]. A theoretical model accounting for the ionic and electronic currents during anodization and supporting the role of oxygen in the growth of the nanotubes (identified as "oxygen

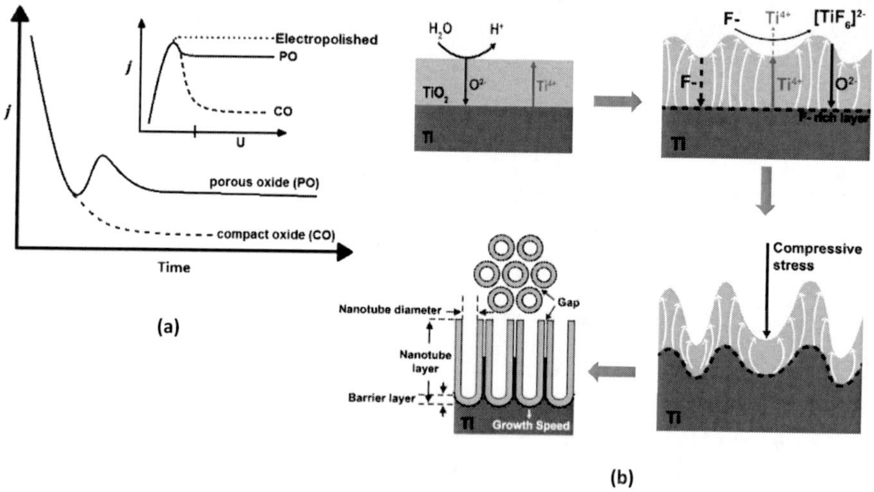

Figure 8. a) Typical current–time (j–t) characteristics after a voltage step in the absence (- - -) and presence (___) of fluoride ions in the electrolyte. The inset shows typical linear sweep voltammograms (j–U curves) for different fluoride concentrations resulting in either electropolished metal (high fluoride concentration), compact oxide (very low fluoride concentration), or tube formation (intermediate fluoride concentration). Adapted from Ref. [34]. b) Illustration of step to formation of nanotubes. Adapted from Ref. [70].

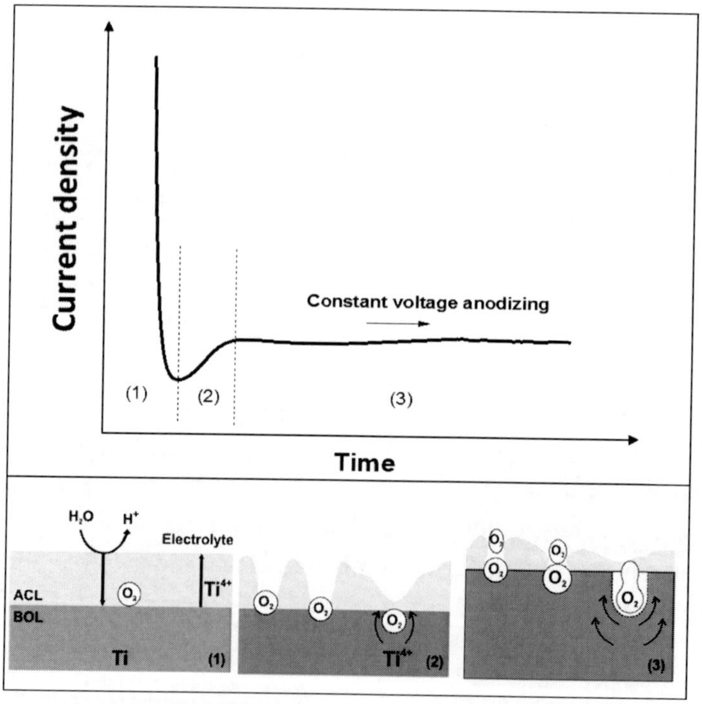

Figure 9: Current density – time curve of potenciostatic anodization and schematic picture of pore formation of nanotube based the oxygen bubble mould. Adapted from Ref. [88].

bubble mould") has been explored [83-87]. It is suggested that the compressive stress in the oxide may result from the expansion of oxygen bubbles. One mechanism to explain the evolution of the TiO_2 nanotubes morphology was proposed by Lu et al. [88] based the oxygen bubble mould [84, 89, 90] in which the stress-driven ion migration and the oxide flow are caused by continuous oxygen evolution from the pore bottom. In the first stage, the oxygen bubbles are formed at the interface between the anion contaminated layers (ACL), F^- and OH^-, and a barrier oxide layer (BOL) (Figure 9b1). Due to expansion of the oxygen bubbles, oxide in BOL/Ti interface will bend and the deformation of BOL will produce hemispherical bottoms in the TiO_2 nanotubes. Due to the high pressure and obstruction of the oxygen bubbles, the oxide flow tends to grow around the oxygen bubbles. So an embryo pore with hemispherical bottom is formed in stage 2 of Figure 9b2. Due to a thin ACL, this layer is broken by oxygen bubbles and the electrolyte penetrates into the embryo pores and the distance between the electrolyte and Ti substrate becomes shorter. When most of the area of the oxygen bubbles is exposed to the electrolyte, the distance between the electrolyte and Ti substrate no longer changes, thus the total anodizing current reaches the equilibrium state (stage 3 of Figure 9a). Due to the presence of ACL, oxygen bubbles cannot be released out the porous freely. Then, the newly-formed TiO_2 near the electrolyte interface moves upward with a channel wall. In this situation, the shape of the bottom will be affected by the oxygen bubbles mould because the newly-formed TiO_2 expands from the bottom to form the nanotube walls, which is consistent with plastic deformation flow of oxide theory [72, 69, 85, 91]. The porous structure of TiO_2 nanotubes can be changed if the gas evolution rate is accelerated, resulting in different morphologies [88].

The Factors That Affects the Growth and Morphology of Nanotubes

The properties of the electrolyte that affect the growth and morphology of the nanotubes are: pH, fluoride concentration, age, water content, applied potential (ΔU), current density, anodization time and the temperature. They all play an important role for improving the tube layer thickness.

At low pH values, longer tubes could be grown [48, 49] because higher oxide dissolution rate occurs in an acidic environment. This is one of the reasons why the dissolution of anodic oxides is lowest using fluoride salts instead of hydrofluoric acid. The concentration of fluoride ions needs to be maintained in a sufficient level to keep a low dissolution rate and to ensure the nanotube growth. Normally, F^- concentrations should be in the range of 0.3 – 0.5 wt.% in order to obtain self-organized TiO_2 nanotubes [92-94].

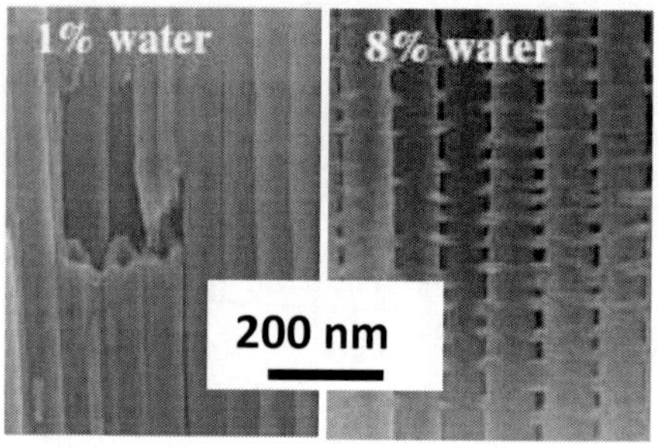

Figure 10. Smooth to rib -tube transition induced by variation of H_2O content in the electrolyte. Adapted from Ref. [96].

In organic electrolytes, the ageing of the solution helps to obtain highly defined nanotubes. The main reasons are the combined increase of the TiF^{+6} content built-up during previous anodization cycles (reduced TiO_2 dissolution) and a higher water content, which reaches saturation due to moisture absorption (for example, ethylene glycol is relatively hygroscopic and tends to take up considerable amount of water from the air) [52, 95].

Water content in the electrolyte affects the growth rate (of oxide formation at the bottom of the tube) and the etching speed (chemical dissolution rate) of the nanotubes. Smooth tube walls are obtained for low water content, whereas side wall ripples are formed at higher water content (see Figure 10) [54, 93, 96, 97]. If the nanotube growth is faster than the etching speed, smooth tube walls are obtained, whereas in the opposite case ribbed tube walls are obtained [98].

While anodization time and etching rate define the tube length, the diameter of the nanotubes is linearly controlled by the applied voltage [93, 99-101]. The anodization time is related to parameters such as electrolyte, potential and concentration. In organic electrolytes, the kinetics are slow and the nanotubes layer grow slowly and the dissolution process is less significant in comparison to aqueous solutions but the diameter of the nanotube is not affected [34].

In long anodization experiments using organic electrolytes, the formation of needle- or grass- like morphologies (*nanograss*) occurs due to the etching of the top of nanotubes. Their walls became very thin to support their own weight, collapsing and bending the tops of the tubes. There are various ways to protect the top of the nanotubes against etching; one is the addition lactic acid during the anodization. The chelating agent is adsorbed on the surface of TiO_2, reducing the dissolution of Ti^{+4} ions by the F^- ions [102]. In addition, there are procedures to remove the *nanograss* after anodization, for example by ultrasonic cleaning combined with acetone or bath in ethanol for 24 hours [103].

The temperature of the electrolyte also affects the dissolution rate of the nanotubes layer, the higher the temperature the faster the charge transfer of these processes [103].

The growth mechanism of the nanotube arrays occur through the competition between field-assisted oxide dissolution and the diffusion rate of the ionic species.

Room temperature conditions are ideal for growing nanotubes in aqueous or organic solutions. However, whereas in organic electrolytes the temperature range most favorable is up to 40ºC. In aqueous solution, the growth of nanotubes is inhibited at lower temperatures. In organic electrolytes, the temperature has a direct influence on the diameter of the nanotubes unlike what occurs in aqueous solutions, however it does not influence on the thickness of the oxide layer in both electrolytes [35, 103, 104].

The TiO_2 nanotubes layer formed at room temperature have large surface area and the shape of the nanotubes are more uniform when compared with those nanotubes produced at lower temperature because the defects in their structures are not shown. Above room temperature, the wall of the nanotubes increases and the diameter decreases due to the higher transport rate of ions as the temperature increases.

Up to 80ºC, there is no formation of nanotubular structures and the anodic oxide consist of an irregular porous layer [103]. At such high temperatures, the chemical etching exceeds the anodic oxidation and the balance required for self-ordering is diminished.

The applied potential determines the electric field strength across oxide, thus affecting the migration of ions and consequently the nanotube diameter. The diameter of the nanotubes is directly related to the number of porous formed during in the early stage of the process. When high potential is employed the pits density etched in the substrate surface increases. The voltage created by volume expansion when converting metal with oxides is also additionally enhanced by potential induced electrostrictive forces [76, 105].

The length of nanotubes is also influenced by the applied potential. When the anodization voltage is very high, the rate of electric field dissolution at the barrier layer inside the nanotubes is higher and longer nanotubes are produced. When the anodization potential is low, less electric field dissolution occurs, forming TiO_2 with smaller diameter pores.

High order degree of nanotubes is obtained when the growth is fast [106]. At high concentrations of fluoride or at higher voltages (higher than the optimized value) the nanotube morphology is lost and the process becomes similar to an electro polishing with a sponge formation [107].

MODIFICATION OF TiO_2 NANOTUBES

In order to enhance the interfacial charge transfer reactions and to reduce of the band gap energy of TiO_2, modifications of TiO_2 nanotubes have been carried out by heat treatment, by introducing other elements (doping) or by tube wall decoration. The aim of these modifications is to make the semiconductor more suitable for various applications, improving its electric conductivity, and its optical and chemical properties.

Considering that the band gap of 3.2 eV of TiO_2 nanotubes allows only the absorption of UV light, which means that only about 7% of the solar spectrum can be utilized, the decoration and doping techniques have been thoroughly investigated for photocatalytic applications, in order to enable photo response in visible light.

Doping

The typical methods to prepare doped TiO_2 nanotubes are: a) growing TiO_2 nanotubes in a solution or melt of the doping species; b) thermal treatments or synthesis in gas atmosphere containing the doping species; c) co-sputtering or sputtering in an atmosphere of doping species; d) high energy implantation; and e) growing TiO_2 structures on a suitable alloy by electrochemical oxidation.

The first report on a successful band-structure modification was reported by Asahi et al. in 2001 [108] using nitrogen as a doping agent. In the following years others elements were investigated, such as carbon, sulfur and a variety of transition metals, including V, Cr, Mn, Nb and Fe [34]. Studies with TiO_2 doped with Fe^{+3}, Mo^{+5}, Ru^{+3} and V^{+4} have demonstrated that the TiO_2 reactivity depends on the concentration of the dopant, energy level of the dopant in TiO_2 lattice, distribution of dopant, donor-receptor relation/concentration and the intensity of light [7, 32]. In the case of titanium alloys, small amounts of the alloying element can drastically affect the properties of TiO_2 while the unique tubular morphology is obtained. For example, TiW (0.2 atom %) alloy showed strong improved photocatalytic properties [109].

In order to obtain an ordered and homogeneous array of nanotubes over a large surface area, the titanium alloy should be single-phase, avoiding different etching/anodization rates on different phases. Besides, the growth is facilitated if the metals involved are valve metals and the morphology follows the specific oxide morphology of the major element in the alloy [34, 110].

The growth principles on different metals and alloys seem to follow the same mechanism for TiO_2 nanotubes. The diameter of the tube is a function of the applied potential and the length of the nanotubes depends of the chemical resistance of the oxide against fluoride. The sidewall ripple formation is also due to the water present in electrolyte which plays an important role for providing the oxygen source for the nanotubes to grow and splits pores into tubes.

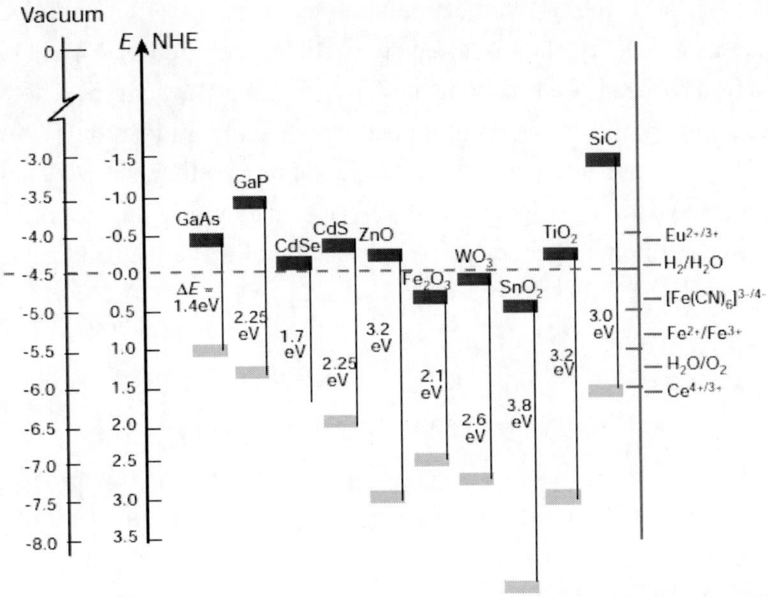

Figure 11. Band gap comparative diagram of some semiconductors. Reproduced from Ref. [111] with permission from Nature.

In general, the doping technique has been efficient in photocatalytic process because slows down the recombination rate of the photo generated charges by transferring the charges between the oxides. However, the selection of the dopant element is the key to a successful improvement of the TiO_2 properties. The position of the valence band (VB) and the conduction band (CB) of the dopant element in relation to the position of the TiO_2 bands, needs to be taken into account. Figure 11 presents a comparative band gap diagram of some semiconductors (dopant elements) in relation to TiO_2 [111].

Decoration

Decoration of TiO_2 nanotubes can be carried out using different methods, such as precipitation reaction of $Me(OH)_x$ followed by a suitable thermal treatment [112], physical dispersion of metal nanoparticles [113, 114], electrodeposition [115], UHV evaporation or chemical reduction [116, 117]. Among these methods, the electrodeposition technique stands out, as it allows for full control of the charge transferred and the ratio between the nucleation and the growth rates, thereby controlling the sizes of the nanoparticles.

The modified TiO_2 nanotubes with transition metals have revealed enhancement of the photocatalytic activity when compared to pure TiO_2. Decoration with electrodeposited Pt substantially reduced the TiO_2 band gap energy and shifted the absorption peak to the visible light region [118]. An enhancement of the photocatalytic activity has been observed in photodegradation experiments of an acid dye, when TiO_2 nanotubes are decorated by photoreduced Ag particles and sputtered Au [119]. Cu and Fe elements have also been used to decorate TiO_2 for obtaining *p*-type photocathodes for water photo-splitting in *pn*-junctions, under visible light irradiation [120, 121]. RuO_2 loaded TiO_2 composite nanotube arrays using a dipping method showed an enhancement of the photocatalytic properties over pure TiO_2 nanotubes due to reduction of the recombination rate of photo-generated electrons and holes and the band gap energy [122].

Heat Treatment and Crystallinity of Anodized Nanotubular Oxides

The nanotubular oxide layer grown on pure metals and alloys by the anodization process of an amorphous area can be transformed into a crystalline structure by an adequate annealing treatment.

TiO_2 has different polymorphs: anatase (tetragonal), brookite (orthorhombic) and rutile (tetragonal). Anatase is the most photoactive phase, whereas brookite is the rarest form of mineral and it is not easily obtained by synthesis and rutile is the lowest free energy and stable phase, but it is not photoactive [123]. Anatase and brookite are metastable polymorphs and can be transformed into rutile phase over a range of temperatures from 400 to 1200°C, depending on several factors such as the presence of impurities, particle size, the subtract where the oxide is supported and the nature of the annealing atmosphere [41, 123-125].

Varghese et al. in 2003 reported the first study on the effect of annealing on TiO_2 nanotubes, using Glancing Angle X-Ray Diffraction (GAXRD) [40]. The presence of anatase (A) crystals begin to appear at 280°C, whereas rutile (R) crystals appeared at 480°C, so that a mixture of (A) and (R) was detected.

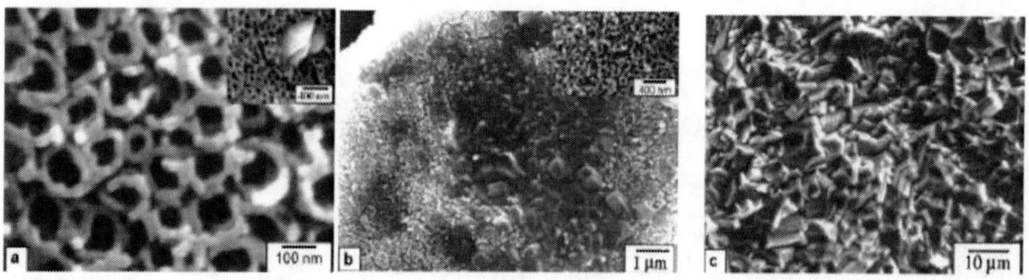

Figure 12. Effect of heat treatment in oxygen ambient on surface morphology at different temperatures: (a) 580°C, the protrusions emerging from the substrate is shown in the inset, (b) 680°C, the distorted tubular structure is show in the inset, and (c) 880°C. Reprinted from reference [41] with permission from Materials Research Society.

When the temperature increased, only the rutile phase is observed and the nanotubes collapse as shown in Figure 12.c. In this case any benefits associated with an ordered nanostructure are lost. The main reason of collapse of nanotubes is thermal oxidation and protrusions emerging from the metal at 550-580°C. Other studies have suggested that the formation of rutile phase at the metal/metal oxide interface in TiO_2 nanotubes is associated with oxygen vacancies generated at interface due to reaction between Ti and tiO_2 on base of nanotubes [126].

In the case of mixed oxide nanotubes, the anatase-rutile transition takes place at a higher temperature than in pure TiO_2. In the amorphous mixed TiO_2-Nb_2O_5 structure, an additional peak attributed to the rutile phase appears only after annealing at 650°C, but the major TiO_2 phase remains anatase [127]. Similar behavior can be observed for the $Ti_{35}Nb$ and $Ti_{35}Nb_4Sn$ alloys [128]. Under annealing heat treatment, the crystallization of TiO_2 formed on CP-Ti (commercially pure titanium) occurred at 225°C, while the TiO_2 formed on $Ti_{35}Nb$ and $Ti_{35}Nb_4Sn$ substrates show crystallization at temperatures of 295°C and 392°C, respectively. The morphology of the layer on $Ti_{35}Nb_4Sn$ at 925°C remained similar to the one before the annealing heat treatment. In this situation the rutile volume fraction was around 60%. According to the XRD patterns shown in **Figure** 13, the rutile structure is predominantly observed in the nanotube layers formed on CP-Ti, while the anatase remained stable in the layers formed on $Ti_{35}N$ and $Ti_{35}Nb_4Sn$.

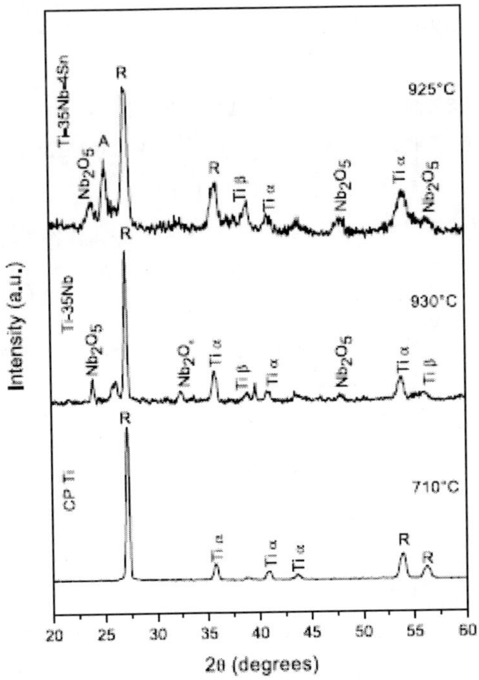

Figure 13. XRD patterns of nanotube layers after annealing heat treatment at higher temperatures on CP-Ti, $Ti_{35}N$ and $Ti_{35}Nb_4Sn$ alloys. Reproduced from Ref. [128] with permission from Applied Surface Science.

APPLICATION OF NANOSTRUCTURED OXIDE

Due to the great biocompatibility, electronic and ionic activity, high surface area and self-organized structure properties of the nanotubular oxides grown on Ti and Ti alloys, they find applications in a large number of areas that includes photocatalysis, solar cells and biomedical sciences.

Photocatalysis

TiO_2 is the most attractive photocatalytic semiconductor for the degradation of organic compounds. The suitable band-edge positions in relation to water is the reason for its high photocatalytic activity, promoting the generation of highly reactive species at valence band ($H_2O + h^+ \rightarrow HO^\bullet$) and at conduction band ($O_2 + e^- \rightarrow O_2^-$) or via direct oxidation of adsorbed species at the valence band. The treatment of different organic compounds contained in wastewaters (textile dyes, pharmaceuticals, pesticides, hormones, to name a few) via photocatalysis using TiO_2 nanotubes was investigated. The performance of the nanotubes was superior to commercial TiO_2 particles and nanoporous materials [37, 129-132]. Better results were obtained with nanotubes with larger wall thicknesses and longer lengths.

Although TiO_2 has high photocatalytic activity, its bandgap around 3.2 eV (anatase phase) is an obstacle when used in visible light and it presents high recombination rate of the photogenerated charges (electron/hole). In order to be applied in visible-light-driven processes and to reduce the recombination rates of the charges, TiO_2 nanotubes should be modified by decorating the surface [133, 134] or by doping processes to form heterojunctions, as mentioned in section 3.3. Many metals and metal oxides are used to modify the TiO_2, among of them, W and Mo, improving the photocatalytic activity and reducing the rates of recombination [109, 135]. The N-doping is the most efficient measure to stimulate a visible photocurrent response but it is questionable if this efficiency is a function of the valence band dominated or attributed to O_2^- generated species at the conduction band [34, 136]. In the case of the photocatalytic reduction of CO_2, Varghese et al., reported that TiO_2 nanotubes modified with Cu nanoparticles provide higher generation rates of hydrocarbon fuel while the TiO_2 nanotubes modified with Pt nanoparticles have higher generation rate for H_2 [137].

Another way to increase the efficiency of photocatalytic processes and to reduce the recombination rate of the electron/hole pair is to combine photocatalysis with electrocatalysis, i.e., photoelectrocatalysis. In this process, the semiconductor is irradiated with visible light and kept under a positive potential. As a result, the negative charges flow towards the counter electrode, enriching the semiconductor surface with holes (h^+) [138]. Due to ordered nanotubes arrangements the charge transport is faster than in nanoparticulate assemblies. Photoelectrocatalysis has shown to be very promising in the inactivation of microorganisms, promoting an attack on the cell wall, leading to lysis and then cell death in a short time of treatment. It is also possible to degrade the toxins generated during cell lysis [139, 140].

Dye-Sensitized Solar Cell (DSSC)

DSSC consists of a working electrode containing metal oxide nanoparticles, such as TiO_2 coated with a dye, a counter electrode and an electrolyte with I^-/I_3^- redox. Its working principle [141, 142] consists of a dye that absorbs in the visible light range and thereby excites the HOMO electrons to the LUMO level, followed by a fast injection of electrons in the conduction band of TiO_2. The electrons move through the TiO_2 layer toward the contact at the back, while the oxidized dye on the surface is regenerated by the electrolyte.

Grimes et al. were the first to use anodised TiO_2 nanotubes in DSSC [143, 144]. The nanotubes were 6.2 μm in length, 110 nm in pore diameter and 20 nm in wall thickness. After anodization, the nanotubes were treated with $TiCl_4$ (decoration with TiO_2 nanoparticles) and then submitted to a sintering process and the crystallization of the amorphous nanotubes. The treatment with $TiCl_4$ increased the surface area for loading of the N719 dye by more than 20%. Although a front illumination is common in solar cells, the structure of the first solar cell presented a back illumination due to the solid substrate in which the nanoutubes were produced. The back illumination structure reduces the efficiency of the process because photons can diffuse when they pass through a counter electrode and the electrolyte before reaching the dye absorbed on the TiO_2 nanotubes [143]. Comparing a back illumination structure with a conventional nanoparticles system, the back illumination presented higher electron transference efficiency due to a direct vectorial charge transfer route in the TiO_2 nanotubes compared to the nanoparticle working electrode.

Lengths of nanotubes between 15 and 20 μm showed higher solar cell efficiencies [57, 145]. Nanotubes with long lengths showed little adhesion and disorders on their top surface (*nanograss*) [145]. With respect to the effect of the nanotube diameter on the conversion efficiency of DSSC, Wantawee et al. observed that the energy conversion efficiency of the DSSCs increased with decreasing the pore size of the nanotubes (as a function of the anodization potential). Decreasing the diameter from 200 to 100 nm, the efficiency increased from 4.21% to 6.89%, because the small tubes have high ability to absorb the dye while promoting electron transport [146].

In general, solar cell conversion efficiency is around 5% for pure TiO_2 nanotubes [145] whereas for TiO_2 nanotubes mixed with TiO_2 nanoparticles is 7% [147]. Another way to increase the efficiency is by modifying the wall of nanotubes by bamboo-type structures or double-walled nanotubes, which present higher specific surface area compared to standard TiO_2 nanotubes [148, 149].

In order to solve the problem of the back structure illumination for nanotubes, Park et al. growth nanotubes with 7-35 μm length and detached them as a film to transfer them onto a FTO glass substrate, using 100 mM Ti-isopropoxide in isopropanol to improve the adhesion. Using front-side illuminated dye-sensitized solar cells (DSSCs), the best performance was obtained with nanotubes of 35 μm length, with around 7.6% efficiency [148, 150]. Under a similar DSSC, Mohammadpour et al. showed that the cell efficiency might strongly relate to the structure of the TiO_2 scaffold. The heating rate used during the crystallization of the photo-anodes largely influenced the structural features of the tubes and their charge carrier transport properties. Fast electron transport was observed for rapidly annealed tubes. A heating rate of 30°C min^{-1} lead to a significant improvement of the solar light conversion efficiency up to 9.79% compared to those treated at 10°C min^{-1}. Such enhanced electron transport of rapidly-annealed tubes may be related to the robust, crack-free and ''single-walled'' morphology [151].

Biomedical Application

The physical connection between the implant and the surface of the bone depends strongly on the surface morphology of the orthopedic implant. Smooth surfaces tend to favor the growth of fibroblasts, undesirable substances. On the other hand, irregular surfaces favor the growth and fixation of osteoblasts, allowing a strong and well-established bone-implant interface [152, 153]. Ti implants, containing a layer of nanotubular TiO_2 allows an effective bone-integration, enabling the development of stem cells [154].

The wettability is an important characteristic of surfaces for biomedical application. Studies carried out by Jin et al. [155] on oxide nanotubular layer on Ti-xNb alloys (x = 5 wt.% - 40 wt. %) by electrochemical anodization treatment shows that the surface wettability varies with the nanotubular diameter. Using the same thickness of nanotube layer, the higher the nanotube diameter the smaller the capillary wetting effect of the nanotubes resulting in less hydrophilic surface. After thermal annealing treatment, the nanotubes show higher contact angle than before thermal treatment due to the removal of hydroxyl groups and residue fluoride containing species. Verissimo et al. also observed that in oxide nanotubular layer on $Ti_{35}Nb$ and $Ti_{35}Nb_4Sn$ alloys the anatase phase shows high values of contact angle, while the rutile phase exhibits lower values and hydrophilic characteristics [128]. However, the combination anatase-rutile is the most suitable for biomedical applications and cell growth.

Another form of application is as a platform for the distribution of antibacterial drugs, which results in the control of harmful bacteria and reduce the possibility of infections [156].

CONCLUSION

This chapter highlighted the importance of using self-organized TiO_2 nanotubes semiconductors in photocatalysis, dye-sensitized solar cells and biomedical applications. As photocatalyst, TiO_2 nanotubes are attractive for waste water treatment due to their suitable band-edge in relation to water since they promote the formation of radicals that are able to oxidise organic compounds. Titanium nanotubes can also be doped with W, N or Mo to modify their photocatalytic activity and have been used for the photocatalytic reduction of CO_2. In DSSC, TiO_2 nanotubes promote the transfer of electrons in their conduction band when the dye absorbs in the visible light. The efficiency of solar cells increases from 5% to 7% when TiO_2 nanoparticles are used and the efficiency could increase further by modifying the wall of nanotubes. TiO_2 nanotubes also have important applications in orthopedic implants thanks to the fact that the cells secreted by the bones attach to the irregular surfaces of the TiO_2 nanotubes and form strong and effective bone-integration able to develop stem cells. TiO_2 nanotubes are also able to act as an antibacterial to deter infections. This chapter also analysed some of the fundamentals aspect of the anodization of titanium and the factors that affect the formation and growth mechanism of TiO_2 when different electrolytes, conditions and dopants are used. A historic perspective of the development of the TiO_2 morphology from the first to the fourth generation was highlighted.

REFERENCES

[1] Pinheiro, L.; Pelegrini, R.; Bertazzoli, R.; De Motheo, A. *Applied Catalysis B: Environmental* 2005, *Vol.* 57, p. 75.

[2] Pelegrini, R.; Freire, R. S.; Duran, N.; Bertazzoli, R. *Environmental Science & Technology* 2001, *Vol.* 35, p. 2845.

[3] Zanoni, M. V. B.; Sene, J. J.; Anderson, M. A. *J. Photochemistry and Photobiology A* 2003, *Vol.* 157, p. 55.

[4] Robert, D.; Malato, S. *Science of Total Environment* 2002, *Vol.* 291, p. 85.

[5] Zhang, L.; Zhu, Y.; He, Y.; Li, W.; Sun, H. *App. Applied Catalysis B: Environmental* 2003, *Vol.* 40, p. 287.

[6] He, C.; Xiong, Y.; Chien, J.; Zha, C.; Zhu, X. *J. Photochemistry and Photobiology A* 2003, *Vol.* 157, p. 71.

[7] Beer, H. B. *U.S. Patent* 1972, 3, 632, 498.
[8] De Nora, O.; Nidola, A.; Trisoglio, G.; Bianchi, G. *Br. Patent* 1973, 1, 399, p. 576.
[9] Friedrich, J.M.; Ponce-De-León, C.; Reade, G.W.; Walsh, F.C. J. *Electroanalytical Chemistry* 2004, *Vol.* 561, p. 203.
[10] Badellino, C.; Rodrigues, C. A.; Bertazzoli, R. *Journal of Hazardous Materials* B 2006, *Vol.* 137, p. 856.
[11] Badellino, C.; Rodrigues, C. A.; Bertazzoli, R. *Journal of Applied Electrochemistry* 2007, *Vol.* 37(4), p. 451.
[12] Brillas, E.; Sirés, I; Oturan, M.A. *Chemical Reviews* 2009, *Vol.* 109, p. 6570.
[13] Panizza, M.; Cerisola, G. *Chemical Reviews* 2009, *Vol.* 109, p. 6541.
[14] Schäpers, T. *Semiconductor Spintronics* - Walter de Gruyter GmbH & Co KG, Germany, 2016. ISBN 9783110425444.
[15] Hoffman, M. R.; Martin, S. T.; Choi, W.; Bahnemann, D.W. *Chemical Reviews* 1995, *Vol.* 95, p. 69.
[16] Sato, N., *in Electrochemistry at Metal and Semiconductor Electrodes, Semiconductor photoelectrodes, Fisrt Edition.* ISBN: 9780444828064. 1998, p. 412.
[17] Jiang, D. et al. *Journal of Photochemistry and Photobiology A: Chemistry* 2001, *Vol.* 144 (2-3), p. 197.
[18] Takahashi, K.; Yoshikawa, A.; Sandhu, A., *in Wide Bandgap Semiconductors: Fundamental Properties and Modern Photonic and Electronic Devices, Mechanical and Thermal Properties of Wide Band Gap Semiconductor Optical, Fisrt Edition.* ISBN: 9783540472353. 2007, p. 52-55.
[19] Bolts, J.M.; Wrighton, M.S. *The Journal of Physical Chemistry* 1976, *Vol.* 80 (24) p. 2641.
[20] López, R.; Gómez, R., *Journal of Sol-Gel Science and Technology* 2012, *Vol.* 61(1), p. 1–7.
[21] Nowak M.; Kauch B.; Szperlich, P., *The Review of Scientific Instruments* 2009, *Vol.* 80, p. 046107.
[22] Philips-Invernizzi, B., Dupont D., Caze C., *Optical Engineering* 2001, *Vol.* 40, p.1082.
[23] Blanco-Galvez, J.; Fernandez-Ibañez, P.; Malato-rodriguez, S., *Journal of Solar Energy Engineering* 2007, *Vol.* 129, p. 4.
[24] Hoffmann, M.R.; Martin, S.T.; Choi, W.; Bahnemann, D.W., *Chemical Reviews* 1995, *Vol.* 95 (1), p. 69.
[25] Hoyer, P., *Langmuir* 1996, *Vol.* 12, p. 1411.
[26] Imai, H.; Takei, Y.; Shimizu, K.; Matsuda, M.; Hirashima, H., *Journal Materials Chemistry* 1999, *Vol.* 9, p. 2971.
[27] Michailowski, D.; Aimaawlawi, D.; Cheng, G.; Moskovits, M., *Chemical Physics. Letters* 2001, *Vol.* 1, p. 349.

[28] Kobayashi, S.; Hamasaki, N.; Suzuki, M.; Kimura, M.; Shirai, H.; Hanabusa, K., *Journal of the American Chemical Society* 2002, *Vol.* 124, p. 6550.

[29] Tian, Z. R. R.; Voigt, J. A.; Liu, J.; Mckenzie B.; Xu, H. F., *Journal of the America Chemical. Society* 2003, *Vol.* 125, p. 12384.

[30] Yao, B. D.; Chan, Y. F.; Zhang, X. Y.; Zhang, W. F.; Yang, Z. Y.; Wang, N., *Chemical Physics. Letters* 2001, *Vol.* 281, p. 82.

[31] Liua, N.; Chen, X.; Zhang, J.; Schwank, J.W., *Catalysis Today* 2014, *Vol.* 225, p. 34-51.

[32] Mor, G.K.; Varghese, O. K.; Paulose, M.; Shankar, K.; Grimes, C. A., *Solar Energy Materials & Solar Cells* 2006, *Vol.* 90, p. 2011.

[33] Chanmanee, W.; Watcharenwong, A.; Chenthamarakshan, C. R.; Kajitvichyanukul, P.; Tacconi, N. R.; Rajeshwar, K., *Journal of the American Chemical Society* 2008, *Vol.* 130, p. 965.

[34] Roy,P.; Berger, S.; Schumki, P., *Angewandte Chemie International Edition* 2011, *Vol.* 50, p. 2904-2939.

[35] Regonini, D.; Bowen, C.R.; Jaroenworaluck, A.; Stevens, R., *Materials Science and Engineering R* 2013, *Vol.* 74, p.377-406.

[36] Chu, S.; Inoue, S.; Wada, K.; Hishita, S.; Kurashima, K., *Advanced Function Materials* 2005, *Vol.* 15, p. 1343.

[37] Cardoso, J.C.; Zanoni, M.V.B., *Separation Science and Technology* 2010, *Vol.* 45, p.1628-1636.

[38] Assefpour-Dezfuly, M.; Vlachos, C.; Andrews, E.H., *Journal Materials Science* 1984, *Vol.* 19, p. 3626– 3639.

[39] Zwilling, V.; Darque-Ceretti, E.; Boutry-Forveille, A.; David, D.; Perrin, M.Y.; Aucouturier, M., *Surface and Interface Analysis* 1999, *Vol.* 27, p. 629–637.

[40] Gong, D.; Grimes, C.G.; Varghese, O.K.; Hu, W.; Singh, R.S.; Chen, Z.; Dickey, E.C., *Journal of Materials Research* 2001, *Vol.* 16, p. 3331–3334.

[41] Varghese, O.K.; Gong, D.; Paulose, M.; Grimes, C.G., Dickey, E. C., *Journal of Materials Research* 2003, *Vol.*18, p. 156–165.

[42] Mor, J.K.; Varghese, O.K.; Paulose, M.; Mukherjee, N.; Grimes, C.G., *Journal of Materials Research* 2003, *Vol.* 18, p. 2588–2593.

[43] Beranek, R.; Hildebrand, H.; Schmuki, P., *Electrochemical and. Solid-State Letters* 2003, *Vol.* 6, p. B12–B14.

[44] Zhao, J.; Wang, X.; Chen, R.; Li, L., *Solid State Communications* 2005, *Vol.* 134, p. 705–710.

[45] Raja, K.; Misra, M.; Paramguru, K., *Electrochimica Acta* 2005, *Vol.* 51, p. 154–165.

[46] Cai, Q.; Paulose, M.; Varghese, O.K.; Grimes, C.A., *Journal of Materials Research* 2005, *Vol.*20, p. 230–236.

[47] Macak, J.M.; Tsuchiya, H.; Schmuki, P.; *Angewandte Chemie International Edition* 2005a, *Vol.* 44, p. 2100–2102.

[48] Macak, J.M.; Sirotna, K.; Schmuki, P.; *Electrochimica Acta* 2005b, *Vol.* 50, p. 3679–3684.

[49] Macak, J.M.; Taveira, L.V.; Tsuchiya, H.; Sirotna, K.; Schmuki, P., *Journal Electroceramics* 2006, *Vol.*16, p. 29–34.

[50] Taveira, L.V.; Macak, J.M.; Tsuchiya, H.; Dick, L.F.P.; Schmuki, P., *Journal of the Electrochemical Society* 2005, *Vol.*152; p. B405.

[51] Ghicov, A.; Tsuchiya, H.; Macak, J.M.; Schmuki, P., *Electro-chemistry Communications* 2005, *Vol. 7*, p. 505.

[52] Raja, K.S.; Gandhi, T.; Misra, M., *Electrochemistry Commun-ications* 2007, *Vol.* 9, p. 1069–1076.

[53] Ruan, C.; Paulose, M.; Varghese, O.K.; Mor, G.K.; Grimes, C.G., *Journal Physical Chemistry. B* 2005, *Vol.*109, p 15754–15759.

[54] Macak, J.M.; Tsuchiya, H.; Taveira, L.V.; Aldabergerova, S.; Schmuki, P., *Angewandte Chemie International Edition* 2005, *Vol* 44, p. 7463–7465.

[55] Paulose, J.M.; Shankar, K.; Yoriya, S.; Prakasam, H.E.; Varghese, O.K.; Mor, G.K.; Latempa, T.A.; Fitzgerald, A.; Grimes, C.G., Journal Physical Chemistry. B 2006, *Vol.* 110, p. 16179–16184.

[56] Yoriya, S.; Prakasam, H.E.; Varghese, O.K.; Shankar, K.; Paulose, M.; Mor, G.K.; Latempa, T.A.; Grimes, C.A.; *Sensors Letters* 2006, *Vol.* 4, p. 334–339.

[57] Shankar, K.; Mor, G.K. Prakasam, H.E.; Yoriya, S.; Paulose, M.; Varghese, O.K.; Grimes C.A., *Nanotechnology* 2007, *Vol.*18, p. 065707.

[58] Prakasam, H.E.; Shankar, K.; Paulose, M., Varghese, O.K.; Grimes, C.G., *Journal Physical Chemistry C* 2007, *Vol.* 111, p. 7235–7241.

[59] Paulose, J.M.; Prakasam, H.E.; Varghese, O.K.; Peng, L.; Popat, K.C.; Mor, G.K.; Desai, T.A.; Grimes, C.G., *Journal Physical Chemistry C* 2007, *Vol.*111, p. 14992–14997.

[60] Albu, S.P.; Schmuki, P., *Phys. Status Solidi – Rapid Reaseach Letters* 2010, *Vol.* 4, p. 215.

[61] Yoo, J.E.; Lee, K.; Altomare, M.; Selli, E.; Schmuki, P., *Angewandte Chemie International Edition* 2013, *Vol.* 52, p. 7514.

[62] Paramasivam, I.; Macak, J.M.; Selvam, T.; Schmuki, P., *Electrochimica Acta* 2008, *Vol.* 54, p. 643.

[63] Tsuchiya, H.; Macak, J.M.; Taveira, L.; Balaur, E.; Ghicov, A.; Sirotna, K.; Schmuki, P., *Electrochemistry Communications* 2005, *Vol.* 7, p. 576.

[64] Watcharenwong, A.; Chanmanee, W.; Tacconi, N. R.; Chentha-marakshan, C. R.; Kajitvichyanukul, P.; Rajeshwar, K., *Journal of Materials Research* 2007, *Vol.* 22, p. 3186.

[65] Macak, J.M., Albu, S.P., Schmuki, P., *Physica Status Solidi – Rapid Reaseach Letters* 2007, *Vol.* 1, p. 181–183.
[66] Shin, Y.; Lee, S., *Nano Letters* 2008, Vol. 8, p. 3171–3173.
[67] Siejka, J.; Nadai, P.; Amsel, G.; *Journal Electrochemical Society* 1970, *Vol.* 118, p. 727-737
[68] A.J. Bard, R. Parsons, J. Jordan, *Standard Potentials in Aqueous Solution*, CRC Press (Taylor & Francis), New York, 1985.
[69] Hebert, K.R.; Albu, S.; Paramasivam, I.; Schmuki, P., *Nature Materials* 2012, *Vol.* 11, p. 162.
[70] Zhou, X.; Nguyen, N.T.; Özkan, S; Schmuki, P. *Electrochemistry Communications* 2014, *Vol.* 46, p. 157-162.
[71] Skeldon, P.; Thompson, G.E.; Garcia-Vergara, S.J.; Iglesias-Rubianes, L.; BlancoPinzon, C.E., *Electrochemical and. Solid-State Letters* 2006, *Vol.* 9, p. B47–B51.
[72] Garcia-Vergara, S.; Skeldon, P.; Thompson, G.; Habazaki, H., *Electrochimica Acta* 2006, *Vol.*52, p. 681–687.
[73] LeClere, D.J.; Velota, A.; Skeldon, P.; Thompson, G.E.; Berger, S.; Kunze, J.; Schmuki, P.; Habazaki, H.; Nagata, S., *Journal Electro-chemical Society* 2008, *Vol.* 155, p. C487–C494.
[74] Berger, S.; Kunze, J.; Schmuki, P.; LeClere, D.; Valota, A.T.; Skeldon, P.; Thompson, G.E., *Electrochimica Acta* 2009, *Vol.* 54, p. 5942–5948.
[75] Berger, S.; Macak, J.M.; Kunze, J.; Schmuki, P., *Electrochemical and. Solid-State Letters* 2008, *Vol.* 11, p. C37–C40.
[76] Vanhumbeeck, J.F.; Proost, J.; *Electrochimica Acta* 2008, *Vol.* 53, p. 6165.
[77] Albu, S.P.; Schmuki, P., *Electrochimica Acta* 2013, *Vol.* 91, p. 90–95.
[78] Jaroenworaluck, A.; Regonini, D.; Bowen, C.R.; Stevens, R., *Applied Surface Science* 2010, *Vol.* 256, p. 2672–2679.
[79] Habazaki, H.; Fushimi, K.; Shimizu, K.; Skeldon, P.; Thompson, G.E., *Electrochemistry Communications* 2007, *Vol.* 9, p. 1222–1227.
[80] Albu, S. P.; Ghicov, A.; Aldabergenova, S.; Drechsel, P.; LeClere, D.; Thompson, G. E.; Macak, J. M.; Schmuki, P., *Advanced Materials* 2008, *Vol.* 20, p. 4135.
[81] Berger, S.; Albu, S.P.; Schmidt-Stein, F.; Hildebrand, H.; Schmuki, P.; Hammond, J.S.; Paul, D.F.; Reichlmaier, S., *Surface Science* 2011, *Vol.* 605, p. L57–L60.
[82] Wei, W.; Berger, S.; Hauser, C.; Meyer, K.; Yang, M.; Schmuki, P., *Electrochemistry Communications* 2010, *Vol.* 12, p. 1184.
[83] Ruiquan, Y.; Longfei, J.; Xufei, Z.; Ye, S.; Dongliang, Y.; Aijun, H., *RSC Advances* 2012, *Vol.* 2, p. 2474–12481.
[84] Zhu, X.F., Song, Y.; Liu, L.; Wang, C. Y.; Zheng, J.; Jia, H.; Wang, X., *Nanotechnology* 2001, *Vol.* 20, p. 475303 (7pp).

[85] Zhu, X.F.; Song, Y.; Yu, D.L;. Zhang, C.; Yao, W., *Electrochemistry Communications* 2013, *Vol.* 29, p. 71-74.
[86] Zhang, S.Y.; Yu, D.L.; Li, D.D.; Song, Y.; Che, J.; You, S.Y.; Zhu. X., *Journal of the Electrochemical Society* 2014, *Vol.* 161, p. E135-E141.
[87] Chong, B.; Yu, D.L.; Gao, M.Q.; Fan; H.W.; Yang; C.Y.; Ma, W.H.; Zhang, S.Y.; Zhu, X.F., *Journal of the Electrochemical Society* 2015, *Vol.* 162, p.) H244-H250.
[88] Lu, H.; Fan, H.; Jin, R.; Chong, B.; Shen, X.; Yan, S.; Zhu, X. *Electrochimica Acta* 2016, *Vol.* 215, p. 380-387.
[89] Jin, R,; Fan, H.; Liu, Y.; Ma, W.; Lu, H.; Yang, P.; Ma, W.H. *Electrochimica Acta* 2016, *Vol.*188, p. 421-427.
[90] Yang, R.Q.; Jiang, L.F.; Zhu, X.F.; Song, Y.; Yu, D.L.; Han, A.J. *RSC Advances* 2012, *Vol.* 2, p. 12474-12481.
[91] Houser, J.E.; Hebert, K.R. *Nature Materials* 2009, *Vol.* 8, p. 415-420.
[92] Kaczmarek, A.; Klekiel, T.; Krasicka-Cydzik, E., *Surface and Interface Analysis* 2010, *Vol.* 42, p. 510–514.
[93] Macak, J.M.; Hildebrand, H.; Marten-Jahns, U.; Schmuki, P., *Journal of Electroanalytical Chemistry* 2008, *Vol.* 621, p. 254–266.
[94] Chen, C.-C.; Hsieh, S.-J., *Journal Electrochemical Society* 2010, *Vol.* 157, p. K125–K130.
[95] Lee, K.; Kim, J.; Kim, H.; Lee, Y.; Kim, D.; Schmuki, P.; Tak, Y., *Journal of the Korean Physical Society* 2009, *Vol.* 54, p. 1027–1031.
[96] Song, Y-Y.; Schmuki, P., *Electrochemistry Communications* 2010, *Vol.* 12, p. 579–582.
[97] Von Der Mark, K.; Park, J.; Bauer, S.; Schmuki, P., *Cell and Tissue Research* 2010, *Vol.* 339, p. 131.
[98] Berger, S.; Kunze, J.; Schmuki, P.; Valota, A. T.; Leclere, D. J.; Skeldon, P.; Thompson, G. E., *Journal Electrochemical Society* 2010, *Vol.* 157, C18.
[99] Yasuda, K.; Schmuki, P., *Electrochimica Acta* 2007, *Vol.* 52, p. 4053.
[100] Yasuda, K.; Macak, J. M.; Berger, S.; Ghicov, A.; Schmuki, P., *Journal Electrochemical Society* 2007, *Vol.* 154, C472.
[101] Bauer, S.; Kleber, S.; Schmuki, P., *Electrochemistry Commun-ications* 2008, *Vol.* 8, p. 1321.
[102] So, S., Lee, K., Schmuki, P., *Journal of the America Chemical Society* 2012, *Vol.* 134, p. 11316–11318.
[103] Lai, C.W.; Sreekantan, S., *International Journal of Photoenergy* 2012, *Vol.* 2013, p. 1-10.
[104] Macak, J.; Schmuki, P., *Electrochimica Acta* 2006, *Vol.* 52, p. 1258–1264.
[105] Garcia-Vergara, S.J.; Iglesias-Rubianes, L.; Blanco-Pinzon, C.E.; Skeldon, P.; Thompson, G.E.; Campestrini, P., *Proceeding of Royal Society: A* 2006, *Vol.* 462, p 2345.

[106] Albu, S.P.; Ghicov, A.; Macak, J.M.; Schmuki, P., *Physica Status Solidi – Rapid Research Letters* 2007, *Vol.*1, p. R65.

[107] Albu, S.P.; Roy, P.; Virtanen, S.; Schmuki, P., Israel *Journal of Chemistry* 2010, *Vol.* 50, p. 453.

[108] Asahi, R.; Morikawa, T.; Ohwaki, T.; Aoki, K.; Taga, Y., *Science* 2001, *Vol.* 293, p. 269.

[109] Paramasivam, I.; Nah, Y-C; Das, C.; Sherstha, N. K.; Schmuki, P., *Chemistry: A European Journal* 2010, *Vol.* 16, p. 8993-8997.

[110] Ferreira, C.P.; Gonçalves, M.C.; Caram, R.; Bertazzoli, R.; Rodrigues C.A., *Applied Surface Science* 2013, *Vol.* 285P, p. 226– 234.

[111] Gartzel, M., *Nature* 2001, *Vol.* 414, p. 338-344.

[112] Shrestha, N. K.; Yang, M.; Nah, Y. C.; Paramasivam, I.; Schmuki, P., *Electrochemistry Communications* 2010, *Vol.* 12, p. 254.

[113] Macak, J. M.; Barczuk, P. J.; Tsuchiya, H.; Nowakowska, M. Z.; Ghicov, A.; Chojak, M.; Bauer, S.; Virtanen, S.; Kulesza, P. J.; Schmuki, P., *Electrochemistry Communications* 2005, *Vol.* 7, p. 1417.

[114] Mandi, U.; Salam, N.; Kundu, S.K.; Bhaumilk, A.; Islam, S.M., *RSC Advances* 2016, *Vol.* 6, p. 73440-73449.

[115] He, D.; Yang, L.; Kuang, S.; Cai, Q., *Electrochemistry Commun-ications* 2009, *Vol.* 9, p. 2467-2472.

[116] Macak, J. M.; Schmidt-Stein, F.; Schmuki, P., *Electrochemistry Communications* 2007, *Vol.* 9, p.1783.

[117] Paramasivam, I.; Macak, J. M.; Ghicov, A.; Schmuki, P., *Chemical Physics Letters* 2007, *Vol.* 445, p. 233.

[118] Almeida, L.C.; Zanoni, M.V.B., *Journal of the Brazilian Chemical Society.* 2014, *Vol.* 25, p. 579-588.

[119] Paramasivam, I., Macak, J.M.; Schmuki, P., *Electrochemistry Communications* 2008, *Vol.* 10, p. 71-75.

[120] Mor, G.K.; Varghese, O. K.; Wilke, R.H.T.; Sharma, S.; Shankar, K.; Latempa, T.J; Choi, K-S; Grimes, C.A., *Nano Letters* 2008, *Vol.* 8, p.1906-1911.

[121] Mor, G.K.; Prakasam, H.E.; Varghese, O. K.; Shankar, K.; Grimes, C.A., *Nano Letters* 2007, *Vol.* 8, p. 2356-2364.

[122] Wang, Z.; Liu, B.; Xie, Z.; Li, Y.; Shen, Z., *Ceramic International* 2016, *Vol.* 42, p. 13664-13669.

[123] Zhang, H.; Banfield, J. F., *Journal of Physical Chemistry B* 2000, *Vol.* 104, p. 3481-3487.

[124] Kumar, K.-N.P.; Keizer, K.; Burggraaf, A.J., *Journal of Materials Chemistry* 1993a, *Vol.* 3, p. 1141–1149.

[125] Kumar, K.-N.P.; Keizer, K.; Burggraaf, A.J.; Okubo, T.; Nagamoto, H., *Journal of Materials Chemistry* 1993b, *Vol.* 3, p. 1151–1159.

[126] Fang, D.; Luo, Z.; Huang, K.; Lagoudas, D.C., *Applied Surface Science* 2011, *Vol.* 257, p. 6451–6461.

[127] Ghicov, A.; Aldabergerova, S.; Tsuchiya, H.; Schmuki, P., *Angewandte Chemie International Edition* 2006, *Vol.* 45, p. 6993.

[128] Verissimo, N.C.; Cremasco, A.; Rodrigues, C.A.; Bertazzoli, R.; Caram R., *Applied Surface Science* 2014, *Vol.* 307, p. 372-381.

[129] SONG, X.; Wu, J. M.; Yan, M., *Thin Solid Films* 2009, *Vol.* 517, p. 4341-4347.

[130] Cardoso, J. C.; Lizier, T. M.; Zanoni, M. V. B., *Applied Surface Science* 2010, *Vol.* 99, p. 96-102.

[131] Weon, S.; Choi. W., *Environmental Science & Technolology* 2016, *Vol.* 50 (5), p. 2556–2563.

[132] Brugnera, M. F.; Rajeshwar, K.; Cardoso, J.C.; Zanoni, M. V. B., *Chemosphere* 2010, *Vol.* 78, p. 569-575.

[133] Brugnera, M. F.; Miyata, M.; Leite, C. Q. F.; Zanoni, M. V. B., *Journal of Photochemistry and Photobiology A: Chemistry* 2014, *Vol.* 278, p. 1-8.

[134] Yu, Y.; Yu, X.; Yang, S., *Journal of Materials Science: Materials in Electronics* 2015, *Vol.* 26, p. 5715-5723.

[135] Paramasivam, I.; Avhale, A.; Inayat, A.; Bosmann, A.; Schmuki, P; Schwieger, W., *Nanotechnology* 2009, *Vol.* 20, p. 225607.

[136] Hoffman, M. R.; Martin, S.T.; Choi, W.; Bahneman, D.W., *Chemical Reviews* 1995, *Vol.* 95, p. 69.

[137] Varghese, O.K.; Paulose, M.; La Templa T.J.; Grimes, C. A., *Nano Letters* 2009, *Vol.* 9, p.171.

[138] Xie, Y., *Electrochimica Acta* 2006, *Vol.* 51, p 3399–3406.

[139] Baram, N.; Starosvetsky, D.; Starosvetsky, J.; Epshtein, M.; Armon, R.; Ein-Eli, Y., *Applied Catalysis B: Environmental* 2011, *Vol.* 101, p. 212–219.

[140] Ng, J.; Zhang, X.; Zhang, T.; Pan, J.; Alan Du, J.; Sun. D. D., *Journal of Chemical Technology and Biotechnology* 2010; *Vol.* 85, p. 1061–1066.

[141] P. Roy, D. Kim, K. Lee, E. Spiecker, P. Schmuki. *Nanoscal*, 2010, *Vol.* 2, p. 45.

[142] Grätzel, M., *Journal of Photochemistry and Photobiology C* 2003, *Vol.* 4, p. 145.

[143] Paulose, M; Shankar, K.; Varghese, O. K.; Mor, G. K.; Hardim, B.; Grimes, C. A., *Nanotechnology* 2006, *Vol.*17, p.1446.

[144] Paulose, M; Shankar, K.; Varghese, O. K.; Mor, G. K.; Grimes, C. A., *Journal Physics D: Applied Physics* 2006, *Vol.* 39, p. 2498.

[145] Roy, P.; Albu, S. P.; Schmuki, P., *Electrochemistry Communications* 2010, *Vol.* 12, p. 949.

[146] Wantawee, S.; Vongwatthaporn, R.; Tipparach, U., *Integrated Ferroelectrics* 2015, *Vol.* 165, p.153–158.

[147] Park, J. H.; Lee, T.-W.; Kang, M. G., *Chemical Communications* 2008, *Vol.* 25, p. 2867.

[148] Kim, D.; Ghicov, A.; Albu, S. P.; Schmuki P.; *Journal of the America Chemical Society* 2008, *Vol.* 130, p. 16454.

[149] Ghicov, A.; Albu, S. P.; Hahn, R.; Kim, D.; Stergiopoulos, T.; Kunze, J.; Schiller, C-A.; Falaras, P.; Schmuki, P., *Chemistry Asian Journal* 2009, *Vol.* 4, p. 520.

[150] Chen, Q.; Xu, D.; Wu, Z.; Liu. Z., *Nanotechnology* 2008, *Vol.*19, p. 365708.

[151] Mohammadpour, F.; Moradi, M.; Lee, K.; Cha, G.; So, S.; Kahnt, A.; Guldi, D. M.; Altomare M.; Schmuki. P., *Chemical Commun-ications* 2015, *Vol.* 51, p. 1631-1634.

[152] Bai, Y.; Park, I. S.; Park, H. H.; Lee, M. H.; Bae, T. S.; Duncan W.; Swain. M., *Surface and Interface Analytical* 2011, Vol. 43, p. 998 – 1005.

[153] Oh, S.; Jin, S., *Materials Science and Engineering C* 2006, *Vol.* 26, p. 1301 – 1306.

[154] Hori, N.; Iwasa, F.; Ueno, T.; Takeuchi, K.; Tsukimura, N.; Yamada, M.; Hattori, M.; Yamamoto, A.; Ogawa. T., *Dental materials* 2010, *Vol.* 26, p. 275–287.

[155] Jin, M., Lu, X.; Qiao, Y; Wang, L.; Volinsky, A.A., *Rare Metals* 2016, *Vol.* 35(2), p. 140-148.

[156] Feschet-Chassot, E.; Raspal, V.; Sibaud, Y.; Awitor, O.K.; Bonnemoy, F.; Bonnet, J.L.; Bohatier, J., *Thin Solid Films* 2011, *Vol.* 519, p.2564-2568.

In: Advanced Materials and Systems … ISBN: 978-1-53613-852-8
Editor: Mauro Coelho dos Santos

Chapter 8

APPLIED ELECTROCHEMICAL MATERIALS FOR TREATMENT OF SYNTHETIC AND REAL WASTE WATER

L. R. Aveiro[1], E. C. Paz[1], F. M. Souza[1], V. S. Pinheiro[1], T. S. Lima[2] and M. C. Santos*

[1]Laboratory of Electrochemistry and Nanostructured Materials, Centre of Natural and Human Sciences, UFABC - Santo André, SP, Brazil

[2]Laboratory of electrochemistry and Electroanalytical, Department of Chemistry, UNIFESP - Diadema, SP, Brazil

ABSTRACT

In this chapter, we first discuss advanced oxidation processes (AOPs). The main focus is on Electrochemical Advanced Oxidation Processes (EAOPs) and their applications. Next, hydrogen peroxide electrogeneration is described mainly for the production of radical hydroxyls to treat synthetic and simulated wastewater. The major nanomaterials and/or materials for gas diffusion cathodes are discussed, distinct from those used for the treatment of water, with (oxide materials/composites and reduced graphene oxide) or without (nanotubes and zeolites) the formation of hydrogen peroxide from the oxygen reduction reaction. The controversy over the formation of both hydroxyl radicals and ferryl ions, which can coexist in Fenton chemistry, is elucidated. Future trends will enable the use of combined treatments with various materials to treat wastewater.

* Corresponding Author Email: mauro.santos@ufabc.edu.br.

Keywords: advanced oxidation processes, electrochemical advanced oxidation processes, wastewater treatment, simulated wastewater treatment, hydrogen peroxide electrogeneration

1. Electrochemical Advanced Oxidation Process

Earth's surface is composed of approximately 70% water [1]. However, much of this water is salt water in oceans. Only 3% of all water on Earth is freshwater, much of which is stored in very distant polar areas; solely 1% of freshwater is located on the surface of the Earth [2]. Due to this small amount of available freshwater, it is fundamental to treat contaminated water using available physico-chemical and biological processes [3, 4].

Increased water pollution has arisen due to organic or other toxic compounds from many industrial, agricultural and urban human activities. For this reason, a major current worldwide concern is wastewater treatment [5]. Most treatment of these compounds is related to persistent organic pollutants, primarily due to substantial conventional treatments, such as coagulation, biological oxidation, adsorption, ion exchange and chemical oxidation, to which these pollutants are resistant.

These pollutants are detected in rivers, lakes, oceans and fresh water all over the globe, causing important environmental health problems due to their toxicity and potential hazardous health effects (carcinogenicity, mutagenicity and bactericidal activity) to living organisms, including human beings [6-9].

At the end of the last century, both physico-chemical and biological treatment technologies reached their maximum efficiency. The importance of new technologies in addition to those using physico-chemical and microorganisms in biological processes in the treatment of wastewater are produced in the industry, providing new requirements for the treatment processes due to the complexity of the waste generated in anthropogenic activities. However, some organic water pollutants pose great risks and challenges, as conventional treatments are not efficient in their removal and/or degradation [10-11].

In this context, so-called Advanced Oxidation Processes (AOPs) are based predominantly (but not exclusively) on the production of very reactive hydroxyl radicals, which are related to similar processes but not the same technologies [5, 12]. These have received much attention [13]. The *In situ* production of highly reactive hydroxyl radicals (•OH) is non-selective and reacts with most organics [14]. The second strongest oxidant after fluorine is this radical, with a high standard redox potential of $E^0(\bullet OH/H_2O) = 2.80$ V/SHE [15]; rate constants for the reaction with several contaminants is in the order of 10^6 to 10^{10} $M^{-1}s^{-1}$ [16]. •OH has a short life-time of only a few nanoseconds in water [17], leading to its automatic elimination from the treatment system.

AOPs include heterogeneous and homogeneous photocatalysis, Fenton and Fenton-like processes, ozonation, the use of ultrasound, microwaves and γ-irradiation, electrochemical processes and wet oxidation processes [10]. More specifically, the most common AOPs are H_2O_2 with UVC radiation (H_2O_2/UVC), ozone and ozone-based processes (O_3, O_3/UVC, O_3/H_2O_2 and O_3/H_2O_2/UVC), titanium dioxide-based processes (TiO_2/UV and TiO_2/H_2O_2/UV) and Fenton's reaction-based methods (Fenton – Fe^{2+}/H_2O_2 and photo-Fenton (PF) – Fe^{2+}/H_2O_2/UV) [5].

Over the last two decades, electrochemical advanced oxidation processes (EAOPs) have increased in relevance as a promising class of AOPs [18–21]. These methods are based on the *In situ* generation of ·OH, which reacts with most organics leading to the mineralization of CO_2, water and inorganic ions due to its high standard redox, as already described. The most powerful EAOPs generate H_2O_2 from the two-electron reduction of injected O_2 [22]. In this case, direct oxidation using anodes such as boron-doped diamond (BDD) are not used—only cathodes in which the main focus is peroxide generation with further ·OH formation.

$$O_2 + 2\,H^+ + 2e^- \rightarrow 2\,H_2O_2 \tag{1}$$

The advanced H_2O_2/UVC process comes from the generation of ·OH from the homolytic cleavage of peroxide (-O-O-) by UVC light Eq. (2) [23, 24], in addition to the UVC radiation oxidation power.

$$H_2O_2 + h\upsilon \rightarrow 2\,OH\ (\lambda < \text{ca. } 280\text{ nm}) \tag{2}$$

The electro-Fenton (EF) process concerns the addition of Fe^{2+} ion in small amounts to the solution containing H_2O_2 to catalyze the conversion reaction into homogeneous •OH, along with Fe^{3+} ion by Fenton's reaction (3) under a pH optimum of ~ 3 [25-27].

$$Fe^{2+} + H_2O_2 \rightarrow Fe^{3+} + {}^\bullet OH + OH^- \tag{3}$$

Reaction (3) can be propagated primordially by the cathodic reduction of Fe^{3+} to Fe^{2+} ion [28, 29]. For EF, an important point is the formation of Fe(III) complexes with final short-linear carboxylic acids, which are difficult to destroy with the generated hydroxyl radicals [25]. Photoelectro-Fenton (PEF) in which the treated solution is simultaneously irradiated with an artificial UVA light can surmount this problem [30-35]. The so-called solar photo-electro-Fenton (SPEF) process is an alternative use of sunlight as a free energy UV source [35] avoiding the excessive cost of artificial UVA lamps. In these processes, mineralization of organic compounds is facilitated by UVA radiation, since more Fe^{2+} from photoreduction of $Fe(OH)^{2+}$ species produces more •OH in the bulk solution. For this reason, Fe(III)-carboxylate species are photolyzed according to reactions (3) and (4) [27,36].

$$Fe(OOCR)^{2+} + h\upsilon \rightarrow Fe^{2+} + CO_2 + R^{\bullet} \quad (4)$$

Above, we discussed the main topics related to the production of radical hydroxyls from H_2O_2 peroxide electrogeneration. Below, we present some important materials used for wastewater treatment using simulated wastewater, considering EAOPs using hydrogen peroxide as precursors of radical hydroxyls.

2. Treatment of Synthetic and Real Wastewater

Numerous reviews and articles on wastewater and simulated wastewater treatments describe advanced materials toward these goals using electrochemical advanced oxidation processes, focusing on the use of hydrogen peroxide as a precursor of hydroxyl radicals. We discuss the waste treatment related to foods and beverages [37-40], pesticides [41, 42], herbicides [43, 44], heavy metals [45, 46], dyes [47-49], pharmaceutical/ hospitals [50-59], domestic waste and landfill leachates [60-63], bacteria wastes [64] and both persistent organic and organic pollutants [65-72]. Below we describe some examples of these treatments.

In the work of Vilar et al. [40], winery wastewater was first subjected to biological oxidation to mineralize the biodegradable fraction, after which an EAOP was applied to mineralize the refractory molecules or transform them into simpler ones that can be further biodegraded. The biological oxidation led to greater than 97% removal of dissolved organic carbon (DOC), chemical oxygen demand (COD) and 5-day biochemical oxygen demand (BOD5) but was inefficient in the degradation of a bioresistant fraction corresponding to 130 mg L^{-1} of DOC. The relative oxidative capability of EAOPs increased in the order AO-H_2O_2 < EF < PEF ≤ SPEF, with DOC removals of biologically treated effluent of 36%, 54%, 84% and 86%, respectively, after 240 min of reaction. The authors attributed the poor DOC removal attained in AO-H_2O_2 due to the low ability of BDD(·OH) generated at the anode surface to react with recalcitrant. However, they do not mention the cathode effect using a 10 cm^2 carbon-PTFE air-diffusion cathode (conventional).

The pesticide atrazine (ATZ) is one of the most common pesticides detected in surface water in Quebec (Canada). Lafrance et al. [41] studied the degradation of ATZ and its by-products using electrochemical advanced oxidation processes such as PEF, EF and anodic oxidation reactions, with simultaneous H_2O_2 formation (AO - H_2O_2). Despite the presence of radical scavenger in surface water (surface water effluent (SWE) was sampled from agricultural areas near intensive crop production), the abatement of TOC was 81.25% after 420 min of treatment. Likewise, the high mineralization current efficiency and low chemical reagents consumption proved that the PEF process was environmentally friendly. H_2O_2 production used graphite carbon plates, a common material.

In the degradation of the herbicide diuron, Brillas et al. [43] worked with solutions of 0.185 mM herbicide at pH 3.0. This reagent was treated by EAOPs such as electrochemical oxidation with electrogenerated H_2O_2 (EO-H_2O_2), EF and UVA PEF or SPEF. The measurements were performed in stirred tank reactors of 100 mL and in a recirculation flow plant of 2.5 L using a filter-press reactor with a Pt or BDD anode and an air-diffusion cathode for H_2O_2 electrogeneration. Importantly, the cathode material is common (carbon-PTFE air-diffusion cathode from E-TEK), i.e., a commercial electrode. The results showed that 100 mL of 0.185 mM diuron at pH 3.0 can be mineralized up to 93% after 360 min of PEF using a stirred BDD/air-diffusion tank reactor at 100 mA cm^{-2} due to the combined action of BDD(·OH), OH and UVA radiation. A lower oxidation ability was found for EF, whereas EO-H_2O_2 was less powerful because only BDD(OH) acted as the main oxidant. The reaction of organics with BDD(·OH) was faster than with Pt(·OH), and the comparative EAOPs in a stirred Pt/air-diffusion cell yielded poorer degradation. Thus, the air diffusion cell contains good materials for hydrogen peroxide production.

Brillas et al. [49] also studied the degradation of Evans Blue diazo dye by electrochemical processes based on Fenton reaction chemistry. A solution with 0.245 mM of the diazo dye Evans Blue and 0.50 mM Fe^{2+} as a catalyst at pH 3.0 was comparatively degraded by electrochemical processes based on Fenton reactions such as EF, PEF with 6 W UVA light and SPEF. Electrolytic trials were made in a 100-mL stirred tank reactor with a BDD anode and an air-diffusion cathode (conventional) at a constant current density. As main conclusions, the authors propose that the relative mineralization rate of the EAOPs rose in the order EF < PEF < SPEF. Mineralization was rapidly attained for the most potent SPEF process at j P 66.7 mA cm^{-2}. Higher j promoted dye decay, decolorization and mineralization in all EAOPs, but with a loss of current efficiency. An initial reaction sequence for Evans Blue degradation was proposed from 19 aromatic intermediates, including diazo, monoazo, biphenylic, benzenic, naphthalenic and phthalic acid compounds, as identified by LC–MS. The SPEF degradation of the Evans Blue solution was performed in a 10-L flow plant with a Pt/air-diffusion cell and a CPC photoreactor. A good mineralization of 88% with MCE = 42% and EDOC = 2.13 kWh kg^{-1} DOC were obtained after 300 min at 55.4 mA cm^{-2}, showing the viability of SPEF to treat this diazo dye at the industrial scale, and only conventional cathode materials were used.

Brillas et al. [50] studied the degradation of 100 mL of 0.245 mM of the antibiotic ciprofloxacin in 0.05 M Na_2SO_4 at pH 3.0 using electrochemical oxidation with electrogenerated H_2O_2 (EO-H_2O_2), EF, UVA PEF and SPEF. Electrolyses were performed with a stirred tank reactor using either a BDD or Pt anode and an air-diffusion cathode. In EF, PEF and SPEF, ciprofloxacin was rapidly removed due to its oxidation with ·OH formed from Fenton's reaction between added iron 2+ and H_2O_2, which was generated at the cathode (conventional). The larger electrochemical incineration of the antibiotic was achieved by SPEF with BDD with 95% mineralization, owing to the additional attack of hydroxyl radicals formed from water oxidation at the BDD anode surface and the photolysis of final Fe(III)-oxalate and Fe(III)-oxamate species from sunlight. Different behavior was found when the same antibiotic concentration was oxidized in a synthetic urine matrix with high urea content and a mixture of PO_3^{4-}, SO_4^{2-} and Cl ions. Since the Fenton reaction was inhibited in this medium, only EO and EO-H_2O_2 processes were useful for mineralization, as the organics primarily degraded by HClO formed from Cl oxidation. At this point, it is important to stress that H_2O_2 production at the cathode is a factor worthy of study, since the Fenton reaction was inhibited. The EO process with a BDD/stainless steel cell was found to be the most powerful treatment for the urine solution, yielding 96% ciprofloxacin removal and 98% mineralization after 360 min of electrolysis at an optimum pH of 3.0 and current density of 66.6 mA cm^{-2}.

Santos et al. [51] studied for the first time the degradation of dipyrone via electrochemical processes and electro-Fenton reaction using a 4% CeO_2/C gas diffusion electrode (GDE) prepared via a modified polymeric precursor method, which is a completely different approach to this type of material. This material was used to electrochemically generate H_2O_2 through oxygen reduction. The authors showed that the incorporation of cerium oxide into the GDE enhances the 2-electron oxygen reduction reaction (ORR) pathway in acidic media, as demonstrated by comparing H_2O_2 production during exhaustive electrolysis using a GDE containing either 4% CeO_2 or Vulcan carbon. The 4% CeO_2 electrode produced 330 mg L^{-1} of H_2O_2, while the Vulcan carbon only produced 108 mg L^{-1}. The increased H_2O_2 production of 4% CeO_2/C might be explained by the formation of two ceria phases, CeO_2 and CeO_{2-x}, with similar concentrations for the Ce(III) and Ce(IV) species, which leads to increased local oxygen concentration and facilitates H_2O_2 formation. Vulcan XC 72R carbon showed that the dipyrone was not removed during the two-hour electrolysis in all applied potentials by electro-degradation. In contrast, using CeO_2/C GDE, in 20 min all of the dipyrone was degraded, with 26% mineralization at −1.3 V. The electro-Fenton process using cerium GDE was most effective among all those employed for dipyrone degradation: all of the dipyrone was removed after 5 min of electrolysis at −1.1 V with 57% mineralization, as verified by TOC analysis. The authors concluded that 4% CeO_2/C is an efficient electrocatalyst for H_2O_2 electrogeneration in acidic media and a very promising material for the degradation of pharmaceuticals such as dipyrone.

In a review paper, Rodrigues et al. [60] noted that the production of domestic waste is increasing so much that waste disposal has become a serious problem in terms of concern around hazards to humans and the environments. In the specific case of Brazil, landfill is the most common method of municipal solid waste disposal. The authors claim that the study aims to evaluate the efficiency of various advanced oxidation processes (ozonation (alkaline pH), electrochemical oxidation + ozonation (alkaline pH) and photo-electrochemical oxidation) and biological systems as a treatment for leachate to remove acute-cyto-genotoxicity in municipal landfills. The authors verified that all investigated advanced oxidation processes higher removal rates than biological treatment. Nonetheless, only the treatment with EO + O_3 provided an effluent of a quality that complies with current legislation. In addition, EO + O_3 have shown the best energetic efficiency, with lower values for EEM than other methods. In contrast, the combinations of various AOPs showed better toxicity removal. The EO +O_3 and PECO treatments were highly efficient in reducing genotoxicity. The authors concluded that EO + O_3 was best, but no study has considered the use of modified cathodes that incorporate advanced materials, such as nanomaterials.

The incorporation of electrochemical advanced oxidation processes in a multistage treatment system for sanitary landfill leachate was performed in the work of Vilar et al. [63]. The authors claim that the use of a current density of 200 mA cm^{-2} was best for a PEF process using $[TDI]_0$ of 60 mg L^{-1}, pH 2.8 and 20 mg. The EAOPs' relative oxidative capability increased in the order EF with 12 mg $[TDI]_0$ L^{-1} < EF with 60 mg [TDI]0 L^{-1} < PEF with 60 mg $[TDI]_0$ L^{-1} SPEF with 60 mg $[TDI]_0$ L^{-1} with a BDD anode, 200 mA cm^{-2}, pH 2.8 and 20 °C were applied. The comparison between EAOPs and their analogous AOPs revealed a high superiority of EF over Fenton, less pronounced superiority of PEF over PF and a similarity between SPEF and SPF in terms of oxidation ability. Conventional cathode electrodes were used.

Staphylococcus aureus inactivation in a simulated wastewater treatment plant effluent by different electrochemical techniques was studied by Torres-Palma et al. [64], including the photo-electro-Fenton process. The authors reported similar results H_2O_2 accumulation, bacterial inactivation and DOC mineralization with either IrO_2 or Ti anodes. Consequently, the influence of anodic oxidation had a minor impact on the PEF, while the cathode was the electrode producing the disinfecting agent H_2O_2 necessary to conduct the Fenton reaction. Hence, a modified cathode rather than conventional carbon, e.g., with nanomaterials, can dramatically enhance the disinfection processes.

A comparative study using different mass proportions of WO_3/C (1%, 5%, 10% and 15%) for H_2O_2 electrogeneration and subsequent phenol degradation was performed by Santos et al. [65]. All synthesis parameters were studied, including the influence of the carbon substrate and the preparation methods. The increase in H_2O_2 production by the addition of a small amount of tungsten was explained by the fact that this material exhibits surface acidity due to the presence of Lewis–Brønsted acidic sites related to W^{6+} species. Additionally, the acid surface has a higher affinity for species with OH−, which provides more hydrophilicity and consequently higher activity in the ORR for H_2O_2 production. These results are associated with a higher content of oxygenated acid species, as identified by the XPS analysis. With respect to phenol degradation, the photo-electro-

Fenton method was most efficient, as it degrades approximately 65% and mineralizes approximately 10% of the phenol in 120 min. In 12 h exhaustive electrolysis total phenol degradation was observed in 4 h, and mineralization was maximized (~75%) after 12 h. Substantial degradation was observed with the photo-electro-Fenton method, which was expected because the UV and Fenton processes increase the hydroxyl radical content, and hydroxyl radicals are more oxidizing than H_2O_2. This is the second work in this chapter to demonstrate the effect of advanced nanomaterials in the cathode for applied organic pollutant degradation.

In fact, the current status of work in the area of hybrid nanomaterials as heterogeneous catalysts combined with EAOPs and hybrid advanced oxidation processes have emerged as a research area of great interest. Even though several processes involve the use of EAOPs with conventional materials for cathodes, research into the use of advanced and nanomaterials for peroxide electrogeneration has increased mainly around the formation of large amounts of this reagent, which is further used to produce radical hydroxyls, primarily to degrade organic pollutants.

3. Advanced Materials for Hydrogen Peroxide Electrogeneration and Its Applications

3.1. General Aspects

Although it involves multiple steps, the anthraquinone oxidation (AO) method is widely used in the large-scale production of hydrogen peroxide, unfortunately with high energy consumption and waste discharge [73, 74]. Recently, *In situ* H_2O_2 generation has gained increasing attention, reducing costs and potential risks in the transportation, storage, and handling of concentrated H_2O_2 [75]. H_2O_2 is a friendly environmental product that decomposes into non-contaminating products, such as water and oxygen [74, 76].

In situ generation of H_2O_2 in aqueous medium at concentrations reaching hundreds of milligrams per liter can be obtained by electrochemical technology [77, 78]. In contrast, the ORR can be conducted at the cathode and can generate various products via two pathways: (i) the two-electron pathway, forming hydrogen peroxide or hydroperoxide ion (HO_2^-) (which is the H_2O_2 conjugated base) and (ii) the four-electron pathway, which forms water or hydroxyl groups, depending on the solution pH.

To produce H_2O_2, stability depends on factors such as cell configuration, cathode properties and operational conditions [79]. Even using undivided or divided electrochemical cells, parasitic reactions occur, reducing H_2O_2 accumulation in the process [80], e.g., in the formation of water and oxygen from hydrogen peroxide.

An undivided cell can provide H_2O_2 oxidation to oxygen at the anode, producing the weak oxidant hydroperoxyl radical as a byproduct [81].

Carbon-based cathodes have been reported to be effective in H_2O_2 electrogeneration Supported metallic nanoparticles have been used to improve the catalytic activity of carbon for hydrogen peroxide electrogeneration, with some important examples discussed below, including metallic oxides and composites, reduced graphene oxide, nanotubes and zeolites.

3.2. Metallic Oxides and Composites

Metals and metallic oxides have been studied with their supports, which form composites; in most cases, these supports contain some type of carbon. These compounds are used to study the electrogeneration of hydrogen peroxide (H_2O_2) by the oxygen reduction reaction (ORR) following the two-electron mechanism, for the subsequent preparation of gas diffusion electrodes (GDE) and application in the degradation of organic pollutants by EAOPs. Some examples are titanium dioxide (TiO_2), cerium dioxide (CeO_2), ruthenium dioxide (RuO_2), tin dioxide (SnO_2), tungsten oxide (WO_3), manganese oxides (MnO_2 and Mn_3O_4), cobalt phthalocyanine and cobalt tetramethoxyphenyl porphyrin (CoPc and CoTMPP), cobalt oxides (CoO_2, Co_2O_4 and Co_3O_4), nickel-tin alloys (SnNi), vanadium oxides (VO_2 and V_2O_5), iron oxide in the magnetite phase (Fe_3O_4), iron (II) phthalocyanine (FePc), iridium dioxide (IrO_2), tantalum oxide (Ta_2O_5), zirconium dioxide (ZrO_2), tungsten-gold core-shell (W@Au), TiO_2-Au and niobium oxide (Nb_2O_5) [82-101]. The nanostructures of metals and metallic oxides are studied due to their high surface area and unique properties that differ from bulk materials to enhance H_2O_2 electrogeneration.

Several materials have been studied and used as a support for transition metals, such as silica, alumina, zeolites, ion exchange resins, and clays, among others [101-104]. However, the advantages of using carbon materials as support in metal and metallic oxide catalysts are significant [99-102, 105-108]; some examples of carbon currently being studied as supports for metals and metallic oxides are carbon black (Printex and Vulcan), graphite felt, carbon nanotubes and graphene [84-86, 88-93, 95, 97-100, 108]. For many years, carbon has been used as a catalyst support due to its properties, such as high stability in basic and acid media, high surface area (which contributes to an excellent dispersion of metals and metallic oxides on its surface), stability at high temperatures, possible control of surface functionalization (which allows greater dispersion of the metal) and control of polarity and hydrophilicity of the synthesized catalyst [100-101, 104, 108].

Supported metal and metal oxide-based electrocatalysts can be synthesized by various methods, such as the impregnation method, the polymer precursor method (PPM) (also known as the Pechini method), thermal decomposition of a polymeric precursor solution (DPP), the sol-gel method (SGM), the hydrothermal method, precipitation or co-precipitation method, chemical vapor deposition, and liquid phase reduction, among other methods [89, 97-100, 104-105, 109-111].

The development of electrodes chemically modified with metallic oxides is an interesting and efficient route to improve carbon properties, and the catalytic activity in ORR and H_2O_2 electrogeneration via the two-electron mechanism [97]. The following recent works describe the potential use of electrocatalysts composed of metals and metallic oxides supported on various types of carbon for H_2O_2 electrogeneration.

J. F. Carneiro et al. [90-91] used carbon black (Printex 6L) modified with 5% of nanoparticles of Ta_2O_5 (w/w). The catalysts based on Ta_2O_5 exhibited high stability in corrosive environments and a high starting potential in the ORR (~0.95 V vs. SHE). This electrocatalyst was prepared by DPP and was characterized morphologically and chemically by transmission electron microscopy (TEM) and X-ray photoelectron spectroscopy (XPS); it was also characterized electrochemically by ORR using the rotating ring-disc electrode (RRDE) technique and a GDE for the study of H_2O_2 electrogeneration. In Figure 1 (A) and (B) we observe nanoparticles of Ta_2O_5 supported on Printex 6L, and in Figure 1 (C) we estimated the nanoparticle size [89]. The electrocatalyst containing nanoparticles of Ta_2O_5 exhibited an efficiency of 83.2% in the conversion of O_2 (g) to H_2O_2, while the pure Printex 6L and treated obtained material, respectively, showed an efficiency of 65.3% and 74.5% in this conversion. This higher selectivity for H_2O_2 electrogeneration is attributed to the synergistic effect between Ta_2O_5 nanoparticles and the functional groups of oxygen on the material surface. The electrogeneration of H_2O_2 using GDE indicated that the electrocatalyst modified with Ta_2O_5 nanoparticles produced 27.9 mg L^{-1} of H_2O_2, while Printex 6L produced only 19.1 mg L^{-1} of H_2O_2, indicating an increase of 46.1% when adding the nanoparticles of Ta_2O_5 to Printex 6L, with lower energy consumption (15 kW h kg^{-1} vs. 18.8 kW h kg^{-1}) [89]. Thus, this GDE composed of Ta_2O_5 nanoparticles supported on Printex 6L proved viable for use as a cathode in EAOPs.

M. H. M. T. Assumpcão et al. [96] used carbon black (Vulcan XC-72R) for its high surface area, and modified this carbon with CeO_2 nanoparticles for their superior properties, such as corrosion resistance and wear resistance; this material is efficient in several applications, such as fuel cell electrodes. Electrocatalysts composed of CeO_2 nanoparticles supported on Vulcan XC-72R were prepared in different proportions by PPM and were characterized by X-ray diffraction (XRD), TEM, XPS, and an ORR study using the RRDE technique. A GDE of the best electrocatalyst (4% CeO_2) was produced. This electrode was used for the electrogeneration of H_2O_2 with varying potential, with a verified improvement in over pure Vulcan XC-72R. The GDE with 4% CeO_2 produced 330 mg L^{-1} of H_2O_2, whereas the Vulcan XC-72R produced 108 mg L^{-1}. The electrodegradation of the dipyrone was studied, verifying that after 2 h of electrolysis, no degradation was observed using the GDE of Vulcan XC-72R at several potentials. Using the GDE with 4% of CeO_2 supported on Vulcan XC-72R, it was found that after 20 min all dipyrone was degraded, and 26% was mineralized when a -1.3 V potential was applied. The electro-Fenton process was also tested; it degraded the dipyrone in 5 min and obtained 57% mineralization in electrolysis when a -1.1 V potential was applied [99, 105]. Therefore, we confirm the potential of metals and metallic oxides supported on materials with relatively lower cost, such as carbon, increasing its efficiency in H_2O_2 electrogeneration and consequently increasing its potential in the degradation of organic pollutants by EAOPs.

E. E. S. Souza et al. [97] used graphite felt for its good electronic conductivity and large number of reaction sites. However, because this material has a hydrophobic surface, its wettability and electrochemical activity must be improved and to this end SnO_2, a semiconductor with a range of 3.0-3.2 eV band at room temperature and conductivity based on O_2 vacancies, was added. It can be deactivated by adding O_2 to its oxide layer when anode potentials are applied, resulting in a less conductive material. To overcome this obstacle, RuO_2 was added as a dopant, commonly used as an active catalyst, to increase the conductivity of this electrocatalyst. In this way, electrodes based on graphite felt (RVG 4000) covered by $(RuO_2)_{0.9}(SnO_2)_{0.1}$ were developed. This electrocatalyst was synthesized by the Pechini method [97]. The modified graphite felt with RuO_2 and SnO_2 yielded an electroactive area 250 times larger than the bare graphite felt and was found to be electrochemically stable. This electrocatalyst had a 72.2% increase in H_2O_2 electrogeneration, so the electrocatalyst is promising for use in PEOA, e.g., in the Electro-Fenton process [97].

W. R. P. Barros et al. [85] used carbon black (Printex 6L) and graphene as supports for Fe_3O_4 (5 wt%) nanoparticles. Printex 6L was chosen due to its low cost and has already been reported in papers as an excellent electrode material for the ORR for electrogeneration of H_2O_2 in alkaline medium [112]. The graphene has high electrical conductivity, high surface area and excellent chemical stability, and ORR using graphene in alkaline medium has already been reported to occur via the 2-electron mechanism [108]. The electrocatalysts were synthesized by the precipitation method, in which sodium borohydride ($NaBH_4$) was used as the reducing agent to form the magnetite phase. These electrocatalysts were characterized by XRD and TEM and the ORR was studied by the RRDE technique. The electrocatalytic activity of the electrocatalysts for the ORR, expressed as the current density -0.3 V vs saturated calomel electrode (SCE), increased in the electrocatalysts containing the Fe_3O_4 nanoparticles relative to the pure supports. In order of increase the electrocatalytic activities were for the ORR for the Printex 6L of 0.29 mA cm^{-2}, Fe_3O_4/Printex 6L of 0.38 mA cm^{-2}, graphene of 0.85 mA cm^{-2} and Fe_3O_4/graphene of 1.12 mA cm^{-2}. The electrocatalysts containing Fe_3O_4 nanoparticles demonstrated a number of electrons transferred from approximately 2.7 electrons and in the range of -0.2 and 0.7 V (SCE), yielding a percentage of H_2O_2

electrogenerated above 60%. The Fe_3O_4/Printex and Fe_3O_4/graphene electrocatalysts presented excellent durability, with H_2O_2 electrogeneration utilizing a GDE. The Fe_3O_4/graphene electrocatalyst showed higher catalytic activity for the ORR, which includes a more positive start potential, indicating lower energy expenditure and a higher current density. The Fe_3O_4/Printex electrocatalyst showed greater selectivity in H_2O_2 electrogeneration [85]. This work showed that these two electrocatalysts are promising for study in the form of GDE for H_2O_2 electrogeneration and later for use as cathodes in EAOPs.

F. V. E dos Reis et al. [95] used carbon black (Vulcan XC-72R) as a support for the synthesis of promising electrocatalysts for H_2O_2 electrogeneration by ORR in alkaline medium, which was modified with TiO_2-Au hybrid material. TiO_2 colloidal spheres decorated by Au [113] and TiO_2 wires were synthesized by the hydrothermal method and decorated with Au, and subsequently these hybrids were supported on Vulcan XC-72R in various proportions by the impregnation method. These electrocatalysts were characterized by SEM, UV-VIS spectra, XRD and XPS, and the ORR was studied by the RRDE technique. Figure 4 shows SEM images of various studied TiO_2 morphologies [95] and Figure 5 shows SEM images of TiO_2-Au hybrid materials in various different morphologies.

The use of TiO_2 colloidal spheres led to higher activities for H_2O_2 electrogeneration by ORR than pure Vulcan XC-72R, and the electrocatalyst containing 3% TiO_2-Au was most promising for the 2-electron mechanism, with obtained 75% of H_2O_2 electrogenerated. In contrast, the TiO_2 wires led to detrimental effects to catalytic performance for all hybrid materials. These results can be explained based on the morphological, chemical and conductivity differences between TiO_2-Au/C electrocatalysts [95].

These studies showed how various metals and metallic oxides are promising for H_2O_2 electrogeneration when supported on different carbon types, indicating that these materials are promising for application to GDE in PEOA to mineralize organic pollutants.

3.3. Reduced Graphene Oxide

Graphene has interesting properties, such as high electrical conductivity, high surface area and many electrochemically favorable edge carbons per mass of graphene. Such properties favor the transfer of electrons between molecules and the substrate with a low overpotential [114-115].

Graphene-based materials have potential for various applications: oxidation of alcohols and/or water [116-117], capacitors [118], lithium-ion batteries [119-120], adsorption of aquatic pollutants [121-122], hydrogen peroxide determination [123], oxygen reduction for fuel cell applications [124-125] or H_2O_2 electrogeneration for polluted water remediation [114, 126-127], among others.

In the last decade, graphene-based materials have been heavily studied for environmental remediation due to their unique properties [128-129]. Among these varieties, reduced graphene oxide (rGO) has shown promise for hydrogen peroxide electrogeneration, a powerful tool for the degradation of persistent organic pollutants.

Graphene-based materials are also attractive for their cost, process capability and environmental implications [128]. Compared to carbon nanotubes (CTNs), graphene presents a larger area available for processes. For example, graphene oxide compared to pure graphene presents more realistic possibilities for application to environmental remediation due to its lower cost of production [128]. Graphene oxide (GO) and reduced graphene oxide can be easily synthesized by chemical exfoliation of graphite without a complex apparatus or metal catalysts. GO already has a large number of functional groups containing oxygen, and no additional acid treatments are required to confer hydrophilic character and reactivity to the GO [128]. This is significant, as these functional groups are likely responsible for the best activity in environmental remediation.

Graphene oxide is usually synthesized through the oxidation of graphite based on Hummers' method, which employs concentrated sulfuric acid, nitric acid and potassium permanganate [130]. Some recent advances in the synthesis of graphene oxide have been achieved, e.g., as proposed by Marcano et al. [131]. These researchers obtained more oxidized and hydrophilic GO, under milder oxidation conditions without generating toxic gases [131].

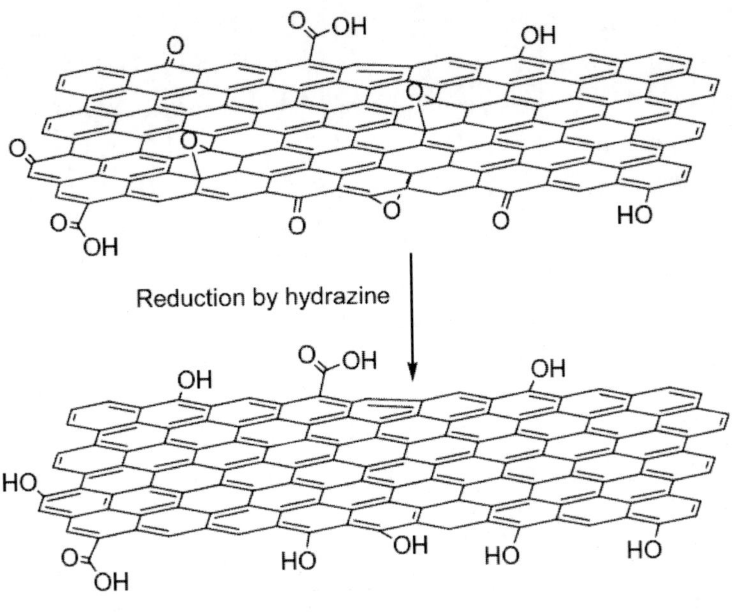

Figure 1. Reduction of graphene oxide to reduced graphene oxide [130] with permission.

The rGO can be obtained from GO by various routes that involve the use of hydrazine, sodium borohydrate, hydroquinone, gaseous hydrogen, strong alkaline solutions or thermal reduction, among other reducing agents [130]. In the hydrazine route, GO sheets dispersed in water are reduced by hydrazine hydrate to form rGO, as shown in Figure 1 [130]. The reduced sheets aggregate and precipitate because rGO is less hydrophilic [130].

Many studies have aimed to electrogenerate H_2O_2 by means of gas diffusion electrodes (GDEs) for use in electrochemical advanced oxidation processes (EAPOs) [132]. Reduced graphene oxide is promising for the reduction of oxygen to H_2O_2 due to its properties, which facilitate the transfer of electrons and accelerate the reaction. Moreover, rGO has good chemical stability. The literature reports that incorporation of metal oxide nanoparticles into rGO [114, 132], electrochemical reduction of rGO [133], doping of rGO with N [134], multilayer composites of rGO [126], among other materials increase the selectivity of rGO for H_2O_2 electrogeneration.

Le et al. [135] studied synthesized graphene oxide prepared using a modified Hummers' method. This GO was deposited on a Carbano Felt (CF) substrate. In the next step the GO was reduced by various methods: electrochemical (CPR-CF), thermal (TR-CF) and chemical (CR-CF). The electrodes of CF modified with rGO prepared by various methods were electrochemically evaluated by cyclic voltammetry (CV) and electrochemical impedance spectrometry (EIS). Figure 2 shows that the GO reduction methodology directly influences the electrochemical activity of the rGO, since material synthesized via the electrochemical reduction of GO showed higher electroactive surface area (137.969, 124.172, 27.594 and 13.797 cm^2 for CPR-CF, TR-CF, CR-CF and rawCF, respectively) and lower resistance to load transfer [135].

Figure 3a shows the effects of graphene modification on electro-catalytic activity for H_2O_2 production. In a saturated solution of O_2, H_2O_2 formation was evidenced by the appearance of a single reduction pseudo-wave related to oxygen reduction between 0 and -0.9 V/SCE. The modified electrode (CPR-CF) in saturated O_2 solution demonstrated a higher current for H_2O_2 production (-22 mA cm^{-2}) compared to crude CF (-4 mA cm^{-2}) [135].

The results of electrochemical analysis were confirmed by comparison of the degradation of Acid Orange 7 (AO7) using rawCF and CPR-CF, which confirmed that degradation with the modified electrode is faster (Figure 3b) [135]. The improved electrocatalytic activity of the modified electrode was attributed to its greater hydrophilicity and electrical conductivity [135].

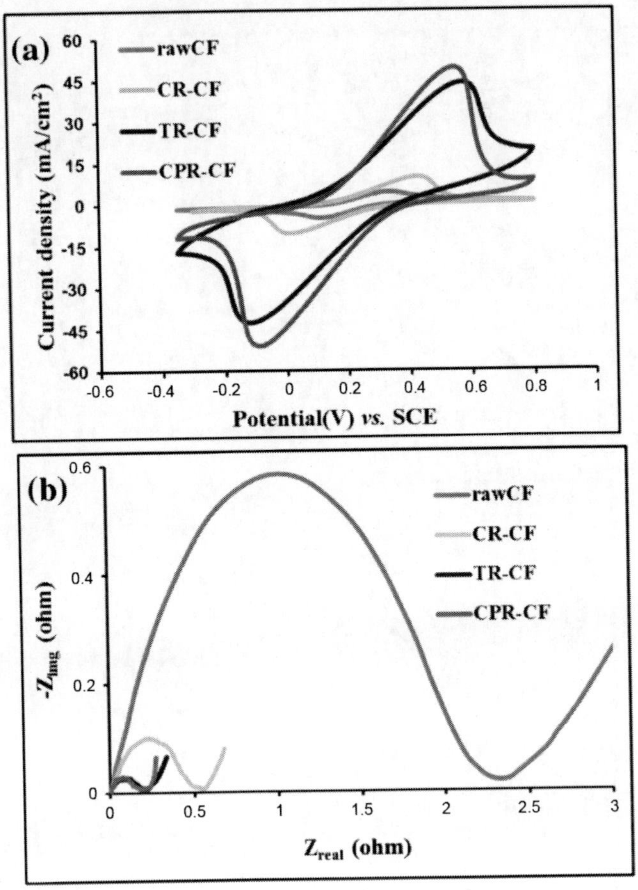

Figure 2. (a) CVs and (b) EISs of modified and unmodified electrodes in 10 mM $K_3[Fe(CN)_6]$ + 1 M KNO_3 solution [135] with permission.

Dong et al. [136] synthesized three GDEs of carbon black modified with rGO. In the process, the GDE [137] was prepared and deposited, and this electrochemically reduced the GO on the surface of the GDE [138], yielding modified carbon black GDE with rGO (ERGO). Dong et al. [136] found that EDGs subjected to reduction for longer times (60 and 120 min) produced more radical species ($^{\bullet}OH$ and $^{\bullet}O_2$) and therefore were more efficient in the degradation of bisphenol A.

From the ESR spectroscopy analysis [136] and evaluation of the electrolysis, the authors proposed a mechanism of ERGO performance in the degradation of bisphenol A in which ORR promoted by ERGOs involved one, two and four electrons per molecule of O_2, resulting in O_2, H_2O_2 and H_2O, respectively [136]. The increased electrochemical reduction time of the GO decreased the basal carbon spacing and increased the concentration of defects at the edges and basal planes of the carbon [136].

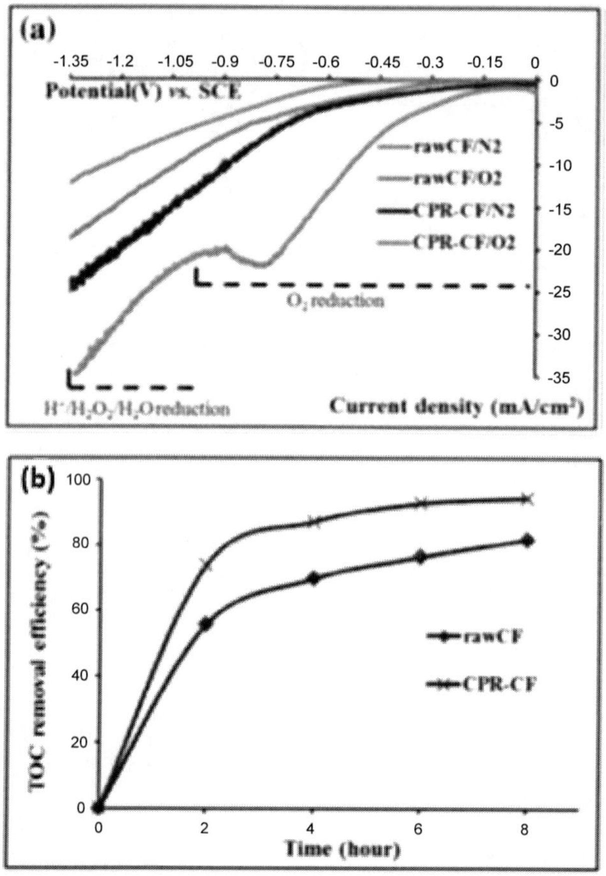

Figure 3. (a) Linear scanning voltammetry of modified and unmodified electrodes in 50 mM Na_2SO_4, N_2/O_2 saturated solution, pH = 3. Scan rate 5 mV/s. (b) TOC removal after 8 h Conditions: V = 30 mL, $[AO_7] = 0.1$ mM; $[Fe^{2+}] = 0.2$ mM; pH = 3; $[Na_2SO_4] = 50$ mM, I = 40 mA, using the Pt anode and raw CF or CPR-CF cathode [135] with permission.

Wu et al. [134] produced N-doped rGO (N-rGO) to control the electron transfer number (n) of ORR and apply this material to the degradation of organic compound by electrochemical methods. Melamine and pyrrole were used as nitrogen sources for the production of N-rGO by microwave-assisted hydrothermal synthesis (MAHS). The electrocatalytic degradation of Orange G (OG) revealed a high degree of decolorization for the azo group (-N = N-) [134].

The study showed that electron transfer number of ORR can be controlled by the synthesis temperature and total concentration of nitrogen groups [134]. The electron transfer number of ORR decreased with increasing synthesis temperature and the content of the quaternary-N structure [134]. The N-dopant structures of these materials were primarily pyridinic-N, pyrrolic-N, quaternary-N [134]. At higher temperatures, the pyridinic-N and pyrrolic-N structures were significantly transformed into the quaternary-N structure. Wu et al. proposed that the quaternary-N structure was the main contributor to the 2-electron pathway ORR [134].

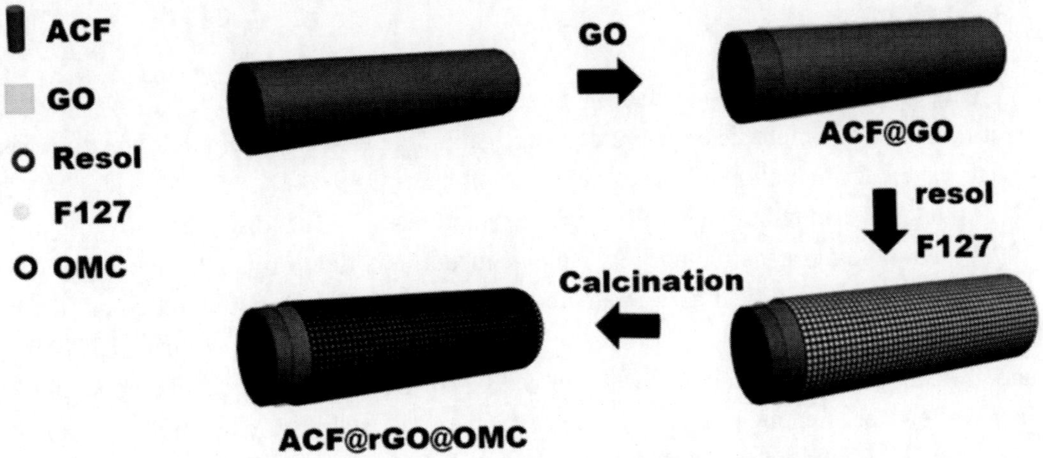

Figure 4. Schematic diagram of the preparation of ACF@rGO@OMC composite [126] with permission.

Ren et al. used a two-step method to synthesize a new surface-coated ACF@rGO@OMC multilayer cathode (ACF = activated carbon fibers, rGO = reduced graphene oxide and OMC = ordered mesoporous carbon) (Figure 4) with a high electrical conductivity and low electrochemical impedance [126]. The ACF was used as the substrate, the rGO as the conducting layer and the OMC as the oxygen diffusion channel [126].

The ACF@rGO@OMC composite compared to the ACF@OMC increased the generation of hydrogen peroxide 31 to 85 mg L^{-1} and the electro-Fenton process current efficiency from 25.0% to 40.0%. It remained stable after 10-fold reaction in an electro-Fenton process to remove phthalic esters [126]. Ren et al. posited that the addition of rGO provided better electrochemical performance because it improved crystallization and electron transport, expanded the electroactive surface area and increased the rate of oxygen diffusion [126].

The properties of reduced graphene oxide depend on the space distribution of functional groups and structural defects [130]. The rGO often contains many defects, such as nanoholes and Stone-Wales defects (heptagon/pentagon bound network of carbon atoms) [135, 138]. Thus, the study of the structure of rGO and the development of synthetic processes capable of modeling the structure of rGO according to the desired property are fundamental for understanding its properties. In this context, prospects for electrogeneration of H_2O_2 by rGO-based materials are promising, since several synthesis methods have been successful in obtaining materials based on rGO with potential application in the degradation of organic pollutants.

3.4. Nanotubes

This section discusses the use of electrochemical materials applied to wastewater treatment with an emphasis on nanotubes, mitigating the worrisome emerging hydrocrisis due to the bioaccumulation of these chemical pollutants [139-142].

Among all nanotubes materials, carbon nanotubes (CNTs), discovered by Ijima in 1991, are a type of nanostructure that have opened a new field of nanotechnology [143]. CNTs have been employed as a strategic material in applications ranging from energy conversion processes [144], the provision of a confined environment for catalysis [145], and as an agent for bacterial and viral removal [146]. CNTs are considered a good option for wastewater treatment due to their mechanical stability, flexibility, chemical resistivity, large surface area, hollow structure, and high electron affinity.

CNTs are hollow cylinders constructed of hexagonal unit cells shaped by carbon atom in sp^2 hybridization, unlike the sp^3 hybridization of carbon diamond [147]. There are two well-known CNT nanostructures: single-walled or multi-walled configurations. The single-walled carbon nanotube (SWCNT) is formed by rolling a single graphene sheet seamlessly to model a cylinder with a diameter of 1 nm and centimeter-scale length [148]. Similar to the basal plane separation in graphite, the multi-walled carbon nanotube (MWCNT) is an array of cylinders formed concentrically and separated by 0.35 nm, with diameters from 2 to 100 nm and lengths of tens of microns [149-150]. Exemplary structures are shown in Figure 5.

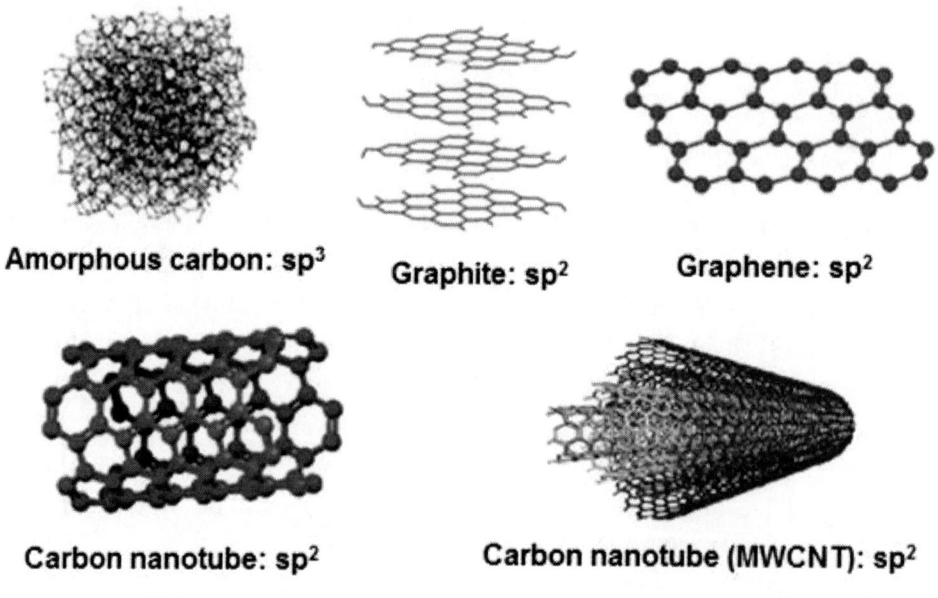

Figure 5. Scheme with carbon nanostructures (our image).

CNTs have unique physical properties that make them very versatile, with a hundred-fold higher tensile strength compared to steel, thermal conductivity superior to other materials except purest diamond, and electrical conductivity similar to copper but higher currents [143]. The chemical reactivity of CNTs is linked to physical characteristics such as (1) an exohedral surface that increases with increasing curvature of the sidewall, due to the distortions of normally planar C–C bonds of sp^2 hybridization and the misalignment of p orbitals; and (2) the high rate of electron transfer on the sidewall, which leads to the high electrical conductivity of CNTs; and (3) especially the lateral wall curvature and structural defects [149].

These carbon nanostructures enable the expansion of many water treatment processes. CNT materials can be used either in anodes (direct oxidation of pollutants) or cathodes (indirect oxidation of pollutants). The combination of electrochemical materials and CNT filters has been applied for wastewater treatment. Recent studies have indicated that CNT filters modified with electrochemical materials effectively removed aqueous organic pollutants, proteins, viruses, azo dyes, pharmaceuticals, perfluorinated chemicals, and phenol [150]. Moreover, the application of electrochemistry could reduce filter fouling rate by *in situ* destruction and biological inactivation [151].

However, these CNT electrochemical materials have a limitation. The anode potential in CNT filters must be higher than the potential redox for a chemical to be oxidized. However, depending on the anode potential, the water-splitting mechanism (Eq. 5) competes with the oxidation of the chemical compound, decreasing the efficiency of the CNT filters. The use of metal nanoparticles coating the CNT filters has improved the efficiency of these electrochemical materials [152].

$$2\,H_2O \rightarrow O_2 + 4H^+ + 4e^-,\ E^0 = +1.03\ V\ vs.\ Ag/AgCl \qquad (5)$$

The surfaces of CNTs can be modified to improve their function and performance through covalent attachment of functional groups or the non-covalent adsorption of functional molecules onto the surface of CNTs [153]. Mishra et al. [154] used MWNT (Fe_3O_4-MWNTs) nanocomposites, and Liu et al. [155] also coated CNT filters with titanium dioxide (TiO_2) for arsenic removal.

The high bacterial adsorption capacity of CNTs has enabled potential applications for pathogen removal and detection [156] and have thus attracted much research interest in developing treatment-based and sensor-based applications in the water and wastewater treatment industry. Recent works showed excellent results in the removal of *Escherichia coli (E. coli)* and *Staphylococcus aureus (S. aureus)* because SWCNTs have high adsorption affinity towards both [157]. It has also been reported in filtration using a poly-(vinylidene fluoride) based microporous membrane covered with SWCNTs to remove pathogens, such as E. coli K12 and the MS2 bacteriophage, a model virus particle [158].

Another class of emerging pollutants are pesticides. The detection of organophosphate pesticides has been achieved using CNTs introduced into an acetylcholinesterase (AChE) based biosensor for pesticide detection [159]. MWCNTs allowed fast and simultaneous extraction and electrochemical detection of organophosphate pesticides, demonstrating one of the myriad applications for CNTs [160].

CNTs can also be used as cathodes in EAOPs to produce hydrogen peroxide (H_2O_2) for further oxidize various organic pollutants [161]. A CNT-modified cathode can be used in the oxygen reduction reaction (ORR) to generate H_2O_2 (Eq. 6) via Fenton reactions (Eq. 7) and to produce oxygen and water as by-products after oxidation [162].

$$O_2 + 2H^+ + 2e^- \rightarrow H_2O_2, E0= +0.45 \text{ V vs. Ag/AgCl} \quad (6)$$

$$Fe^{2+} + H_2O_2 \rightarrow Fe^{3+} + OH^- + \bullet OH \quad (7)$$

Recent studies with wastewater treatments using EAOP with modified CNTs were applied to the following pollutants: basic red 46 azo dye; basic yellow 28 dye; and direct red 23 azo dye, all using cathodes with multi-walled CNT with PTFE [163].

With continuously falling prices for CNTs, electrochemical CNT filters could be widely adopted for wastewater treatment. Surface modification with metallic nanoparticles, as shown in Figure 6, and functionalization of CNT filters will further improve their applications and treatment efficiency [164].

The use of CNTs as a support for catalytic metals is very interesting for electrochemical materials because CNTs provide high electrical conductivity and a specific interaction between the d-orbital of catalytic metal and the carbon surface to promote delocalization of electrons that favor catalytic activity [165].

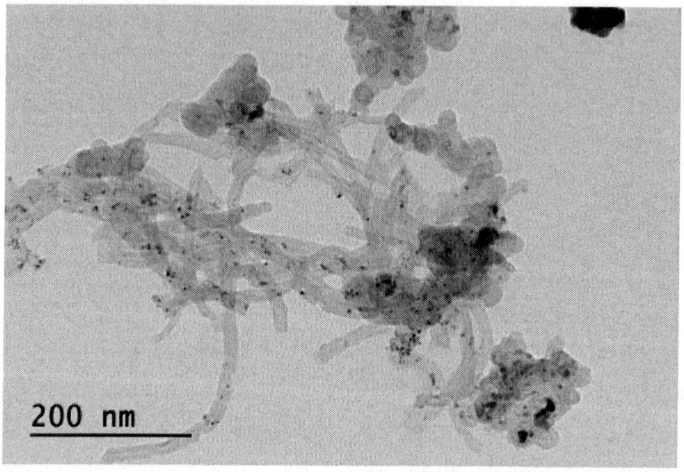

Figure 6. MWCNT modified with Pt metallic nanoparticles (prepared by our group).

Soon after the discovery of CNTs, it became evident that properties achieved with carbon could be expanded to other inorganic materials. Investigations have been initiated to synthesize other compounds such as graphite carbon to achieve a nanotube-like structure.

The most studied non-carbon nanotubes are B_4C carbides [165]; chlorides [165]; boron nitride (BN-NT) [165]; dichalcogenides of transition MeX_2 (where Me = Mo, W, Nb, and X = S, Se, Te [166-169]; oxides (MoO_3, SiO_2, Al_2O_3, V_2O_5, TiO_2, MnO_2, Co_3O_4, WO_3) [170]; halogenides; and nanofibers, for example SiC-SiO_2-BN [171-172].

Some specific properties of this nanotube (NT) allow a wide field of applications. The best physical properties are symmetric nanotubes for research into applications in Raman and Infrared techniques [173]; electronic properties for nanoelectronic research [174]; unique quantum properties, since although they are extended objects, at least two dimensions are smaller than the corresponding wavelengths [175]; thermodynamic and mechanical properties such as stability and flexibility that allow use in industrial processes [176-177]; and superconductivity and semiconductor properties suitable for energy research [178].

Using a simple optimized electrochemical anodization of a titanium plate, the first ordered matrices of TiO_2 NT were obtained in 1999 by Zwilling and co-workers [179]. Since then, there has been intense research activity into the synthesis, properties, modifications, and applications of various TiO_2 nanomaterials [180]. Figure 7 shows a typical formation of NT TiO_2.

The band gaps for TiO_2 NT are relatively smaller than TiO_2 bilayers, resulting in red optical absorption. This change has indicated that the electronic and optical properties of TiO_2 nanostructures can be effectively controlled by changing their symmetry, dimensions, and structure for each specific application [182]. TiO_2 NT has shown successful removal of dyes and pharmaceuticals, as well as deactivation of bacteria in wastewater treatments [183-184]. TiO_2 NT, a stable, non-toxic and abundant semiconductor, is widely studied as a promising material for the remediation of polluted water [185].

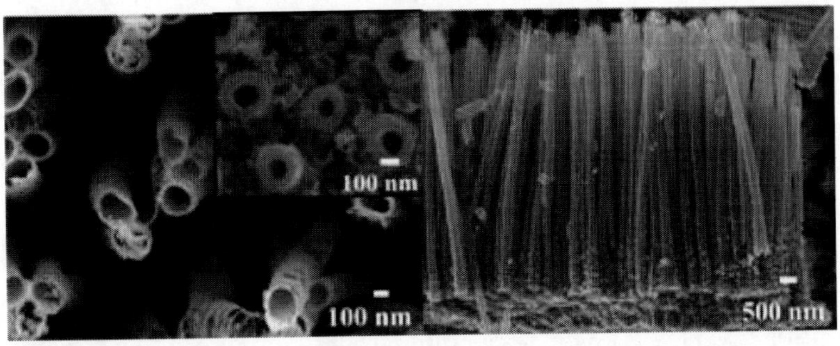

Figure 7. Typical NT TiO_2 formation by anodization [181].

As an important n-type semiconductor, WO_3 NT has received considerable attention due to outstanding properties. It has been used to degrade dyes and as a strong adsorbent for wastewater treatment [186].

In summary, carbon nanotubes and non-carbon nanotubes have been widely applied as electrochemical materials for wastewater treatment. Its uses range from filters to bacterial deactivation due to its versatility in the control of synthesis and modifications. Therefore, due to their vast unique properties, nanotubes will expand and multiply in wastewater treatment technologies through electrochemical processes.

3.5. Zeolites

Zeolites are crystalline solids composed of hydrated aluminosilicate, which are minerals composed of interlinked tetrahedra of alumina (AlO_4) and silica (SiO_4). These solid three-dimensional crystal structures are built from aluminum and alkaline and alkaline-earth metals (such as sodium, potassium, and magnesium). Zeolites have a very regular arrangement, with large open pores, sometimes referred to as cavities or channels. These pores are arranged in a negatively charged honeycomb framework, inside where cations or anions, water or smaller molecules may reside. The zeolites have many different crystalline structures [187-188]. The selective characteristics of zeolites can be used to control pollutants and treat effluents.

The first zeolite, identified as stilbite, was discovered in a copper mine in Sweden in 1756 by the chemist and mineralogist A. Crönstedt. He observed that the mineral exhibits intumescence when heated in a blowpipe flame and was thus termed a zeolite. Zeolite is derived from two Greek words, zeo and lithos, meaning "to boil" and "stone," given its capacity to release water vapor under heating. There are approximately 60 naturally occurring zeolites, formed in both volcanic and sedimentary rocks; according to the US Geological Survey, the most commonly mined forms include chabazite, clinoptilolite, and mordenite. However, synthetic zeolites are more commonly employed commercially due to their higher uniformity in composition, high purity and possibility of optimizing their properties for specific applications [189-191].

The most common zeolite synthesis is the hydrothermal method, proposed by Barrer in 1948 and Milton in 1949. The methodology developed by Barrer was based on the conversion of mineral phases under concentrated salt solutions from 180 to 270°C. Milton employed more reactive reagents, for example, silica gel or sodium silicate, as sources of silicon, with alumina or sodium aluminate as sources of aluminum. Sodium hydroxide was used as a source of sodium and a mineralizing medium. This allowed the use of milder conditions, such as lower temperatures and pressures. These works initiated a series of studies of reaction routes to obtain zeolites [192-193].

Zeolites are currently prepared using hydrothermal conditions because the open structure of these aluminosilicates must be stabilized during growth by occupying the channels and cavities formed by host molecules. The hydrothermal synthesis of a zeolite is generally performed in basic media and consists of mixing silicon and aluminum-containing sources with a compensatory cation. The reaction mixture is heated (above 100 °C) in an autoclave under high pressure. After this temperature induction period, crystals of the zeolitic product can be detected and recovered by filtration, washing and drying [194].

Synthetic zeolites have a highly uniform composition and high purity and are commonly used as catalysts. However, natural zeolites are more commonly used in effluent treatment. The low cost of natural zeolites has stimulated the development of cheaper treatment systems [195].

Zeolites can be used in industrial products (absorbents for oil and spills, gas separations, ion exchange), water treatment (e.g., water filtration, heavy metal removal, swimming pools) and wastewater treatment (e.g., ammonia/ammonium removal in municipal sludge/ wastewater, heavy metal removal, septic leach fields) [196].

Zeolites are extremely useful as catalysts and can promote a diverse range of catalytic reactions involving organic molecules. The reactions can occur in the pores of the zeolite, which allows a greater degree of product control. The most important are cracking, isomerization and hydrocarbon synthesis. Another widely used property of zeolites is in gas separation. The porous structure of zeolites captures molecules of certain dimensions inside pores. In water treatment, the principal property of zeolites is ion exchanges, as cations hydrated within the zeolite pores are bound loosely to the zeolite framework and can readily exchange with other cations in aqueous media [195, 197-198].

Hermassi et al. [199] developed a powdered zeolitic material NaP1-FA zeolitic material synthesized from Narcea FA (fly ash) and its calcium modified form (CaP1-NA), which are capable of high phosphate sorption in neutral to slightly basic conditions for the study of sorbent materials to recover phosphate from treated waste-water effluents. For phosphate-sorption equilibrium, experiments were performed by varying the experimental conditions, including solution pH, phosphate concentration, and the presence of competing ions. The maximum phosphate-sorption capacities were 57 ± 5 and 203 ± 11 mg $P–PO_4$ g^{-1} for NaP1-FA and CaP1-NA, respectively. The sorption capacities of both zeolites in the pH range expected for waste-water effluents (pH 7 to 9) were slightly dependent on pH, exhibiting a maximum at pH 8. Phosphate removal proceeds through two primary mechanisms: (a) surface complexation with ≅AlOH and ≅FeOH groups of the zeolitic structure or unreacted minerals from the FA; and (b) the formation of Ca–phosphate phases, mainly brushite. The removal mechanisms were confirmed by XRD analyses and P speciation. The stabilities of the phosphate-loaded zeolite samples as evaluated by the extraction experiments confirmed their potential availabilities in soil applications. The higher solubility of this zeolitic material is promising as a novel inorganic zeolite/CaP1-NA/brushite fertilizer.

In other work, Humelnicu et al. [200] studied in batch mode for the adsorption of Basic Blue 41 textile dye from aqueous solutions on a zeolite tuff (Nereju-Romania). Basic Blue 41 textile dye adsorption was observed in batch mode on a natural zeolitic volcanic tuff (analcite is predominant zeolite), and activity depended on the selected size of the tuff particle, correlating with their varying composition. There was a higher efficiency of small tuff particles (0.085 mm diameter) with large amounts of zeolites, compared to those with large dimensions (0.9 mm) with low amounts of zeolites. The zeolite content was the main factor controlling the adsorption capacity of the volcanic tuff for Basic Blue-41 dye. The initial pH and dye concentration, contact time between adsorbent and dye, and temperature are factors that influence this adsorption. For equilibrium analysis, the Langmuir, Freundlich, Temkin and Dubinin–Radushkevich adsorption isotherms models were applied. The obtained data fit the Langmuir isotherm model. The monolayer adsorption capacity of the adsorbent was 192.31 mg g^{-1} at 50°C and pH 5 for tuff particles of 0.085 mm. The mechanism of this process is complex and can be discussed in terms of the adsorption of the dye cation on the particle surface due to Si OH groups and an ion exchange equilibrium after diffusion of the dye cation to active sites inside the zeolite network. The increased external surface area and mesopore volume for the 0.085 mm particles with high zeolite content exposes more active sites both on the surface and inside the adsorbent due to its high content of analcite.

In the case of wastewater treatment, we find several works in the literature; recently, Wang et al. [201] reported an alternative method for synthesizing zeolites from coal fly ash via mechanochemical treatment. The raw fly ash was size-reduced to approximately 75 μm and mechanically ground via a wet ball milling system for 6 h in the presence of a 2.0 M sodium hydroxide (NaOH) solution to synthesize zeolite. The synthesized zeolite and the raw fly ash were used to remove Cr(VI) from aqueous solutions. The results show that, under equilibrium conditions, the maximum removal of Cr(VI) increases from 0.803 mg/g raw fly ash to 1.549 mg/g synthesized zeolite. The improved Cr(VI) removal capacity of the synthesized zeolite can be attributed to the formation of zeolite X and to a lesser extent, the improved physical adsorption capacity. Low-valence iron (Fe_3O_4) in the raw fly ash and synthesized zeolite plays a significant role in reducing and removing Cr(VI) from aqueous solution, possibly reducing the toxicity of hexavalent chromium polluted solution, which could cause health problems for human beings.

In other work, Filippidis et al. [202] developed a methodology to treat wastewater from the Sindos industrial area of Thessaloniki (Greece) using HENAZE, a very high-quality zeolitic tuff containing 86 wt.% HEU-type zeolite (clinoptiloliteheulandite), 4 wt.% mica + clay minerals, 5 wt.% quartz + cristobalite + tridymite and 5 wt.% feldspar. HENAZE y cannot be used as a feed additive for animals or in nutrition supplements, since it contains SiO_2-minerals (quartz, cristobalite, tridymite). Chemically, it is suitable for use as a soil conditioner, since the concentrations of trace elements (Cd, Cr, Cu, Hg, Ni, Pb, Zn) are lower than the maximum allowable concentrations for agricultural soils. The treatment of 300 mL of wastewater from the Sindos industrial area of Thessaloniki (initial pH 7.7) with various proportions (0.1, 0.2, 0.4, 0.8, 1.6, 3.2, 6.4 g) of HENAZE with a grain size of < 0.5 mm, resulted in overflowing clear water at pH 7.4 – 7.3, free of odors and improved 69–76%, over 97% for the P_2O_5 content, 33 to above 50% for Pb and 77% to over 88% for Cr.

Wang et al. [203] developed a modified zeolite for adsorption of bisphenol-A (BPA). Specific modifications of zeolite were optimized using an orthogonal experimental design and a range of analysis. The best modified zeolite for the removal of bisphenol-A (BPA) was prepared by 10% cetyltrimethyl ammonium bromide (CATB) with pre-treated zeolites at an operating temperature of 25 °C and processing time of 36 h. Comparing the property characteristics of raw zeolite (PZ) and CTAB-modified zeolite (CTABZ), it is evident that the modifier, CTAB, was bound to zeolite after modification. The adsorption characteristics of bisphenol-A (BPA) onto CTABZ from aqueous solution show that the sorption kinetics can be well described by a pseudosecond-order reaction model with a rate constant (Kv) of 1.07 g mg^{-1} h^{-1}. BPA sorption onto the CTABZ fitted better to the Freundlich isotherm (R^2 = 0.996) than the Langmuir model, and 25 °C was the best adsorption temperature for BPA to be adsorbed by CTABZ. Furthermore, the adsorption capacity of CTABZ increased with increasing solution pH, ranging from 5 to 10, and the optimal adsorption pH was pH 10. In addition, the adsorption capacity of BPA increased with the concentration of HCO^{3-} and decreased with increasing concentrations of NO_3, OH^-, and Al^{3+} in the test solution.

Visa [204] presented a comparative study of two types of zeolite materials prepared through a hydrothermal method, using two fly ash types as raw materials, termed ZCET40 and ZDs40. These zeolites are investigated as adsorbents for the simultaneous removal of heavy metals (Cd^{2+}, Cu^{2+}, Ni^{2+}, Zn^{2+} and Pb^{2+}) from solutions containing up to five cations. The ash converted into zeolite materials were collected from two Combined Heat and Power Plants (CET Brasov and DEVA) from Romania. The zeolite-type substrates were obtained using fly ash modified through hydrothermal processes using a low concentration of NaOH, and was further used for heavy metal removal (Cd^{2+}, Ni^{2+}, Cu^{2+}, Zn^{2+}, Pb^{2+}) from synthetic solutions with mono-, bi-, and pentavalent cations. The zeolite materials were characterized by AFM, XRD, FT-IR, SEM, BET and surface energy measurements to outline the crystalline and morphology modifications. The specific surface area was greater for ZCET40; ionic species of heavy metals are less bulky, and surfaces are richer in micropores, which is essential for efficiency over 80%. This means that ZCET40 was more efficient in the adsorption of heavy metals than ZDs40.

As evidenced, the surface of zeolites has developed new types of active centers (`SiO^-) and (AlO^-); the heavy metal cations must move through the pores, and the channels of the lattice can replace the exchangeable Na^+ and K^+ cations. Adsorption from solutions containing two or five heavy metal cations occurs with reduced efficiency for each cation according to competitive mechanisms at similar active sites (homogeneity substrates). Heavy metals (Pb^{2+}, Cu^{2+}, Cd^{2+}, Zn^{2+} and Ni^{2+}) compete with each other for adsorption sites. The results indicate that adsorption efficiency depends on the pH of the aqueous system, with 7.85 of ZCET40 and 8.42 of ZDs40. For both substrates, the working pH is not favorable for the adsorption of Ni^{2+} cations. The sequence adsorption efficiency of heavy metals on zeolites must be correlated with the hydrolysis constant, the adsorbed volume cation and the electronegative value. According to experimental data, the 60–90 min. contact time appears optimal for all substrates and is technologically feasible. Kinetic studies indicate that the adsorption of heavy metals follows the pseudo-second order equation, indicating chemisorption. The results show that substrates derived from CET Brasov ash has more efficient adsorption properties. Fly ash reuse is a great option for developing efficient, low-cost adsorbents for zeolite-type wastewater treatment [204].

Waste management and water quality are two main problems that humanity faces. Increased urbanization and industrialization lead to excess release of waste into the environment. New studies on wastewater treatment are of great interest to minimize these environmental problems, and thus, studies of this topic are accumulating in the literature. In this chapter, we describe recent papers published recently, indicating the huge range of studies on this topic.

3.6. Controversial Problems

3.6.1. Controversies Involving the Fenton Reaction Mechanism

Pollution of water by various toxic contaminants has become one of the most serious problems worldwide [205-207]. In recent decades, nanotechnology has received great attention, and several nanomaterials have been developed for water remediation [208-209].

Many nanomaterials serve as a potent alternative to conventional treatment methods, due to the increased adsorption and/or photocatalytic activity and specificity of the substance. However, most applications are not yet ready for the market, due to challenges such as scale and system configuration, environmental concerns and cost-benefit analyses. Therefore, only a few commercial products are available on the market [210].

In this scenario, the use of AOPs stand out, especially those involving the Fenton reaction. Fenton reactions occur in chemical and biochemical systems and in the natural environment [211-212]. The importance of Fenton chemistry has long been recognized among others in food chemistry, material aging and environmental engineering. The oxidant species obtained in the Fenton reaction are still controversial [213-215]. It is intriguing that a simple reaction (Fe^{2+} ions with H_2O_2) observed by H. J. H. Fenton more than a hundred years ago [216], continues to be very difficult to describe and understand. Paradoxically, the Fenton reaction is successful when used in environmental protection (for example, in the treatment and remediation of groundwater) and is thought to be a factor that damages biomolecules and plays an important role in the aging process and a variety of diseases [213].

The Fenton processes (reaction (8) is the classical Fenton reaction) consists of 2 reaction paths, a fast reaction with ferrous ion species (reactions (8) and (9)), which initiate the reaction of Fe(II) + H_2O_2, and a slow reaction with the ferric ion (reactions (10), (11) and (12)), dominated by the reduction of Fe(III). The ferrous regime was studied with the stopped-flow technique using yellow methyl (MY) as substrate. The kinetic model considered for all observed results requires that the reaction between Fe(II) and H_2O_2 produces both OH (~ 60% yield) and other species, possibly the ferryl (FeO^{2+}) ion (~ 40% yield) in a parallel process, with the ferryl reaction with MY at a negligible rate. The experimental data are consistent with the degradation of MY by reacting with OH, with a constant rate $k_{Da} = (1.3 \pm 0.1) \times 10^{10}\ M^{-1}s^{-1}$ at 25°C. The hypotheses for the formation of both OH and FeO^{2+}, with OH being the only reagent, also occur for the degradation of dichloroanilines (2,4-DCA and 3,4-DCA) in the ferrous ion stage. The kinetic model proposed by Minero et al. [215] may be a long-term departure from controversies about the Fenton reaction, allowing the quantification of the formation of reactive species.

$$Fe^{2+} + H_2O_2 \rightarrow Fe^{3+} + OH^- + \bullet OH \qquad (8)$$

$$\bullet OH + H_2O_2 \rightarrow HO_2^{\bullet} + H_2O \qquad (9)$$

$$Fe^{3+} + HO_2^{\bullet} \rightarrow Fe^{2+} + H^+ + O_2 \qquad (10)$$

$$Fe^{2+} + HO_2^{\bullet} \rightarrow Fe^{3+} + HO_2^- \qquad (11)$$

$$Fe^{2+} + \bullet OH \rightarrow Fe^{3+} + OH^- \qquad (12)$$

While several studies have suggested that OH is the oxidant species involved in the oxidative processes connected to the Fenton reaction, other possibilities have not been ruled out. For example, a study of the reduction of dissolved iron species by humic acids has suggested that in the addition of the radical •OH, another oxidant must be involved in Fenton reactions in marine water with neutral pH (7.0 - 7.5) [217]. Studies on the oxidation of arsenic [218] indicated that •OH is involved in the oxidation of arsenic (III) to arsenic (IV), which occurs readily at low pH, but that high valence metal species must also be formed at high pH, which does not readily oxidize arsenic (III). Other studies [219] suggest that at Fe(II) nanomolar levels, the oxidation of Fe(II) by H_2O_2 in marine water predominantly involves $FeOH^+$ species at pH 6-8 [213].

3.6.2. Possible Mechanisms of Fenton's Reaction

Many studies examining the nature of the reactive oxidant species in the Fenton reaction have been driven by the possibility that many reaction mechanisms are present, some of which presented below. For example, the scheme with the simple free radical pathway (Figure 1) can be shown, but the mechanism of the Fenton's reaction has been suggested to be more complicated than that shown in Figure 8. Since iron has a variable valence, its oxidation by H_2O_2 can occur by transferring one or two electrons (reactions (13) and (14), respectively) [213].

$$Fe(II) + H_2O_2 \rightarrow Fe(III) + OH^- + \bullet OH \quad (13)$$

$$Fe(II) + H_2O_2 \rightarrow Fe(IV) + 2OH^- \quad (14)$$

Some studies suggest, therefore, that the classical Fenton reaction occurs using only Fe(II) as an electron donor for H_2O_2. This would be an external sphere electron transfer reaction without direct bonding interactions between the electron donor and acceptor (Mechanism I, Figure 9). Recent studies have shown and favored the electron transfer mechanisms of the inner sphere, which involve a direct connection between iron and H_2O_2. This interaction can produce a metal-peroxo complex, Fe(II)OOH, which must react may react more to generate HO (one-electron oxidant) or Fe(IV)O radicals (two-electron oxidant). (Mechanism II, Figure 9). The key issue is therefore which species is the main oxidant in these reactions [213].

Several pathways have been proposed [220-221], including non-radical mechanisms, radical mechanisms involving radicals centralized in oxygen and reactions of high-valence metal species (Figure 10).

Kremer [223-224] proposed a non-radical mechanism for the Fenton reaction (Figure 11), suggesting that the reaction begins with the reversible formation of a Fe^{2+} and H_2O_2 primary intermediate (Fe^{2+} and H_2O_2) (exchanging a water molecule in the hydration layer of the Fe^{2+} ions for H_2O_2.) This species is the key intermediate in the reaction, which reacts with Fe^{2+} ions to produce Fe^{3+} (k5) or H_2O_2 to produce O_2 (k4). FeO^{2+} can react further with Fe^{3+} to form a binuclear $[FeOFe]^{5+}$ (k6) species. (k7) or re-decomposing into FeO^{2+} and Fe^{3+} (k8). Kremer indicated an error in the analysis of Barb et al. [225] because they assumed that a stable state is reached in $[Fe^{2+}]$, while in fact $[Fe^{2+}]$ goes to zero [213].

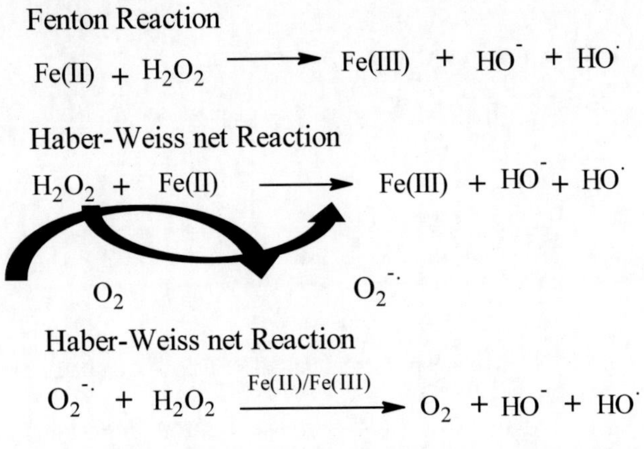

Figure 8. Basic free radical mechanisms for Fenton and Haber-Weiss reactions. Adapted from Barbusinski et al.

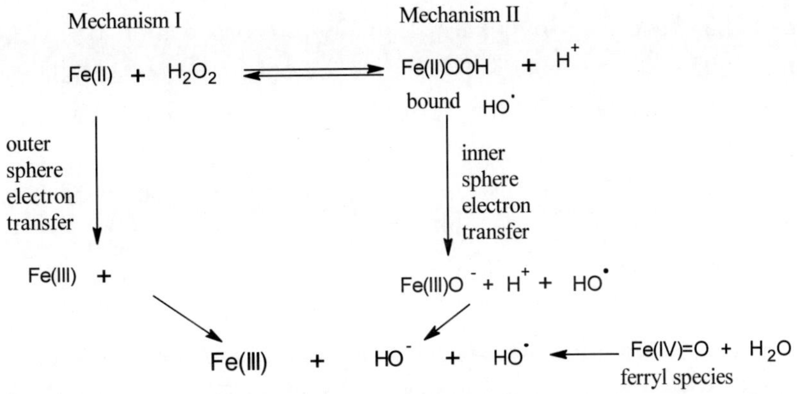

Figure 9. Basic reactions and intermediates involved in classical reactions of Fenton and central metal. Adapted from Barbusinski et al.

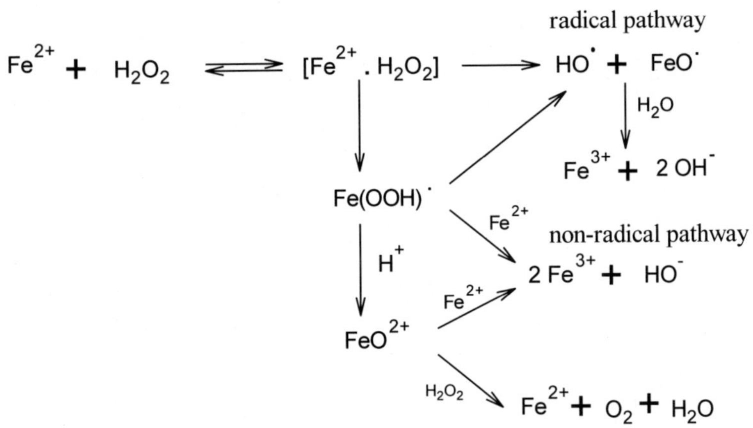

Figure 10. Possible reaction paths for Fenton reaction in the absence of organic substrates. Adapted from Barbusinski et al.

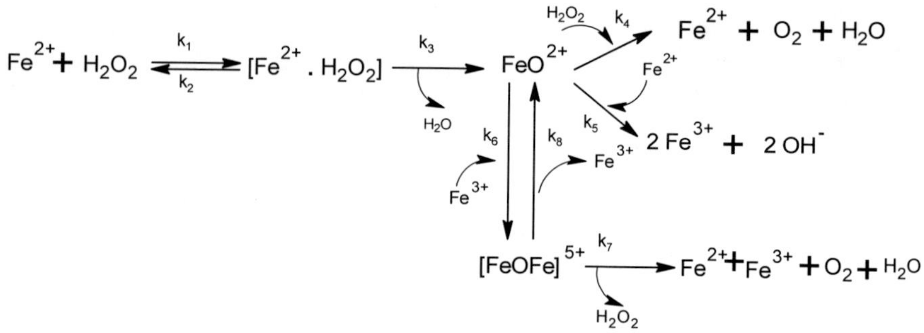

Figure 11. Non-radical proposed mechanism for the Fenton reaction. Adapted from Barbusinski et al.

More recent studies [218, 226] show that the Fenton chemistry reaction cannot be restricted to the mechanism of Barb et al. [225] or Bray and Gorin [226]. In fact, these studies postulate the presence of an active intermediate, which should be a weak acid with a pKa close to 2, providing hydroxyl and iron (III) radical formation at low pH or ferryl ions at high pH. This hypothesis explains the observation of hydroxyl radical production at equimolar concentrations of dilute reagents in water and values below pH 2 [218, 226, 228]. Therefore, according to the pH, the active intermediate is OH (radical path) or the ferryl ion (non-radical path) [213-214, 221, 229].

Ensing et al. [230-231] demonstrated the spontaneous formation of ferryl ($Fe^{IV}O^{2+}$) ion in aqueous iron (II) and hydrogen peroxide solutions in molecular dynamics simulations of first principles, confirming the model proposed by Bray and Gorin [213, 227, 232].

Their simulations are unfavorable but do not completely rule out the radical mechanism OH of Haber and Weiss (which is, especially in biochemistry, often taken as synonymous with Fenton chemistry). In the initial step of dissociation of iron-catalyzed hydrogen peroxide, the short lifetime of the OH radical and the L-Fe^{III}-OH complex always appear first. This radical has no independent existence, since it immediately abstracts a hydrogen or a short transfer through one or two solvent molecules of the binder water to form the dihydroxoiron (IV) complex, or may act directly from the hydroxyl linker to form the ferryl ion. In both cases, a molecule of water is neutralized. When other linkers are used in addition to water molecules, such as chelating agents, the radical can eliminate these linkers [213]. As the Fenton reaction is common in chemical, biological, and environmental systems where conditions can be very diverse, there is a high likelihood of more than one universal Fenton mechanism. It is possible that both hydroxyl radicals and ferryl ions can coexist in Fenton chemistry (Fenton reactions and such as Fenton) and depending on the environmental conditions or operational parameters, one species will predominate. Given the above doubts, it is desirable to conduct further research to prove the hypothesis or prove that there is only one mechanism (radical or non-radical) of the Fenton reaction [208-209, 211, 219].

4. Future Trends

Earlier research focused on conventional or traditional systems for wastewater treatment, but the field has shifted to AOPs and/or EAOPs such as photocatalysis and ozonation. The current trend of research on EAOPs conjunction with nanotechnology for water and wastewater treatment is increasing around the globe. The nanomaterials' unique properties and use with EAOPs present great advantages, with the possibility of a revolution in water and wastewater treatment. Although many nanomaterials can be assessed mainly in EAOPs in the gas diffusion cathodes, some materials have already been used at the pilot scale. The amount of ·OH produced is an important factor, and thus, more work is needed with respect to nanomaterials.

Other materials have growing environmental applications such as metal-organic frameworks (MOFs), which are highly porous inorganic-organic hybrid crystalline materials, prepared from metal ion/clusters and multidentate organic ligands. More recent examples of the photocatalytic degradation of dyes have been published.

Different materials for the wastewater treatment of other water samples are being studied, such as Net-zero graywater (NZGW) reuse or a nearly closed-loop recycle of graywater for all original uses, which can recover both water and its attendant hot-water thermal energy while avoiding the installation and maintenance of a separate graywater sewer in residential areas. In this case, specifically TiO_2-UV, UV-hydrogen peroxide, hydrogen peroxide-ozone, ozone-UV advanced oxidation processes, UV, ozone, hydrogen peroxide, filtration, and chlorine disinfection processes were reviewed for performance, energy demand, environmental impact, and operational simplicity.

When energy is not the problem, UV-hydrogen peroxide and UV-TiO_2 can be used to mineralize organics, but may require a secondary process to prevent microbial regrowth when working with real samples. Chlorine and ozone may be useful for maintenance of disinfection residues. The choice of combined processes to degrade pollutants is a key factor, including the use of hydrogen peroxide Fenton processes, boron-doped diamond, heterogeneous Fenton processes and photo-eletrochemical processes, in addition to solar energy.

References

[1] Victor S. Pinheiro, Edson C. Paz, Luci R. Aveiro, Luanna S. Parreira, Felipe M. Souza, Pedro H. C. Camargo, Mauro C. Santos. Ceria high aspect ratio nanostructures supported on carbon for hydrogen peroxide electrogeneration, *Electrochimica Acta* 259 (2018) 865-872.

[2] U. D. Hinrichsen, D. Robey, B. Upadhyay, Solutions for a water-short world, *Popul. Rep. XXVI* (1998) 1-31.

[3] T. A. Kurniawan, G. Y. S. Chan, W. H. Lo, S. Babel. Physico-chemical treatment techniques for wastewater laden with heavy metals, *Chem. Eng. J.* 118 (2006) 83-98.

[4] M. S. Nawaz, M. Ahsan. Comparison of physico-chemical, advanced oxidation and biological techniques for the textile wastewater treatment, *Alex. Eng. J.* 53 (2014) 717-722.

[5] Francisca C. Moreira, Rui A. R. Boaventura, Enric Brillas, Vítor J. P. Vilar. Electrochemical advanced oxidation processes: A review on their application to synthetic and real wastewaters. *Applied Catalysis B: Environmental* 202 (2017) 217–261.

[6] H. Zollinger. Color Chemistry: *Syntheses, Properties, and Applications of Organic Dyes and Pigments*, 3rd revised ed., VHCA and Wiley-VCH, Zurich, Switzerland/Weinheim, Germany, 2003.

[7] K. P. Sharma, S. Sharma, S. Sharma, P. K. Singh, S. Kumar, R. Grover, P. K. Sharma. A comparative study on characterization of textile wastewaters (untreated and treated) toxicity by chemical and biological tests. *Chemosphere* 69 (2007) 48–54.

[8] K. Kümmerer. The presence of pharmaceuticals in the environment due to human use–present knowledge and future challenges. *J. Environ. Manage.* 90 (2009) 2354–2366.

[9] C. A. Damalas, I. G. Eleftherohorinos. Pesticide exposure, safety issues, and risk assessment indicators. *Int. J. Env. Res. Public Health* 8 (2011)1402–1419.

[10] Raf Dewil, Dionissios Mantzavinos, Ioannis Poulios, Manuel A. Rodrigo. New perspectives for advanced oxidation processes. *Journal of Environmental Management* 195 (2017) 93-99.

[11] Klavarioti, M., Mantzavinos, D., Kassinos, D. Removal of residual pharmaceuticals from aqueous systems by advanced oxidation processes. *Environ. Int.* 35 (2009), 402-417.

[12] Comninellis, C., Kapalka, A., Malato, S., Parsons, S. A., Poulios, I., Mantzavinos, D. Advanced oxidation processes for water treatment: advances and trends for R&D. *J. Chem. Technol. Biotechnol.* 83, 2008, 769-776.

[13] M. A. Oturan, J. J. Aaron. Advanced oxidation processes in water/wastewater treatment: principles and applications. *Crit. Rev. Environ. Sci. Technol.* 44 (2014) 2577–2641.

[14] W. H. Glaze, J.-W. Kang, D. H. Chapin. The chemistry of water treatment processes involving ozone, hydrogen peroxide and ultraviolet radiation. *Ozone Sci. Eng.* 9 (1987) 335–352.

[15] W. M. Latimer, *Oxidation Potentials, 2nd ed.,* Prentice-Hall, New York, UnitedStates, 1952.

[16] G. V. Buxton, C. L. Greenstock, W. P. Helman, A. B. Ross, *J. Phys. Critical review of rate constants for reactions of hydrated electronschemical kinetic data base for combustion chemistry. Part 3: Propane. Chem. Ref. Data*17 (1988) 513–886.

[17] E. G., Janzen, Y. Kotake, R. D. Hinton. Stabilities of hydroxyl radical spin adducts of PBN-type spin traps. *Free Radic. Biol. Med.* 12 (1992) 169–173.

[18] C. Comninellis, A. Kapalka, S. Malato, S.A. Parsons, I. Poulios, D. Mantzavinos. Advanced oxidation processes for water treatment: advances and trends for R&D. *J. Chem. Technol. Biotechnol*. 83 (2008) 769–776.

[19] E. Brillas, I. Sirés, M. A. Electro-Fenton process and related electrochemical technologies based on Fenton's reaction chemistry. Oturan, *Chem. Rev*. 109 (2009) 6570–6631.

[20] C. Barrera-Díaz, P. Cañizares, F. J. Fernández, R. Natividad, M. A. Rodrigo. Electrochemical advanced oxidation processes: an overview of the current applications to actual industrial effluents. *J. Mex. Chem. Soc.* 58 (2014) 256–275.

[21] B. P. Chaplin. Critical review of electrochemical advanced oxidation processes for water treatment applications. *Environ. Sci. Process. Impacts* 16 (2014) 1182–1203.

[22] Angelo R. F. Pipi, Ignasi Sirés, Adalgisa R. De Andrade, Enric Brillas. Application of electrochemical advanced oxidation processes to the mineralization of the herbicide diuron, *Chemosphere* 109 (2014) 49–55.

[23] Baxendale, J. H., Wilson, J. A. The photolysis of hydrogen peroxide at high light intensities. *Trans. Faraday Soc*. 53, (1957) 344-356.

[24] Francisca C. Moreira, J. Soler, M. F. Alpendurada, Rui A. R. Boaventura, Enric Brillas, Vítor J. P. Vilar. Tertiary treatment of a municipal wastewater toward pharmaceuticals removal by chemical and electrochemical advanced oxidation processes, *Water Research* 105 (2016) 251-263.

[25] Enric Brillas, Abdoulaye Thiam, Sergi Garcia-Segura. Incineration of acidic aqueous solutions of dopamine by electrochemical advanced oxidation processes with Pt and BDD anodes, *Journal of Electro-analytical Chemistry* 775 (2016) 189–197.

[26] I. Sirés, E. Brillas. Remediation of water pollution caused by pharmaceutical residues based on electrochemical separation and degradation technologies: a review, *Environ. Int*. 40 (2012) 212–229.

[27] I. Sirés, E. Brillas, M. A. Oturan, M. A. Rodrigo, M. Panizza. Electrochemical advanced oxidation processes: today and tomorrow. A review, *Environ. Sci. Pollut. Res*. 21 (2014) 8336–8367.

[28] A. Özcan, M. A. Oturan, N. Oturan, Y. Şahin. Removal of acid Orange 7 from water by electrochemically generated Fenton's reagent, *J. Hazard. Mater*. 163 (2009) 1213–1220.

[29] E. B. Cavalcanti, S. Garcia-Segura, F. Centellas, E. Brillas. Electrochemical incineration of omeprazole in neutral aqueous medium using a platinum or boron-

doped diamond anode: Degradation kinetics and oxidation products, *Water Res.* 47 (2013) 1803–1815.

[30] E. Guinea, J. A. Garrido, R. M. Rodríguez, P. L. Cabot, C. Arias, F. Centellas, E. Brillas. Degradation of the fluoroquinolone enro-floxacin by electrochemical advanced oxidation processes based on hydrogen peroxide electrogeneration, *Electrochim. Acta* 55 (2010) 2101–2115.

[31] S. Garcia-Segura, J. A. Garrido, R. M. Rodríguez, P. L. Cabot, F. Centellas, C. Arias, E. Brillas. Mineralization of flumequine in acidic medium by electro-Fenton and photoelectro-Fenton processes, *Water Res.* 46 (2012) 2067–2076.

[32] A. El-Ghenymy, R. M. Rodríguez, C. Arias, F. Centellas, J. A. Garrido, P. L. Cabot, E. Brillas. Electro-Fenton and photoelectro-Fenton degradation of the antimicrobial sulfamethazine using a boron-doped diamond anode and an air-diffusion cathode, *J. Electroanal. Chem.* 701 (2013) 7–13.

[33] F. C. Moreira, S. Garcia-Segura, R. A. R. Boaventura, E. Brillas, V. J. P. Vilar. Degradation of the antibiotic trimethoprim by electro-chemical advanced oxidation processes using a carbon-PTFE air-diffusion cathode and a boron-doped diamond or platinum anode, *Appl. Catal. B Environ.* 160-161 (2014) 492–505.

[34] L. C. Almeida, S. Garcia-Segura, N. Bocchi, E. Brillas. Solar photoelectro-Fenton degradation of paracetamol using a flow plant with a Pt/air-diffusion cell coupled with a compound parabolic collector: Process optimization by response surface methodology, *Appl. Catal. B Environ.* 103 (2011) 21–30.

[35] S. Garcia-Segura, E. B. Cavalcanti, E. Brillas. Mineralization of the antibiotic chloramphenicol by solar photoelectro-Fenton. From stirred tank reactor to solar pre-pilot plant, *Appl. Catal. B Environ.* 144 (2014) 588–598.

[36] E. Brillas, C. A. Martínez-Huitle. Decontamination ofwastewaters containing synthetic organic dyes by electrochemical methods. An updated review, *Appl. Catal. B Environ.* 166–167 (2015) 603–643.

[37] Aziza Hadj Ltaïef, Adriana D'Angelo, Salah Ammar, Abdellatif Gadri, Alessandro Galiab, Onofrio Scialdoneb. Electrochemical treatment of aqueous solutions of catechol by various electrochemical advanced oxidation processes: *Effect of the process and of operating* parameters.

[38] J. M. Ochando-Pulido, S. Pimentel-Moral, V. Verardo, A. Martinez-Ferez. A focus on advanced physico-chemical processes for olive mill wastewater treatment, *Separation and Purification Technology* 179 (2017) 161–174.

[39] Heng Lin, Nihal Oturan, Jie Wu, Virender K. Sharma, Hui Zhang, Mehmet A. Oturan. Removal of artificial sweetener aspartame from aqueous media by electrochemical advanced oxidation processes, *Chemosphere* 167 (2017) 220-227.

[40] Francisca C. Moreira, Rui A. R. Boaventura, Enric Brillas, Vıtor J. P. Vilar. Remediation of a winery wastewater combining aerobic biological oxidation and electrochemical advanced oxidation processes, *Water Research,* 75 (2015) 95 -108.

[41] Simon Komtchou, Ahmad Dirany, Patrick Drogui, Didier Robert, Pierre Lafrance. Removal of atrazine and its by-products from water using electrochemical advanced oxidation processes, *Water Research* 125 (2017) 91-103.

[42] Anita Klancar, Jurij Trontelj, Albin Kristl, Andrej Meglic, Tinkara Rozinac, Maja Zupanci, Justin, Robert Roskar. An advanced oxidation process for wastewater treatment to reduce the ecological burden from pharmacotherapy and the agricultural use of pesticides, *Ecological Engineering* 97 (2016) 186–195.

[43] Angelo R. F. Pipi, Ignasi Sirés, Adalgisa R. De Andrade, Enric Brillas. Application of electrochemical advanced oxidation processes to the mineralization of the herbicide diuron, *Chemosphere* 109 (2014) 49–55.

[44] Worachate Khongthon, Goran Jovanovic, Alex Yokochi, Polkit Sangvanich, Varong Pavarajarn. Degradation of diuron via an electrochemical advanced oxidation process in a microscale-based reactor, *Chemical Engineering Journal* 292 (2016) 298–307.

[45] Yudong Xue, Shili Zheng, Zhi Sun, Yi Zhang, Wei Jin. Alkaline electrochemical advanced oxidation process for chromium oxidation at graphitized multi-walled carbon nanotubes, *Chemosphere* 183 (2017) 156-163.

[46] Yudong Xue, Shili Zheng, Hao Du, Yi Zhang, Wei Jin. Cr(III)-induced electrochemical advanced oxidation processes for the V_2O_3 dissolution in alkaline media, *Chemical Engineering Journal* 307 (2017) 518–525.

[47] Ricardo E. Palma-Goyes, Javier Silva-Agredo, Ignacio González, Ricardo A. Torres-Palma. Comparative degradation of indigo carmine by electrochemicaloxidation and advanced oxidation processes, *Electrochimica Acta* 140 (2014) 427–433.

[48] Abdoulaye Thiam, Ignasi Sirés, José A. Garrido, Rosa M. Rodríguez, Enric Brillas. Decolorization and mineralization of Allura Red AC aqueous solutions by electrochemical advanced oxidation processes, *Journal of Hazardous Materials* 290 (2015) 34–42.

[49] Vanessa S. Antonin, Sergi Garcia-Segura, Mauro C. Santos, Enric Brillas. Degradation of Evans Blue diazo dye by electrochemical processes based on Fenton's reaction chemistry, *Journal of Electroanalytical Chemistry* 747 (2015) 1–11.

[50] Vanessa S. Antonin, Mauro C. Santos, Sergi Garcia-Segura, Enric Brillas. Electrochemical incineration of the antibiotic ciprofloxacin in sulfate medium and synthetic urine matrix) *Water Research* 83 (2015) 31-41.

[51] M. H. M. T. Assumpção, A. Moraes, R. F. B. De Souza, R. M. Reis, R. S. Rocha, I. Gaubeur, M. L. Calegaro, P. Hammer, M. R. V. Lanza, M. C. Santos. Degradation of dipyrone via advanced oxidation processes using a cerium nanostructured electrocatalyst material, *Applied Catalysis A: General* 462–463 (2013) 256–261.

[52] Francisca C. Moreira, J. Soler, M. F. Alpendurada, Rui A. R. Boaventura, Enric Brillas, Vítor J. P. Vilar. Tertiary treatment of a municipal wastewater toward pharmaceuticals removal by chemical and electrochemical advanced oxidation processes, *Water Research* 105 (2016) 251-263.

[53] Enric Brillas, Abdoulaye Thiam, Sergi Garcia-Segura. Incineration of acidic aqueous solutions of dopamine by electrochemical advanced oxidation processes with Pt and BDD anodes, *Journal of Electroanalytical Chemistry* 775 (2016) 189–197.

[54] Ling Feng, Eric D. van Hullebusch, Manuel A. Rodrigo, Giovanni Esposito, Mehmet A. Oturan. Removal of residual anti-inflammatory and analgesic pharmaceuticals from aqueous systems by electro-chemical advanced oxidation processes. *A review, Chemical Engineering Journal* 228 (2013) 944–964.

[55] Nihal Oturan, Jie Wu, Hui Zhang, Virender K. Sharma, Mehmet A. Oturan. Electrocatalytic destruction of the antibiotic tetracycline in aqueous medium by electrochemical advanced oxidation processes: Effect of electrode materials, *Applied Catalysis B: Environmental* 140– 141 (2013) 92– 97.

[56] Aline Maria Sales Solano, Carlos Alberto Martínez-Huitle, Sergi Garcia-Segura, Abdellatif El-Ghenymy, Enric Brillas. Application of electrochemical advanced oxidation processes with a boron-doped diamond anode to degrade acidic solutions of Reactive Blue 15 (Turqueoise Blue) dye, *Electrochimica Acta* 197 (2016) 210–220.

[57] Elvira Bocos, Elena Alfaya, Olalla Iglesias, Marta Pazos, M. Ángeles Sanromán. Application of a new sandwich of granular activated and fiber carbon as cathode in the electrochemical advanced oxidation treatment of pharmaceutical effluents, *Separation and Purification Technology* 151 (2015) 243–250.

[58] Soliu O. Ganiyu, Eric D. van Hullebusch, Marc Cretin, Giovanni Esposito, Mehmet A. Oturan. Coupling of membrane filtration and advanced oxidation processes for removal of pharmaceutical residues: A critical review, *Separation and Purification Technology* 156 (2015) 891–914.

[59] Bo Yang, Jianping Deng, Gang Yu, Shubo Deng, Juying Li, Caizhen Zhu, Qiongfang Zhuo, Huabo Duan, Tao Guo. Effective degradation of carbamazepine using a novel electro-peroxone process involving simultaneous electrochemical generation of ozone and hydrogen peroxide, *Electrochemistry Communications*, 86 (2018) 26-29.

[60] Cláudia Regina Klauck, Alexandre Giacobbo, Erlon Diego Lorenz de Oliveira, Luciano Basso da Silva, Marco Antônio Siqueira Rodrigues. Evaluation of acute toxicity, cytotoxicity and genotoxicity of landfill leachate treated by biological lagoon and advanced oxidation processes, *Journal of Environmental Chemical Engin-eering* 5 (2017) 6188–6193.

[61] Cláudia Regina Klauck, Alexandre Giacobbo, Christian Gabriel Altenhofen, Luciano Basso Silva, Alvaro Meneguzzi, Andréa Moura Bernardes, Marco Antônio Siqueira Rodrigues. Toxicity elimination of landfill leachate by hybrid processing of advanced oxidation process and adsorption, *Environmental Technology & Innovation* 8 (2017) 246–255.

[62] Francisca C. Moreira, J. Soler, Amélia Fonseca, Isabel Saraiva, Rui A. R. Boaventura, Enric Brillas, Vítor J. P. Vilara. Electro-chemical advanced oxidation processes for sanitary landfillleachate remediation: Evaluation of operational variables, *Applied Catalysis B: Environmental* 182 (2016) 161–171.

[63] Francisca C. Moreira, J. Soler, Amélia Fonseca, Isabel Saraiva, Rui A. R. Boaventura, Enric Brillas, Vítor J. P. Vilar. Incorporation of electrochemical advanced oxidation processes in a multistage treatment system for sanitary landfill leachate, *Water Research* 81 (2015) 375-387.

[64] Pilar Valero, Martha Verbel, Javier Silva-Agredo, Rosa Mosteo, Maria P. Ormad, Ricardo A. Torres-Palma. Electrochemical advanced oxidation processes for *Staphylococcus. aureus* dis-infection in municipal WWTP effluents, *Journal of Environmental Management* 198 (2017) 256-265.

[65] M. H. M. T. Assumpção, R. F. B. De Souza, R. M. Reis, R. S. Rocha, J. R. Steter, P. Hammer, I. Gaubeur, M. L. Calegaro, M. R. V. Lanza, M. C. Santos. Low tungsten content of nanostructured material supported on carbon for the degradation of phenol, *Applied Catalysis B: Environmental* 142–143 (2013) 479–486).

[66] Mohammad Reza Delsouz Khaki, Mohammad Saleh Shafeeyan, Abdul Aziz Abdul Raman, Wan Mohd Ashri Wan Daud. Application of doped photocatalysts for organic pollutant degradation - A review, *Journal of Environmental Management* 198 (2017) 78-94.

[67] Mariana Lanzarini-Lopes, Sergi Garcia-Segura, Kiril Hristovski, Paul Westerhoff. Electrical energy per order and current efficiency for electrochemical oxidation of p-chlorobenzoic acid with boron-doped diamond anode, *Chemosphere* 188 (2017) 304-311.

[68] Gholamreza Moussavi, Mohsen Rezaei. Exploring the advanced oxidation/reduction processes in the VUV photoreactor for dechlorination and mineralization of trichloroacetic acid: Parametric experiments, degradation pathway and bioassessment, *Chemical Engineering Journal* 328 (2017) 331–342.

[69] Gholamreza Moussavi, Sakine Shekoohiyan. Simultaneous nitrate reduction and acetaminophen oxidation usingthe continuous-flow chemical-less VUV process as an integratedadvanced oxidation and reduction process, *Journal of Hazardous Materials* 318 (2016) 329–338.

[70] Matz Dietricha, Marcus Frankea, Michael Steltera, Patrick Braeutigama. Degradation of endocrine disruptor bisphenol A by ultrasound-assisted electrochemical oxidation in water *Ultrasonics - Sonochemistry* 39 (2017) 741–749.

[71] Marek Trojanowicz, Krzysztof Bobrowski, Bogdan Szostek. A survey of analytical methods employed for monitoring of Advanced Oxidation/Reduction Processes for decomposition of selected perfluorinated environmental pollutants, *Talanta* 177 (2018) 122–141.

[72] Marek Trojanowicz, Anna Bojanowska-Czajka, Iwona Bartosiewicz, Krzysztof Kulisa. Advanced Oxidation/Reduction Processes treatment for aqueous perfluorooctanoate (PFOA) and perfluoro-octanesulfonate (PFOS) – A review of recent advances, *Chemical Engineering Journal* 336 (2018) 170–199.

[73] I. Yamanaka, T. Hashimoto, R. Ichihashi, K. Otsuka. Direct synthesis of H_2O_2 acid solutions on carbon cathode prepared from activated carbon and vapor-growing-carbon-fiber by a H_2/O_2 fuel cell. *Electrochim. Acta* 53 (2008) 4824.

[74] M. Campos-Martin, G. Blanco-Brieva, J. L. G. Fierro. Hydrogen peroxide synthesis: an outlook beyond the anthraquinone process. *Angew. Chem. Int. Ed.* 45 (2006) 6962.

[75] G. Xia, Y. Lu, H. Xu. Electrogeneration of hydrogen peroxide for electro-Fenton via oxygen reduction using polyacrylonitrile-based carbon fiber brush cathode. *Electrochim. Acta* 158 (2015) 390.

[76] Y. Sheng, S. Song, X. Wang, L. Song, C. Wang, H. Sun, X. Niu. Electrogeneration of hydrogen peroxide on a novel highly effective acetylene black-PTFE cathode with PTFE film. *Electrochim. Acta* 56 (2011) 8651.

[77] M. H. M. T. Assumpção, A. Moraes, R. F. B. De Souza, I. Gaubeur, R. T. S. Oliveira, V. S. Antonin, G. R. P. Malpass, R. S. Rocha, M. L. Calegaro, M. R. V. Lanza, M. C. Santos. Low content cerium oxide nanoparticles on carbon for hydrogen peroxide electro-synthesis. *Appl. Catal. A-Gen.* 411–412 (2012) 1.

[78] F. L. Silva, R. M. Reis, W. R. P. Barros, R. S. Rocha, M. R. V. Lanza. Electrogeneration of hydrogen peroxide in gas diffusion electrodes: Application of iron (II) phthalocyanine as a modifier of carbon black. *J. Electroanal. Chem.* 722–723 (2014) 32.

[79] F. C. Moreira, R. A. R. Boaventura, E. Brillas, V. J. P. Vilar. Electrochemical advanced oxidation processes: a review on their application to synthetic and real wastewaters. *Appl. Catal. B- Environ.* 202 (2017) 217.

[80] A. A. Gallegos, Y. V. García, A. Zamudio. Solar hydrogen peroxide. *Sol. Energy Mater. Sol. Cells* 88 (2005) 157.

[81] E. Brillas, R. M. Bastida, E. Llosa, J. Casado. Electrochemical destruction of aniline and 4- chloroaniline for wastewater treatment using a carbon- PTFE O 2- fed cathode. *J. Electrochem. Soc.* 142 (1995) 1733.

[82] N. Guillet, L. Roué, S. Marcotte, D. Villers, J. P. Dodelet, N. Chhim, S. Trévin. (2006). Electrogeneration of hydrogen peroxide in acid medium using pyrolyzed cobalt-based catalysts: Influence of the cobalt content on the electrode performance, *J. Appl. Electrochem.*, 36: 863–870.

[83] V. S. Antonin, M. H. M. T. Assumpcao, J. C. M. Silva, L. S. Parreira, M. R. V Lanza, M. C. Santos. (2013). Synthesis and characterization of nanostructured electrocatalysts based on nickel and tin for hydrogen peroxide electrogeneration, *Electrochim. Acta.* 109: 245–251.

[84] A. Moraes, M. H. M. T. Assumpção, R. Papai, I. Gaubeur, R. S. Rocha, R. M. Reis, M. L. Calegaro, M. R. V Lanza, M. C. Santos. (2014). Use of a vanadium nanostructured material for hydrogen peroxide electrogeneration, *J. Electroanal. Chem.* 719: 127–132.

[85] W. R. P. Barros, Q. Wei, G. Zhang, S. Sun, M. R. V Lanza, A. C. Tavares. (2015). Oxygen reduction to hydrogen peroxide on Fe_3O_4 nanoparticles supported on Printex carbon and Graphene, *Electrochim. Acta.* 162: 263–270.

[86] F. L. Silva, R. M. Reis, W. R. P. Barros, R. S. Rocha, M. R. V Lanza. (2014). Electrogeneration of hydrogen peroxide in gas diffusion electrodes: Application of iron (II) phthalocyaninc as a modifier of carbon black, *J. Electroanal. Chem.* 722-723: 32–37.

[87] A. Xu, W. Han, J. Li, X. Sun, J. Shen, L. Wang. (2016). Electrogeneration of hydrogen peroxide using Ti/IrO_2-Ta_2O_5 anode in dual tubular membranes Electro-Fenton reactor for the degradation of tricyclazole without aeration, *Chem. Eng. J.* 295: 152–159.

[88] F. Xu, T. Song, Y. Xu, Y. Chen, S. Zhu, S. Shen. (2009). A new cathode using CeO_2/MWNT for hydrogen peroxide synthesis through a fuel cell, *J. Rare Earths.* 27: 128–133.

[89] J. F. Carneiro, R. S. Rocha, P. Hammer, R. Bertazzoli, M. R. V Lanza. (2016). Hydrogen peroxide electrogeneration in gas diffusion electrode nanostructured with Ta_2O_5, *Appl. Catal. A Gen.* 517: 161–167.

[90] J. F. Carneiro, L. C. Trevelin, A. S. Lima, G. N. Meloni, M. Bertotti, P. Hammer, R. Bertazzoli, M. R. V. Lanza. (2017). Synthesis and Characterization of ZrO_2/C as Electrocatalyst for Oxygen Reduction to H_2O_2, *Electrocatalysis.* 0–6.

[91] J. F. Carneiro, M. J. Paulo, M. Siaj, A. C. Tavares, M. R. V Lanza. (2015). Nb_2O_5 nanoparticles supported on reduced graphene oxide sheets as electrocatalyst for the H_2O_2 electrogeneration, *J. Catal.* 332: 51–61.

[92] V. S. Antonin, L. S. Parreira, L. R. Aveiro, F. L. Silva, R. B. Valim, P. Hammer, M. R. V. Lanza, M. C. Santos. (2017). W@Au Nanostructures Modifying Carbon as Materials for Hydrogen Peroxide Electrogeneration, *Electrochim. Acta.* 231: 713–720.

[93] J. F. Carneiro, M. J. Paulo, M. Siaj, A. C. Tavares, M. R. V. Lanza. (2017). Zirconia on Reduced Graphene Oxide Sheets: Synergistic Catalyst with High Selectivity for H_2O_2 Electrogeneration, *Chem Electro Chem.* 4: 508–513.

[94] R. S. Rocha, F. L. Silva, R. B. Valim, W. R. P. Barros, J. R. Steter, R. Bertazzoli, M. R. V. Lanza. (2016). Effect of Fe^{2+} on the degradation of the pesticide profenofos by electrogenerated H_2O_2, *J. Electroanal.* Chem. 783: 100–105.
[95] F. V. E. dos Reis, V. S. Antonin, P. Hammer, M. C. Santos, P. H. C. Camargo. (2015). Carbon-supported TiO_2–Au hybrids as catalysts for the electrogeneration of hydrogen peroxide: Investigating the effect of TiO_2 shape, *J. Catal.* 326: 100–106.
[96] M. H. M. T. Assumpção, D. C. Rascio, J. P. B. Ladeia, R. F. B. De Souza, E. T. Neto, M. L. Calegaro, R. T. S. Oliveira, I. Gaubeur, M. R. V Lanza, M. C. Santos. (2011). Comparative study of different methods for the preparation of CoxOy/C for the electrosynthesis of hydrogen peroxide, *Int. J. Electrochem. Sci.* 6: 1586–1596.
[97] E. E. S. Souza, A. F. T. O. Santos, I. M. D. Gonzaga, N. R. Alves, L.P. Costa, L. S. R. Silva, K. I. B. Eguiluz, G. R. Salazar-Banda, E. B. Cavalcanti. (2015). Ruthenium-tin oxides-coated graphite felt: Enhanced active area and improved efficiency for the electrochemical generation of hydrogen peroxide, *Ceram. Int.* 41: 10293–10297.
[98] M. H. M. T. Assumpcao, A. Moraes, R. F. B. De Souza, M. L. Calegaro, M. R. V Lanza, E. R. Leite, M. A. L. Cordeiro, P. Hammer, M. C. Santos. (2013). Influence of the preparation method and the support on H_2O_2 electrogeneration using cerium oxide nanoparticles, *Electrochim. Acta.* 111: 339–343.
[99] M. H. M. T. Assumpção, a. Moraes, R. F. B. De Souza, I. Gaubeur, R. T. S. Oliveira, V. S. Antonin, G. R. P. Malpass, R. S. Rocha, M. L. Calegaro, M. R. V. Lanza, M. C. Santos. (2012). Low content cerium oxide nanoparticles on carbon for hydrogen peroxide electrosynthesis, *Appl. Catal. A Gen.* 411-412: 1–6.
[100] M. H. M. T. Assumpção, R. F. B. De Souza, R. M. Reis, R. S. Rocha, J. R. Steter, P. Hammer, I. Gaubeur, M. L. Calegaro, M. R. V Lanza, M. C. Santos. (2013). Low tungsten content of nanostructured material supported on carbon for the degradation of phenol, *Appl. Catal. B Environ.* 142-143: 479–486.
[101] C. A. Martínez-Huitle, E. Brillas. (2009). Decontamination of wastewaters containing synthetic organic dyes by electrochemical methods: A general review, *Appl. Catal. B Environ.* 87: 105–145.
[102] M. Munoz, Z. M. de Pedro, J. A. Casas, J. J. Rodriguez. (2015). Preparation of magnetite-based catalysts and their application in heterogeneous Fenton oxidation - A review, *Appl. Catal. B Environ.* 176-177: 249–265.
[103] S. Navalon, M. Alvaro, H. Garcia. (2010). Heterogeneous Fenton catalysts based on clays, silicas and zeolites, *Appl. Catal. B Environ.* 99: 1–26.
[104] R. S. Ribeiro, A. M. T. Silva, J. L. Figueiredo, J. L. Faria, H.T. Gomes. (2016). Catalytic wet peroxide oxidation: A route towards the application of hybrid magnetic carbon nanocomposites for the degradation of organic pollutants. A review, *Appl. Catal. B Environ.* 187: 428–460.

[105] M. H. M. T. Assumpção, A. Moraes, R. F. B. De Souza, R. M. Reis, R. S. Rocha, I. Gaubeur, M. L. Calegaro, P. Hammer, M. R. V Lanza, M. C. Santos. (2013). Degradation of dipyrone via advanced oxidation processes using a cerium nanostructured electrocatalyst material, *Appl. Catal. A Gen*. 462-463: 256–261.

[106] Y. Yang, C. Zhang, Z. Hu. (2013). Impact of metallic and metal oxide nanoparticles on wastewater treatment and anaerobic digestion, *Environ. Sci. Process. Impacts*. 15: 39–48.

[107] M. C. Pereira, L. C. a. Oliveira, E. Murad. (2012). Iron oxide catalysts: Fenton and Fentonlike reactions – a review, *Clay Miner*. 47: 285–302.

[108] B. Weng, J. Wu, N. Zhang, Y.-J. Xu. (2014). Observing the Role of Graphene in Boosting the Two-Electron Reduction of Oxygen in Graphene–WO_3 Nanorod Photocatalysts, *Langmuir*. 30: 5574–5584.

[109] J. L. Figueiredo. (2013). Functionalization of porous carbons for catalytic applications, *J. Mater. Chem*. A. 1: 9351.

[110] H. A. Asmaly, B. Abussaud, Ihsanullah, T. A. Saleh, V. K. Gupta, M. A. Atieh. (2015). Ferric oxide nanoparticles decorated carbon nanotubes and carbon nanofibers: From synthesis to enhanced removal of phenol, *J. Saudi Chem. Soc*. 19: 511–520.

[111] M. A. Andrade, R. J. Carmona, A. S. Mestre, J. Matos, A. P. Carvalho, C. O. Ania. (2014). Visible light driven photooxidation of phenol on TiO_2/Cu-loaded carbon catalysts, *Carbon N. Y.* 76: 183–192.

[112] M. H. M. T. Assumpção, R. F. B. De Souza, D. C. Rascio, J. C. M. Silva, M. L. Calegaro, I. Gaubeur, T. R. L. C. Paixão, P. Hammer, M. R. V Lanza, M. C. Santos. (2011). A comparative study of the electrogeneration of hydrogen peroxide using Vulcan and Printex carbon supports, *Carbon N. Y.* 49: 2842–2851.

[113] T. C. Damato, C. C. S. De Oliveira, R. A. Ando, P. H. C. Camargo. (2013). A facile approach to TiO_2 colloidal spheres decorated with Au nanoparticles displaying well-defined sizes and uniform dispersion, *Langmuir* 29.5: 1642-1649.

[114] J. F. Carneiro, M. J. Paulo, M. Siaj, A. C. Tavares, e M. R. V Lanza. Nb_2O_5 nanoparticles supported on reduced graphene oxide sheets as electrocatalyst for the H_2O_2 electrogeneration," *J. Catal*., vol. 332, p. 51–61, 2015.

[115] J. Wang, S. Yang, D. Guo, P. Yu, D. Li, J. Ye, e L. Mao. Comparative studies on electrochemical activity of graphene nanosheets and carbon nanotubes. *Electrochem. commun*., vol. 11, no 10, p. 1892–1895, 2009.

[116] H. Ashassi-Sorkhabi, B. Rezaei-Moghadam, E. Asghari, R. Bagheri, e R. Kabiri, Sonoelectrosynthesized polypyrrole-graphene oxide nanocomposite modified by carbon nanotube and Cu2O nano-particles on copper electrode for electrocatalytic oxidation of methanol., *J. Taiwan Inst. Chem. Eng*., vol. 69, p. 118–130, 2016.

[117] E. Umeshbabu e G. Ranga Rao, $NiCo_2O_4$ hexagonal nanoplates anchored on reduced graphene oxide sheets with enhanced electrocatalytic activity and stability for methanol and water oxidation, *Electrochim. Acta*, vol. 213, p. 717–729, 2016.

[118] F. Li, J. Song, H. Yang, S. Gan, Q. Zhang, D. Han, A. Ivaska, e L. Niu, One-step synthesis of graphene/SnO_2 nanocomposites and its application in electrochemical supercapacitors., *Nanotechnology,* vol. 20, no 45, p. 455602, 2009.

[119] D. Yoon, J. Hwang, D. H. Kim, W. Chang, K. Y. Chung, e J. Kim, One-pot route for uniform anchoring of TiO_2 nanoparticles on reduced graphene oxides and their anode performance for lithium-ion batteries, *J. Supercrit. Fluids*, vol. 125, p. 66–78, 2017.

[120] S.-C. Weng, S. Brahma, C.-C. Chang, e J.-L. Huang, Synthesis of MnOx/reduced graphene oxide nanocomposite as an anode electrode for lithium-ion batteries, *Ceram. Int.*, vol. 43, no 6, p. 4873–4879, 2017.

[121] W. Konicki, M. Aleksandrzak, D. Moszyński, e E. Mijowska, Adsorption of anionic azo-dyes from aqueous solutions onto graphene oxide: Equilibrium, kinetic and thermodynamic studies, *J. Colloid Interface Sci.,* vol. 496, p. 188–200, 2017.

[122] H. Raghubanshi, S. M. Ngobeni, A. O. Osikoya, N. D. Shooto, C. W. Dikio, E. B. Naidoo, E. D. Dikio, R. K. Pandey, e R. Prakash, Synthesis of graphene oxide and its application for the adsorption of Pb+2 from aqueous solution, *J. Ind. Eng. Chem.*, vol. 47, p. 169–178, 2016.

[123] S. Muralikrishna, S. Cheunkar, B. Lertanantawong, T. Ramakrishnappa, D. H. Nagaraju, W. Surareungchai, R. G. Balakrishna, e K. R. Reddy, Graphene oxide-Cu(II) composite electrode for non-enzymatic determination of hydrogen peroxide, *J. Electroanal. Chem.,* vol. 776, p. 59–65, 2016.

[124] M. Jahan, Z. Liu, e K. P. Loh, A graphene oxide and copper-centered metal organic framework composite as a tri-functional catalyst for HER, OER, and ORR, *Adv. Funct. Mater.*, vol. 23, no 43, p. 5363–5372, 2013.

[125] Y. Dong, Y. Deng, J. Zeng, H. Song, e S. Liao, A high-performance composite ORR catalyst based on the synergy between binary transition metal nitride and nitrogen-doped reduced graphene oxide, *J. Mater. Chem. A*, vol. 5, no 12, p. 5829–5837, 2017.

[126] W. Ren, D. Tang, X. Lu, J. Sun, M. Li, S. Qiu, e D. Fan, Novel Multilayer ACF@rGO@OMC Cathode Composite with Enhanced Activity for Electro-Fenton Degradation of Phthalic Acid Esters, *Ind. Eng. Chem. Res.*, vol. 55, no 42, p. 11085–11096, 2016.

[127] C.-Y. Chen, C. Tang, H.-F. Wang, C.-M. Chen, X. Zhang, X. Huang, e Q. Zhang, Oxygen Reduction Reaction on Graphene in an Electro-Fenton System: *In situ* Generation of H_2O_2 for the Oxidation of Organic Compounds, *ChemSusChem,* vol. 9, no 2, p. 1194–1199, 2016.

[128] C. Santhosh, V. Velmurugan, G. Jacob, S. K. Jeong, A. N. Grace, e A. Bhatnagar, Role of nanomaterials in water treatment applications: A review, *Chem. Eng. J.*, vol. 306, p. 1116–1137, 2016.

[129] A. Reina, X. Jia, J. Ho, D. Nezich, H. Son, V. Bulovic, M. S. Dresselhaus, J. Kong, K. S. K. S. Kim, Y. Zhao, H. Jang, S. Y. Lee, J. M. Kim, J.-H. Ahn, P. Kim, J.-Y. Choi, e B. H. Hong, Large-scale pattern growth of graphene films for stretchable transparent electrodes. *Nano Lett*., vol. 9, no 1, p. 30–35, 2009.

[130] V. Singh, D. Joung, L. Zhai, S. Das, S. I. Khondaker, e S. Seal, Graphene based materials: Past, present and future, *Prog. Mater. Sci.,* vol. 56, no 8, p. 1178–1271, 2011.

[131] D. Marcano, D. Kosynkin, J. M. Berlin, A. Sinitskii, Z. Z. Sun, A. Slesarev, L. B. Alemany, W. Lu, e J. M. Tour, Improved Synthesis of Graphene Oxide, Acs Nano (2010 *Am. Chem. Soc*., vol. 4, no 8, p. 4806–4814, 2010.

[132] J. F. Carneiro, M. J. Paulo, M. Siaj, A. C. Tavares, M. R. V. Lanza, Zirconia on Reduced Graphene Oxide Sheets: Synergistic Catalyst with High Selectivity for H_2O_2 Electrogeneration, *Chem Electro Chem*, 2017.

[133] T. X. H. Le, M. Bechelany, J. Champavert, e M. Cretin, A highly active based graphene cathode for the electro-fenton reaction, *RSC Adv.,* vol. 5, no 53, p. 42536–42539, 2015.

[134] S. H. Wu, P. C. Li, e C. C. Hu, Electron transfer number control of the oxygen reduction reaction on nitrogen-doped reduced graphene oxides for the air electrodes of zinc-air batteries and organic degradation, *Mater. Chem. Phys*., vol. 183, p. 551–560, 2016.

[135] T. X. H. Le, M. Bechelany, S. Lacour, N. Oturan, M. A. Oturan, e M. Cretin, High removal efficiency of dye pollutants by electron-Fenton process using a graphene based cathode, *Carbon N. Y.,* vol. 94, p. 1003–1011, 2015.

[136] H. Dong, H. Su, Z. Chen, H. Yu, e H. Yu, Fabrication of Electrochemically Reduced Graphene Oxide Modified Gas Diffusion Electrode for In-situ Electrochemical Advanced Oxidation Process under Mild Conditions, *Electrochim. Acta*, vol. 222, p. 1501–1509, 2016.

[137] H. Dong, H. Yu, X. Wang, Q. Zhou, e J. Feng, A novel structure of scalable air-cathode without Nafion and Pt by rolling activated carbon and PTFE as catalyst layer in microbial fuel cells, *Water Res.,* vol. 46, no 17, p. 5777–5787, 2012.

[138] C. Liu, K. Wang, S. Luo, Y. Tang, e L. Chen, Direct electrodeposition of graphene enabling the one-step synthesis of graphene-metal nanocomposite films, *Small,* vol. 7, no 9, p. 1203–1206, 2011.

[139] M. Pumera, Graphene-based nanomaterials for energy storage, *Energy Environ. Sci.,* vol. 4, no 3, p. 668–674, 2011.

[140] N. Singh, N. Sharma, and J. K. Katnoria, Monitoring of Water Pollution and Its Consequences : An Overview, *Int. J. Environ. Chem. Ecol. Geol. Geophys. Eng.*, vol. 8, no. 2, pp. 1–9, 2014.

[141] J. Kim, B. P. Lingaraju, R. Rheaume, J. Y. Lee, and K. F. Siddiqui, Removal of ammonia from wastewater effluent by chlorella vulgaris, *Tsinghua Sci. Technol.*, vol. 15, no. 4, pp. 391–396, 2010.

[142] D. L. Huang et al. Application of molecularly imprinted polymers in wastewater treatment: a review, *Environ. Sci. Pollut. Res.*, vol. 22, no. 2, pp. 963–977, 2015.

[143] R. F. Gibson, E. O. Ayorinde, and Y.-F. Wen, Vibrations of carbon nanotubes and their composites: A review, *Compos. Sci. Technol.*, vol. 67, no. 1, pp. 1–28, 2007.

[144] J. Zhang, D. Su, and R. Schlögl, Commercial carbon nanotubes as heterogeneous catalysts in energy related applications, *Phys. status solidi,* 2009.

[145] X. Pan and X. Bao, The effects of confinement inside carbon nanotubes on catalysis, *Acc. Chem. Res.*, 2011.

[146] C. Vecitis, M. Schnoor, and M. Rahaman, Electrochemical multiwalled carbon nanotube filter for viral and bacterial removal and inactivation, *Sci. Technol.*, 2011.

[147] M. M. J. Treacy, T. W. Ebbesen, and J. M. Gibson, Exceptionally high Young's modulus observed for individual carbon nanotubes, *Nature,* vol. 381, no. 6584, pp. 678–680, 1996.

[148] S. Iijima and T. Ichihashi, Single-shell carbon nanotubes of 1-nm diameter, *Nature,* vol. 363, no. 6430, pp. 603–605, 1993.

[149] S. Imani Yengejeh, S. A. Kazemi, and A. Öchsner, Advances in mechanical analysis of structurally and atomically modified carbon nanotubes and degenerated nanostructures: A review, *Compos. Part B Eng.*, vol. 86, pp. 95–107, 2016.

[150] C. W. Tan, K. H. Tan, Y. T. Ong, A. R. Mohamed, S. H. S. Zein, and S. H. Tan, Energy and environmental applications of carbon nanotubes, *Environmental Chemistry Letters*, vol. 10, no. 3. pp. 265–273, 2012.

[151] S. A. Jame and Z. Zhou, Electrochemical carbon nanotube filters for water and wastewater treatment, *Nanotechnology Reviews*, vol. 5, no. 1. pp. 41–50, 2016.

[152] H. Huang, K. Schwab, and J. G. Jacangelo, "Pretreatment for Low Pressure Membranes in Water Treatment: A Review," *Environ. Sci. Technol.,* vol. 43, no. 9, pp. 3011–3019, 2009.

[153] P. R. Gogate and A. B. Pandit, "A review of imperative technologies for wastewater treatment I: Oxidation technologies at ambient conditions," *Advances in Environmental Research*, vol. 8, no. 3–4. pp. 501–551, 2004.

[154] L. Meng, C. Fu, and Q. Lu, Advanced technology for functionalization of carbon nanotubes, *Prog. Nat. Sci.*, vol. 19, no. 7, pp. 801–810, 2009.

[155] A. K. Mishra and S. Ramaprabhu, Magnetite decorated multiwalled carbon nanotube based supercapacitor for arsenic removal and desalination of seawater, *J. Phys. Chem. C,* vol. 114, no. 6, pp. 2583–2590, 2010.

[156] H. Liu, A. Vajpayee, and C. D. Vecitis, Bismuth-doped tin oxide-coated carbon nanotube network: Improved anode stability and efficiency for flow-through organic electrooxidation, *ACS Appl. Mater. Interfaces*, vol. 5, no. 20, pp. 10054–10066, 2013.

[157] G. Wei, H. Yu, X. Quan, S. Chen, and H. Zhao, Constructing all carbon nanotube hollow fiber membranes with improved performance in separation and antifouling for water treatment, *Environ. Sci.,* 2014.

[158] V. K. K. Upadhyayula, S. Deng, M. C. Mitchell, G. B. Smith, V. K. Nair, and S. Ghoshroy, Adsorption kinetics of *Escherichia coli* and *Staphylococcus aureus* on single-walled carbon nanotube aggregates., *Water Sci. Technol*., vol. 58, no. 1, pp. 179–184, 2008.

[159] A. S. Brady-Estévez, S. Kang, and M. Elimelech, A single-walled-carbon-nanotube filter for removal of viral and bacterial pathogens, *Small,* vol. 4, no. 4, pp. 481–484, 2008.

[160] M. Shim, N. W. S. Kam, R. J. Chen, Y. Li, and H. Dai, Functionalization of Carbon Nanotubes for Biocompatibility and Biomolecular Recognition, *Nano Lett*., vol. 2, no. 4, pp. 285–288, 2002.

[161] D. Du, M. Wang, J. Zhang, J. Cai, H. Tu, and A. Zhang, Application of multiwalled carbon nanotubes for solid-phase extraction of organophosphate pesticide, *Electrochem. commun*., vol. 10, no. 1, pp. 85–89, 2008.

[162] L. Zhou, Z. Hu, C. Zhang, Z. Bi, T. Jin, and M. Zhou, Electrogeneration of hydrogen peroxide for electro-Fenton system by oxygen reduction using chemically modified graphite felt cathode, *Sep. Purif. Technol*., vol. 111, pp. 131–136, 2013.

[163] S. Yuan, Y. Fan, Y. Zhang, M. Tong, and P. Liao, Pd-Catalytic *In situ* Generation of H_2O_2 from H_2 and O_2 Produced by Water Electrolysis for the Efficient Electro-Fenton Degradation of Rhodamine B, *Environ. Sci. Technol*., vol. 45, no. 19, pp. 8514–8520, 2011.

[164] F. C. Moreira, R. A. R. Boaventura, E. Brillas, and V. J. P. Vilar, Electrochemical advanced oxidation processes: A review on their application to synthetic and real wastewaters, *Applied Catalysis B: Environmental,* vol. 202. pp. 217–261, 2017.

[165] S. Jame and Z. Zhou, Electrochemical carbon nanotube filters for water and wastewater treatment, *Nanotechnol. Rev*., 2016.

[166] H. Dai, Carbon nanotubes: opportunities and challenges, *Surf. Sci.,* 2002.

[167] L. Rapoport, Y. Bilik, Y. Feldman, and M. Homyonfer, Hollow nanoparticles of WS2 as potential solid-state lubricants, *Nature,* 1997.

[168] Y. Feldman, E. Wasserman, D. Srolovitz, and R. Tenne, High-rate, gas-phase growth of MoS_2 nested inorganic fullerenes and nanotubes, *Science* (80-.)., 1995.

[169] D. Galvan, J. Kim, and M. Maple, Formation of $NbSe_2$ nanotubes by electron irradiation, *Fuller. Sci*., 2000.

[170] Q. Li, J. Newberg, E. Walter, and J. Hemminger, Polycrystalline molybdenum disulfide (2H−MoS_2) nano-and microribbons by electrochemical/chemical synthesis, *Nano Lett.*, 2004.

[171] T. Kasuga, M. Hiramatsu, A. Hoson, T. Sekino, and K. Niihara, Formation of titanium oxide nanotube, *Langmuir,* 1998.

[172] Y. Hacohen, E. Grunbaum, R. Tenne, and J. Sloan, Cage structures and nanotubes of $NiCl_2$, *Nature,* 1998.

[173] Y. Zhang, K. Suenaga, C. Colliex, and S. Iijima, Coaxial nanocable: silicon carbide and silicon oxide sheathed with boron nitride and carbon, *Science*, 1998.

[174] M. Damnjanovic, T. Vukovic, and I. Milosevic, Super-slippery Carbon Nanotubes: Symmetry Breaking breaks friction, *Eur. Phys. J. B, vol.* 25, p. 4, 2001.

[175] R. A. Evarestov et al., DFT modeling of anatase nanotubes, *Phys. Rev. B,* vol. 68, no. 1, p. 4568, 1992.

[176] J. Hu, T. Odom, and C. Lieber, Chemistry and physics in one dimension: synthesis and properties of nanowires and nanotubes, *Acc. Chem. Res*., 1999.

[177] M. O. Watanabe, T. Sasaki, S. Itoh, and K. Mizushima, Structural and electrical characterization of BC_2N thin films, *Thin Solid Films,* vol. 281–282, no. 1–2, pp. 334–336, 1996.

[178] V. V. Pokropivny et al. Boron Nitride Analogs of Fullerenes (the Fulborenes), Nanotubes, and Fullerites (the Fulborenites), *J. Solid State Chem.,* vol. 154, no. 1, pp. 214–222, 2000.

[179] V. V. Pokropivny, Composite on base of 2D nanotubular lattice as ideal high-Tc superconductor, *Phys. C Supercond. its Appl.,* vol. 351, no. 1, pp. 71–77, 2001.

[180] V. Zwilling, E. Darque-Ceretti, A. Boutry-Forveille, D. David, M. Y. Perrin, and M. Aucouturier, Structure and physicochemistry of anodic oxide films on titanium and TA6V alloy, *Surf. Interface Anal.*, vol. 27, no. 7, pp. 629–637, 1999.

[181] P. Roy, S. Berger, and P. Schmuki,TiO_2 nanotubes: synthesis and applications, *Angew Chem Int Ed Engl,* vol. 50, no. 13, pp. 2904–2939, 2011.

[182] S. Yoriya and C. A. Grimes, Self-assembled anodic TiO_2 nanotube arrays: electrolyte properties and their effect on resulting morphologies, *J. Mater. Chem*., vol. 21, no. 1, pp. 102–108, 2011.

[183] Q. Meng, Z. Guan, J. Huang, Q. Li, and J. Yang, Electronic and optical properties of TiO_2 nanotubes and arrays: a first-principles study, *Phys Chem Chem Phys*, vol. 16, no. 23, pp. 11519–11526, 2014.

[184] M. Lazar, S. Varghese, and S. Nair, Photocatalytic water treatment by titanium dioxide: recent updates, *Catalysts,* 2012.

[185] M. Kiser, P. Westerhoff, T. Benn, and Y. Wang, Titanium nanomaterial removal and release from wastewater treatment plants, *Sci. Technol*., 2009.

[186] Q. Zhang, J. Zhu, Y. Wang, J. Feng, W. Yan, and H. Xu, Electrochemical assisted photocatalytic degradation of salicylic acid with highly ordered TiO_2 nanotube electrodes, *Appl. Surf. Sci.*, 2014.

[187] J. Li, X. Liu, Q. Han, X. Yao, and X. Wang, Formation of WO_3 nanotube-based bundles directed by $NaHSO_4$ and its application in water treatment, *J. Mater. Chem. A*, vol. 1, no. 4, p. 1246, 2013.

[188] Ribeiro, F. R., Rodrigues, A. E., Rollmann, L. D., Naccache, C., Proceedings of the NATO Advanced Science Institutes Series. Series E, Applied sciences. n. 80 - Zeolites: Science and Technology (Alcabideche, Portugal).; *Martinus Nijhoff Publishers in cooperation with NATO Scientific Affairs Division*: Amsterdam, Holanda, 1983.

[189] Flanigen, E. M., Zeolites and molecular sieves. An historical perspective, in Introduction to Zeolite Science and Practice, 2 ed, *Stud. Surf. Sci. Catal*. v. 137, Elsevier, Amsterdam, p. 11 – 35, 2001.

[190] Cronstedt, A. F., Ron och beskriting om en obekant barg ant, som kallas zeolites. *Akad. Handl. Stockh*. v. 18, p. 120–130, 1756. [Ron, and writing about an unknown barg ant, called Zeolites.]

[191] Coombs, D. S., Alberti, S., Armbruster, T., Artioli, G., Colella, C., Galli, E., Grice, J. D., Liebau, F., Mandarino, J. A., Minato, H., Nickel, E. H., Passaglia, E., Peacor, D. R., Quartieri, S., Rinaldi, R., Ross, M., Sheppard, R. A., Tillmasnns, E., Vezzalini, G., Recommended nomenclature for zeolite minerals: Report of the subcommittee on zeolites of the International Mineralogical Association, Commission on New Minerals and Mineral Names. *Canadian Mineralogist*. v. 35, 1571-1606, 1997.

[192] Pérez-Pariente, J., *Em Materiales Zeolíticos: Síntesis, Propriedades y Aplicaciones*; Universidad Alicante: San Vicente, 2002, chap. 1.

[193] Barrer, R. M., baynham, J. W., Bultitude, F. W., Meier, W. M., hydrothermal chemistry of the silicates. Part viii. Low-temperature crystal growth of aluminosilicates, and of some gallium and germanium analogues. *J. Chem. Soc. P*. 195-208, 1959. Doi: 10. 1039/jr9590000195.

[194] Barrer, F. R. S. R. M., Zeolites and their synthesis – review. *Zeolites*. v. 1, p. 130-140, 1981.

[195] Cundy, C. S., Cox, P. A., the hydrothermal synthesis of zeolites: precursors, intermediates and reaction mechanism. *Microporous and mesoporous materials*. V. 82, issues 1–2, p. 1–78, 2005.

[196] Aguiar, M. R. M. P., Novaes, A. C., Guarino, A. W. S., Remoção de metais pesados de efluentes industriais por aluminossilicatos. *Química Nova*. v. 25, p. 1145-1154, 2002. [Removal of heavy metals from industrial effluents by aluminosilicates *New chemistry*. v. 25, p. 1145-1154, 2002].

[197] Rodhes, C. J., properties and applications of zeolites. *Sci prog*. V. 93, p. 223-284, 2010.

[198] Akgul, M., Karabakan, A., Acar, O., Yurum, Y., Removal of silver (I) from aqueous solutions with clinoptilolite. *Microporous and Mesoporous Materials*. v. 94, p. 99–104, 2006.

[199] Yilmaz, B., Müller, U., catalytic applications of zeolites in chemical industry. *Topics in catalysis*. V. 52, issue 6, p. 888-895, 2009.

[200] Hermassi, M., Valderrama, C., Moreno, N., Oriol, F., Querol, X., Batis, N., Cortina, J. L, Powdered Ca-activated zeolite for phosphate removal fromtreated waste-water. *J Chem Technol Biotechnol*. v. 91, p. 1962–1971, 2016.

[201] Humelnicua, I., Baiceanu, A., Ignatc, M. E., Dulman, V., The removal of Basic Blue 41 textile dye fromaqueous solution by adsorption onto naturalzeolitic tuff: Kinetics and thermodynamics. *Process Safety and Environmental Protection*. V. 105, p. 274-287, 2017.

[202] Wang, W. X., Qiao, Y., Li, T., Liu, S., Zhou, J., Yao, H., Yang H., Xu, M., Improved removal of Cr(VI) from aqueous solution using zeolite synthesized from coal fly ash via mechano-chemical treatment. *Asia-Pac. J. Chem. Eng*. 2017, Published online in Wiley Online Library.

[203] Filippidis, A., Tziritis, E., Kantiranis, N., Tzamos, E., Gamaletsos, P., Papastergios, G., Filippidis, S., Application of Hellenic Natural Zeolite in Thessaloniki industrial area wastewater treatment. *Desalination and Water Treatment*. p. 1-11, 2015.

[204] Wang, H., Zhanga, H., Jiang, J. Q., Ma, X., Adsorption of bisphenol A onto cationic-modified zeolite. *Desalination and Water Treatment*. p. 1–8, 2016.

[205] Visa, M., Synthesis and characterization of newzeolite materials obtained fromfly ash for heavy metals removal in advanced wastewater treatment. *Powder Technology* v. 294, p. 338–347, 2016.

[206] Brillas, E., Sirés, I. & Oturan, M. A. Electro-fenton process and related electrochemical technologies based on Fenton's reaction chemistry. *Chem. Rev.* 109, 6570–6631 (2009).

[207] Sirés, I., Brillas, E., Oturan, M. A., Rodrigo, M. A. & Panizza, M. Electrochemical advanced oxidation processes: Today and tomorrow. A review. *Environ. Sci. Pollut. Res.* 21, 8336–8367 (2014).

[208] Brillas, E. & Martínez-Huitle, C. A. Decontamination of wastewaters containing synthetic organic dyes by electrochemical methods. An updated review. *Applied Catalysis B: Environmental* (2015).

[209] Santhosh, C. et al. Role of nanomaterials in water treatment applications: A review. *Chemical Engineering Journal* (2016).

[210] Moreira, F. C., Boaventura, R. A. R., Brillas, E. & Vilar, V. J. P. Electrochemical advanced oxidation processes: A review on their application to synthetic and real wastewaters. *Applied Catalysis B: Environmental* (2017).

[211] Groves, J. T. High-valent iron in chemical and biological oxidations. *J. Inorg. Biochem.* 100, 434–447 (2006).

[212] Barbusiński, K. Fenton Reaction - Controversy Concerning the Chemistry. *Ecol. Chem. Eng.* 16, 347–358 (2009).

[213] Logager, T., Holcman, J., Sehested, K. & Pedersenf, T. Oxidation of Ferrous Ions by Ozone in Acidic solutions. *Inorg. Chem.* 31, 3523–3529 (1992).

[214] Minero, C., Lucchiari, M., Maurino, V. & Vione, D. A quantitative assessment of the production of ˙OH and additional oxidants in the dark Fenton reaction: Fenton degradation of aromatic amines. *RSC Adv.* (2013).

[215] Walling, C. Fenton's Reagent Revisited. *Water Res.* 8, 125–131 (1975).

[216] Paciolla, M. D., Kolla, S. & Jansen, S. A. The reduction of dissolved iron species by humic acid and subsequent production of reactive oxygen species. *Adv. Environ. Res.* (2002).

[217] Hug, S. J. & Leupin, O. Iron-catalyzed oxidation of Arsenic(III) by oxygen and by hydrogen peroxide: pH-dependent formation of oxidants in the Fenton reaction. *Environ. Sci. Technol.* (2003).

[218] González-Davila, M., Santana-Casiano, J. M. & Millero, F. J. Oxidation of iron (II) nanomolar with H_2O_2 in seawater. *Geochim. Cosmochim. Acta* (2005).

[219] Goldstein S. & Meyerstein D. Comments on the mechanism of the 'Fenton-like' reaction. *Acc. Chem. Res.* 32, 547–550 (1999).

[220] Enami, S., Sakamoto, Y. & Colussi, A. J. Fenton chemistry at aqueous interfaces. *Proc. Natl. Acad. Sci.* (2014).

[221] Bossmann, S. H. et al. New Evidence against Hydroxyl Radicals as Reactive Intermediates in the Thermal and Photochemically Enhanced Fenton Reactions. *J. Phys. Chem. A* (1998).

[222] Kremer, M. L. & Stein, G. The Catalytic Decomposition of Hydrogen Peroxide by Ferric Perchlorate. *Trans. Faraday Soc.* 55, 595–560 (1959).

[223] Kremer, M. L. Mechanism of the Fenton reaction. Evidence for a new intermediate. *Phys. Chem. Phys.* 1, 3595–3605 (1999).

[224] Barb, W. G., Baxendale, J. H., George, P. & Hargrave, K. R. Reactions of ferrous and ferric ions with hydrogen peroxide. Part II.—The ferric ion reaction. *Trans. Faraday Soc.* (1951).

[225] Gozzo, F. Radical and non-radical chemistry of the Fenton-like systems in the presence of organic substrates. *Journal of Molecular Catalysis A: Chemical* (2001).

[226] Bray. William C. & Gorin, M. H. Ferryl Ion, a Compound of Tetravalent Iron. *J. Am. Chem. Soc.* 54, 2124–2125 (1932).

[227] Chevallier, E., Jolibois, R. D., Meunier, N., Carlier, P. & Monod, A. 'Fenton-like' reactions of methylhydroperoxide and ethylhydro-peroxide with Fe^{2+} in liquid aerosols under tropospheric conditions. *Atmos. Environ.* (2004).

[228] Keenan, C. R. & Sedlak, D. L. Factors Affecting the Yield of Oxidants from the Reaction of Nanoparticulate Zero-Valent Iron and Oxygen. *Environ. Sci. Technol.* 42, 1262–1267 (2008).

[229] Ensing, B., Buda, F., Blöchl, P. E. & Baerends, E. J. A Car–Parrinello study of the formation of oxidizing intermediates from Fenton's reagent in aqueous solution. *Phys. Chem. Chem. Phys.* (2002).

[230] Ensing, B., Buda, F. & Baerends, E. J. Fenton-like chemistry in water: Oxidation catalysis by Fe(III) and H_2O_2. *J. Phys. Chem. A* (2003).

[231] Bokare, A. D. & Choi, W. Review of iron-free Fenton-like systems for activating H_2O_2 in advanced oxidation processes. *Journal of Hazardous Materials* 275 (2014): 121-135.

INDEX

#

A

B

C

D

E

F

G

H

I

K

L

M

N

O

P

Q

R

S

T

U

V

W

X